Critical Studies of Education

Volume 19

We live in an era where forms of education designed to win the consent of students, teachers, and the public to the inevitability of a neoliberal, market-driven process of globalization are being developed around the world. In these hegemonic modes of pedagogy, questions about issues of race, class, gender, sexuality, colonialism, religion, and other social dynamics are simply not asked. Indeed, questions about the social spaces where pedagogy takes place—in schools, media, corporate think tanks, etc.—are not raised. When these concerns are connected with queries such as the following, we begin to move into a serious study of pedagogy: What knowledge is of the most worth? Whose knowledge should be taught? What role does power play in the educational process? How are new media reshaping as well as perpetuating what happens in education? How is knowledge produced in a corporatized politics of knowledge? What sociopolitical role do schools play in the twenty-first century? What is an educated person? What is intelligence? How important are sociocultural contextual factors in shaping what goes on in education? Can schools be more than a tool of the new American (and its Western allies') twenty-first century empire? How do we educate well-informed, creative teachers? What roles should schools play in a democratic society? What roles should media play in a democratic society? Is education in a democratic society different than in a totalitarian society? What is a democratic society? How is globalization affecting education? How does our view of mind shape the way we think of education? How does affect and emotion shape the educational process? What are the forces that shape educational purpose in different societies? These, of course, are just a few examples of the questions that need to be asked in relation to our exploration of educational purpose. This series of books can help establish a renewed interest in such questions and their centrality in the larger study of education and the preparation of teachers and other educational professionals.

Geraldine L. Cochran • Jennifer D. Adams
Editors

Equity-Oriented, Discipline-Based STEM Education Research

Springer

Editors
Geraldine L. Cochran
Department of Physics
The Ohio State University
Columbus, OH, USA

Jennifer D. Adams
Department of Chemistry
Werklund School of Education
University of Calgary
Calgary, AB, Canada

ISSN 2543-0467 ISSN 2543-0475 (electronic)
Critical Studies of Education
ISBN 978-981-96-9135-7 ISBN 978-981-96-9136-4 (eBook)
https://doi.org/10.1007/978-981-96-9136-4

This Springer imprint is published by the registered company Springer Nature Singapore Pte Ltd.
The registered company address is: 152 Beach Road, #21-01/04 Gateway East, Singapore 189721, Singapore

Foreword

Warning—this handbook is for the "adult" STEM reformer—the reformer who has already tried, failed, learned, and partly succeeded in making the culture of STEM innovative, equitable, and humane. If that is you, then it's likely you've selected this book because, like many other adults doing such work, you know that time is a formidable enemy of those in the fight for achieving equity in STEM. Or perhaps you selected it because you are fed up with the stubborn paralysis of the proverbial "needle," which never seems to want to boldly move in a positive direction. Or maybe you have been wishing for someone who really understood your anxiety and frustration to do something to reassure you that there is still reason to hope. Whatever your reason, the decision to pick up this "handbook" was the best one you could have made. But you should know—this is no ordinary handbook. It's so much more. The chapters in this book fill a critical information gap and provide vital insight into the constructs of discipline-based education research—both constructs that have served us well and those that haven't. Specifically, the authors take an audacious step and unpack how this highly acclaimed, well-documented approach to examining and understanding what leads to undergraduate STEM learning falls short in prioritizing the ways in which STEM teaching and learning happens for students who have been historically marginalized from these disciplines.

The origins of discipline-based education research, or DBER (pronounced dee' bər), can be traced as far back as the early 1900s when science faculty raised serious questions about the "quality" of undergraduate science teaching and learning. However, in answering those questions, they apparently looked no further than their own disciplinary views, attitudes, and research methods to determine what quality in science teaching should look like. And they, either consciously or unconsciously, discarded and disregarded the methods, perspectives, and ways of knowing of other fields—namely, the field of education research. What resulted were limited ideas and innovations in undergraduate STEM education. In the years that followed, and even now, DBER has corrected this misstep in very significant ways. This robust area of education research allows us to blend experimental and exploratory methodologies related to education with the peculiar customs and cultures of various STEM disciplines. In short, DBER allows us to form and make use of an arsenal of scientific investigative methods to better understand, document, and explain what works in undergraduate STEM education.

But methods minus meaning won't suffice for the kind of reform we need in today's STEM academy. Indeed, if we know what works in undergraduate STEM teaching, do we not also need to know for whom it works best, or doesn't work at all? As global economies rapidly create opportunities for new jobs in emerging scientific and technology-based fields and as higher education continues to overlook and undervalue the STEM talent of historically marginalized groups, the need to fully understand what works and grapple with what doesn't work is of the essence, especially now.

Admittedly, higher education is changing, though some would argue not fast enough. We have experienced unconscionable chaos as a result of a global pandemic that was set amid public outcries and unrest, which were rooted in deep and historic sociopolitical divides, embedded in racism, and fueled by economic class differences. The horrific images of police brutality, protests, and incivility that we have seen of late are not that far removed from the everyday encounters of microaggressions, indignities, and denigrations that minoritized STEM faculty and students face in their respective disciplines, departments, and campuses. As of the writing of this foreword, the lasting chaos of COVID-19 is now mired in recent US Supreme Court decision-making that has upended basic principles of democracy and broken the promises of education that guarantee access to a life of liberty for all. Now more than ever, we will have to rely on the collective conscience of higher education institutions, and the individuals therein, to guide us in doing "the right thing" absent the measures of accountability that we have come to rely on. Clearly, such a state of affairs as this will disadvantage STEM higher education reform efforts—leaving us ill-prepared to fully support the next generation of discoverers and innovators, ill-equipped to prepare citizenries for the ways that technological advances will impact our daily lives and, ultimately, vulnerable to the next global disaster.

The process of questioning and interrogating STEM teaching and learning methods and determining what counts as "quality" undergraduate STEM education is critical for redressing this trend. But discipline-based education research inquiry, in and of itself, while adequate, is insufficient in uncovering the complete truths about how science teaching and learning happens. This we learned from STEM faculty of the last century who ignored the potential for other disciplines to make meaningful contributions to STEM reform discourse and plans of action. Simply put, we have to be open to questioning and interrogating our questions and interrogations. In other words, we cannot continue to rely solely on Eurocentric frameworks and conventional forms of evidence to inform what is meant by "quality" STEM teaching and learning. Nor can we afford to continue to submit to dominant Western research paradigms that call for rigor but conveniently neglect relevancy, cultural responsiveness, and authenticity.

Without a collective commitment to discontinue these practices and hold ourselves accountable for doing so, we can never hope to dissolve, or even slightly diminish, the myriad ways in which dominant cultures within undergraduate STEM education have and continue to systematically deny access, delay change, or slow progress. Nor can we ever expect to rid ourselves of the old assumptions that have given rise to anti-inclusive STEM teaching practices or overcome the instinctive impulse we have as scientists to want to solve the "equity problem" in STEM teaching before understanding the problem of equity itself.

Maya Angelou was once quoted as saying, "When we know better, do better." This handbook masterfully offers a platform to not only know and do better but also to *be* better. We are indeed scientists and educators, but we can also be social justice activists, lifelong learners, and conscious humans who desire and require a different approach to discipline-based education research. Here, the editors and authors have called us in to be a part of what has been sorely missing from STEM higher education discussions and debates. And they lead us to the important conclusion that discipline-based education research, if it is to meaningfully contribute to the twenty-first-century STEM higher education reform, must be focused on and grounded in critical theories and frameworks that will ensure we not only know what works but also for whom "what works" works and under what conditions "what works" works best.

To that end, the accounts in this handbook lean heavily toward integrating justice into discipline-based education research. While many reformers skirt around the topic, these authors put faculty first! And they elegantly employ a broad range of culturally responsive methodologies including storytelling, personal reflection, and community-based participatory action research. Collectively, they are resistant to the use of Western conventions in data interpretation and reporting related to the experiences of historically marginalized groups. Their resistance means that marginalized students are seen as individuals, not a singular monolith that can be easily served with broad brushstrokes of reform that miss the mark and all but foreclose on our prospects for a diverse and competitive STEM workforce.

By the time you finish this book, it will not be lost on you that there are important voices of STEM education researchers here, all with a unique flavor for what education research should look like. They represent a broad range of cultural diasporas and all professorial and educational levels, including graduate students and early career faculty who, arguably, are closest to the locus of innovation in STEM education research. By including their perspectives, which are untouched by the decades of exposure to oppressive STEM cultures that often color the lenses of more seasoned researchers, this book offers us a much-needed space to revisit and reclaim our hopes that an equitable undergraduate STEM enterprise is not only possible, but probable.

American Association of Colleges and Universities
Washington, DC, USA

Kelly Mack

Contents

1 **Introduction: The Imperative of Equity-Oriented, Discipline-Based STEM Education Research** 1
Jennifer D. Adams and Geraldine L. Cochran

2 **Liberal Ideologies in Academic STEM: How Faculty Can (Re)produce Racial Inequities** 9
Tatiane Russo-Tait and Aireale J. Rodgers

3 **Critical Race Theory with Applications to DBER** 37
Miguel Rodriguez

4 **QuantCrit as a Methodological Framework for Equity-Centered Discipline-Based Education Research** 55
Paulette Vincent-Ruz, Kathryn N. Hosbein, and Vanessa Rosa

5 **Moving Beyond Deficit: DisCrit and Disability Justice in STEM Postsecondary Education** 91
Lisette E. Torres, Cathery Yeh, Paulo Tan, Daniel L. Reinholz, and Roy Payan

6 **Disabled by Design: Dreaming the Future of DBER Through Intersectional Disobedience** 111
Phillip A. Boda

7 **Political Disability Identity and DBER Disability Politics** 131
Liam McDermott

8 **Addressing Biases Through Decoloniality and Belonging in STEM: A Cross-Disciplinary Dialogue** 147
Jamilah Dei-Sharpe, Kirsten Crandall, Jessica Ford, Nathalie Jreidini, Ezgi Ozyonum, and Hadjara Sidibé

9 **(Un)Diluting the Diaspora: Acknowledging Cultural Identity Intersections in Postsecondary STEM Education** 175
Takeshia Pierre and Rose Pringle

10 It's in the Way We Talk: Exploring the Use of Everyday Language in Teaching Technical Content STEM Learning Environments 193
Nicole Pitterson and Benjamin Goldschneider

11 From Here to Equity: Promoting STEM Inclusion and Equity by Engaging in Authentic Field-Based Ecology Experiences 213
Christina Colon and Gillian U. Bayne

12 Promoting a Culture of Inclusion in Graduate Engineering Education: A Case Study 235
Walter C. Lee, Jeremi S. London, Adam S. Masters, Teirra K. Holloman, David B. Knight, and Eunsil Lee

13 Cultural and Systemic Barriers to Graduate Stem Education 255
Diana Sachmpazidi

14 Theoretical and Methodological Innovative Futures for Physics in Postsecondary Contexts 273
Eleanor S. Armstrong, Allison J. Gonsalves, Anna T. Danielsson, and Emma Mckay

Contributors

Jennifer D. Adams is a Tier 2 Canada Research Chair of Creativity, Equity, and STEM and Professor at the University of Calgary in the Faculty of Science. She is the PI of the Creativity, Equity, and STEM Lab, where she leads her team in research on equity in STEM teaching and learning environments with an emphasis on identity-affirming, anti-deficit, desiring, and justice-oriented approaches. Dr. Adams is the President of the NARST, a global organization for improving science education through research and the author of "*Informal Science Education and Teacher Learning*" by Peter Lang Press. Her prior appointments include Brooklyn College and the Graduate Center, City University of New York, the American Museum of Natural History, and the New York City Department of Education.

Eleanor S. Armstrong (she/her) is a Space Research Fellow at the University of Leicester's Institute of Space Research (UK), where she takes a critically queer feminist approach to social studies of outer space. She has held Postdoctoral Fellowship at the University of Delaware (USA, 2021) and was a Postdoctoral Researcher at Stockholm University (Sweden, 2022–24) since finishing her PhD in 2020 at University College London (UK). She has been a visiting scholar at the University of Cambridge (UK) and the Canadian Aviation and Space Museum (Canada) and held honorary research appointments at UCL, University of York, University of Vienna, and New York University. She developed queer science tours in the UK, including the Queering the Science Museum (2018) and Bridging Binaries (2019) programs for the University of Cambridge Museums. She co-organizes the virtual conference Space Science in Context (2020, 2023, 2024).

Gillian U. Bayne is a Professor of Science Education who holds a dual appointment in the Urban Education Department at CUNY's Graduate Center. She has over 30 years of science education experience, which has helped to deepen not only her passion for the science content but, equally important, the ethical underpinnings of its pedagogical enactment. She has worked with and supported beginning and preservice teachers, as well as science education doctoral students. Dr. Bayne's research is grounded in sociocultural theory and encompasses addressing STEM equity issues within communities that have been marginalized by systemic oppression locally, nationally, and internationally, including examining the personal and professional trajectories of underrepresented scientists of color, the leadership development of professional Ethiopian women in STEM, science educational outreach at Genspace, a Brooklyn-based community lab, and the National Society of Black Physicists x Harlem Gallery of Science Mentoring Program. Dr. Bayne has served as the Principal Investigator and Co-PI on numerous NSF-, NASA-, and NOAA-funded projects.

Phillip A. Boda is an Assistant Professor in the Special Education Department at the University of Illinois Chicago (UIC) and received his Ph.D. in Science Education from Teachers College, Columbia University. His work focuses on inclusive and special education in K-12 STEM technology-enhanced learning environments, environmental justice policy, and the development and sustainability of Community-Policy Participatory Partnerships (C-PPPs). He currently works with interdisciplinary scholars across UIC's College of Education, College of Engineering, Computer Science, Law School, and School of Public Health supporting Chicago communities to identify cumulative environmental hazards in their locales and include their voices among the plans to determine what industries are permitted. Combined with his work with the Gregory S. Fehribach Center around disability-centered instrumentation and his role supporting PhDs to leverage item response theory for small-sample analyses in teacher education, he employs an applied research design to all his work so that the impacts and outputs are socially consequential.

Geraldine L. Cochran is an Associate Professor of Physics at the Ohio State University. Cochran is a Fellow of the American Physical Society and a Fellow of the American Association of Physics Teachers, and a member of the International Union of Pure and Applied Physics' Commission on Physics Education. Cochran's research focuses on equity in physics education, including improving access, experiences, and success in physics. She has extensive experience developing and assessing curricula, mentoring programs, and interventions that enhance student learning and experiences in physics. Cochran also leads the Transforming Introductory Physics Sequences to Support all Students (TIPSSS) network. This network of educators and researchers is working to transform introductory physics sequences by developing curricular materials, evaluating diverse teaching models, and incorporating quantitative literacy and mathematical skills development alongside physics instruction.

Christina Colón is a full professor with tenure in the Department of Biological Sciences at Kingsborough Community College where she teaches biology classes for both majors and non-majors. Her approach to teachings is hands-on, writing intensive, civic engagement, student focused, active, and inquiry based. In person, she runs numerous field trips where she steps aside to let students take the lead and earn the status of colleague and citizen science. In an online environment, she engages students in active learning through virtual labs (e.g., Google Scholar,

SimBio, iNaturalist, and Zooniverse) and in-person activities that they do on their own.

After years of research in Asia studying civets, binturongs and sun bears, she shifted to local species in order to involve students in her research. She now works with students each spring to study the spawning activity and reproductive ecology of American horseshoe crab (*Limulus polyphemus*). She works in collaboration with Fordham, Cornell, Audubon, and the IUCN Horseshoe Crab Specialist Group to study, monitor, and protect this 450 MYO species that plays a vital role in the biomedical industry due to the clotting properties of its blue blood.

She also works with Gotham Coyote and AMNH to record acoustic behaviors of species in areas with urban coyotes to better understand the impact of changes (such as the arrival of this mesopredator) on the acoustic behavior of critters and the urban soundscape overall.

Kirsten Crandall is a researcher working in zoonotic disease systems in Canada with specialties in vector-borne diseases. More broadly, she is interested in science communication and equity, diversity, and inclusion issues in the STEMM (Science, Technology, Engineering, Math, Medicine) fields. An ecologist by training, she holds a master's degree in Biology from McGill University and a joint PhD in Biology from McGill University and the University of Ottawa.

Anna T. Danielsson is Professor of Science Education and Head of the Science Education Section at the Department of Teaching and Learning, Stockholm University. Danielsson has previously held appointments at Uppsala University, University of Cambridge, and King's College London. Her research interests are centered on identity and gender perspectives on the teaching and learning of science. Much of her work has concerned identity constitution in higher education physics, but the research has also included science teacher education, engineering education, and power relations in physics and technology school classroom. Her current projects explore students' trajectories to and through higher education physics and mathematics education, with a particular focus on spaces and materialities.

Jamilah Dei-Sharpe is a Canadian sociologist whose research explores the intersections of race, gender, and structural inequality. She is an SSHRC Joseph Armand Bombardier recipient for her study on racialized masculinities, which examines how gendered racism shapes the lives of Black men in Canada and their contributions to gender justice. She also serves as a qualitative process evaluator at the McGill University Health Centre Research Institute (RI-MUHC), advancing patient-centered clinical trials by addressing intersectionality, stakeholder engagement, and structural barriers in trial design and implementation. Dr. Dei Sharpe is committed to reimagining research and pedagogy through community-oriented, decolonial approaches. She is the founder of the *Decolonial Perspectives and Practices Hub*, a nonprofit dedicated to *reworlding* Canadian universities—an initiative whose methods have been published by Alberta Press and Bloomsbury. She also co-chairs the Race and Ethnicity Cluster of the Canadian Sociological Association and leads several initiatives to co-create knowledge and foster inclusive learning environments across disciplines.

Jessica Ford is a faculty member in the Biology Department of Dawson College. She studies the ecological consequences of tadpole loss and how tadpoles influence their aquatic environment. She is equally passionate about herpetology, conservation, diversity and inclusion, and outreach. Jessica is also the co-founder and chair of STEMM Diversity @McGill, the author and illustrator of the STEMM diversity coloring and activity book, and sits on local and international EDI committees.

Benjamin Goldschneider is an Assistant Professor and Founding Member of the First-Year Engineering Center. Through a combination of teaching, advising, curriculum development, and research, Benjamin seeks to foster students' passions for engineering and help them navigate the tumultuous first steps of their college career. Benjamin's previous research has placed a strong emphasis on students' socialization and sense of belonging with the goal of identifying the individuals, organizations, and institutional structures that are most important for students' success and connecting them. Benjamin graduated with a bachelor's degree in Industrial Engineering from Purdue University in 2018 before moving on to further study at Virginia Tech. He completed his Ph.D. in Engineering Education in 2023.

Allison J. Gonsalves is an Associate Professor at McGill University's Faculty of Education. Her research program has involved the tracing of "science identity trajectories" across learning spaces and the intersections of social identities and experiences that shape these movements. She has worked extensively in the areas of gender and physics education research, and informal science learning. Her current research program investigates postsecondary students' affective engagement in informal sustainability and equity initiatives and the consequences this has for their identity work and ongoing participation in science.

Teirra K. Holloman is a postdoctoral associate in engineering education at Virginia Tech. Teirra received her Ph.D. in Engineering Education and an M.Eng. in Industrial and Systems Engineering from Virginia Tech and her B.S. in Industrial Engineering from Clemson University. Her research interests include organizational resilience; organizational change; diversity, equity, and inclusion issues in engineering and global education programs.

Kathryn N. Hosbein is an Assistant Professor in the Chemistry Department at Middle Tennessee State University. Her ongoing research seeks to improve our understanding of how to support first-generation STEM students in developing as scientists throughout their academic careers. Her interest in this research was sparked by her own experiences navigating science throughout her academic career, having ADHD and anxiety. Dr. Hosbein has a Master's degree in Chemistry and a Ph.D. in Chemistry from Portland State University and completed two postdoctoral fellowships in Chemistry Education Research at East Carolina University and the University of Michigan.

Nathalie Jreidinir is the Director of Education at the Ecomuseum Zoo in Montreal, Quebec. They are a biologist with a passion for science communication and environmental education. Their goal as a scientist and educator is to inspire curiosity, not only in the next generation of scientists but in individuals of all ages, by fostering an understanding of ecological stewardship that transcends career paths.

David B. Knight is a Professor in the Department of Engineering Education at Virginia Tech and serves as Special Assistant to the Dean for Strategic Plan Implementation in the College of Engineering. His research tends to be at the macro-scale, focused on a systems-level perspective of how engineering education can become more effective, efficient, and inclusive, and considers the intersection between policy and organizational contexts. Knight currently serves as the Co-Editor-in-Chief of the *Journal of Engineering Education*.

Eunsil Lee is an Assistant Professor at University at Buffalo in the Department of Engineering Education. Her research interests center on the concept of sense of belonging, interpersonal interactions, diversity in citizenship in engineering graduate education. She received B.S. and M.S. in Clothing and Textiles from Yonsei University (South Korea) with the concentration area of Nanomaterials and Biomaterials in Textiles. She began her Ph.D. study in Textile Engineering but shifted her path toward Engineering Education, earning her Ph.D. from Arizona State University, in Engineering Education. After gaining her Ph.D., she worked as a postdoctoral associate at Florida International University in the School of Universal Computing, Construction, and Engineering Education and as Visiting Assistant Professor at Virginia Tech in the Department of Engineering Education.

Walter C. Lee is an Associate Professor in the Department of Engineering Education at Virginia Tech and the Director for Research in the Center for the Enhancement of Engineering Diversity. He is broadly interested in inclusion, diversity, and educational equity, particularly as it relates to students from groups that are historically underrepresented or marginalized in engineering. Lee received his Ph.D. in Engineering Education from Virginia Tech, his M.S. in Industrial and Systems Engineering from Virginia Tech, and his B.S. in Industrial Engineering from Clemson University.

Jeremi S. London is the Assistant Provost for Academic Opportunity and Belonging with a faculty appointment in the Mechanical Engineering Department at Vanderbilt University. London is a mixed methods researcher with interests in research impact, broadening participation, and organizational change promoting equitable outcomes in STEM education. London holds engineering degrees from Purdue University (BS & MS in Industrial Engineering, PhD in Engineering Education) and MBA from the University of Maryland-Global Campus.

Adam S. Masters is a doctoral student and graduate research assistant at Virginia Polytechnic Institute and State University. They received a B.S. in Mechanical Engineering from the University of Delaware and are currently pursuing a Ph.D. in Engineering Education at Virginia Tech. Adam's research interests include access, equity, and social justice in engineering.

Liam McDermott is a postdoctoral fellow at the University of Connecticut studying neurodiversity in physics and the ways in which neurodivergent students learn/

perform physics. He very recently earned his Ph.D. from Rutgers University under the advising of Dr. Geraldine Cochran. Being disabled himself and navigating public schools and higher education while being both hard-of-hearing and neurodivergent, he became fascinated by disability justice and the politics of disability in education while in graduate school. He has since dedicated his studies to making academia more accessible and developing more equitable science classrooms/ curricula.

Emma Mckay Emma studies student organizing in physics departments through affective and anarchist lenses. How is it that students build power to change their educational contexts together? Emma also organizes with the McGill TA and invigilator union, AGSEM. In their spare time, they watch birds, stare at plants, cycle, and weave with leaves.

Ezgi Ozyonum is a Public Scholar, Researcher, & Instructor at Concordia University, where they completed their Ph.D. in Education. As an educator and researcher, they focus on inclusive, student-centered learning and critical pedagogies. Their experience spans teaching, research, educational development, and community-engaged work, particularly in areas of internationalization, social justice, and higher education policy. They've taught undergraduate and graduate courses in education and engineering, developed equity-focused curriculum and guidebooks, and contributed to institutional initiatives on decolonization, EDI, and student support.

Roy Payan identifies as a blind Latino and veteran who champions Disability Rights through his advocacy work in education, workforce, and housing. He earned a bachelor's degree from Cal State Los Angeles, double majoring in Rehabilitation Services and Public Policy. He pursued this track because he wanted to better comprehend the mannerisms, traits, and characteristics of a wide spectrum of disabilities, in an effort to put forth and implement better policies that impact persons with disabilities. Payan also earned a master's degree from the USC Sol Price School of Public Policy. As Vice President of the National Federation of the Blind (NFB) affiliate *Poder Y Movimiento Latino*, Payan has been instrumental in forcing several shelters throughout the state of California to accept homeless, blind DACA students into their shelters. He has also worked on a program analysis of the Los Angeles County's Homeless Shelter Authority to review program misuse and policy failures to accept persons with disabilities into the homeless system simply for being disabled. He has even worked to preserve open land spaces and wildlife corridors for the purpose of land back repatriation to Indigenous nations as Vice President of the Northeast Los Angeles—Save Our Undeveloped Lands (NELA SOUL).

Payan is most known for his lawsuit against the Los Angeles City College District Board (*Payan v LACCD*) to force compliance with Title II Section 504 of the Rehabilitation Act of 1973 and the Americans with Disabilities Act (ADA). He won that fight after 7 long years, benefiting future students with disabilities. His work, dedication, and advocacy on behalf of persons with disabilities have resulted in his winning several national awards, including the California Association of

Postsecondary Education in Disabilities (CAPED) President's Award, the Justin Shapel Award from RespectAbility.org, and recognition by the USC Sol Price School of Public Policy for Advocacy of Disability Rights on the 32nd anniversary of the signing of the ADA.

Takeshia Pierre is the Assistant Professor of Critical STEM Education at Tufts University. Her scholarship is shaped by a twofold goal of disrupting racism in [STEM] education and sharing the stories of marginalized communities. Experienced in K-12 schools and trained as a public health practitioner, she champions interdisciplinary collaboration, culturally responsive curriculum development, and professional learning. Adopting a critical lens, she tackles topics such as STEM engagement, retention, equity, and access in her research. This primarily includes amplifying the voices of STEM learners and professionals across the Black Diaspora, uncovering institutions that support their STEM matriculation and retention, and interrogating educational disparities. In addition, her research agenda focuses on the culturally responsive professional development of STEM teachers domestically and on engaging in global discussions surrounding decolonizing STEM, collaborating with STEM scholars across North America, Africa, and Europe.

Nicole Pitterson is an Associate Professor and Assistant Department Head for Undergraduate Programs in the Department of Engineering Education at Virginia Tech. She holds a Ph.D. in Engineering Education from Purdue University and other background degrees in Manufacturing and Electrical and Electronic Engineering. Her research interests are exploring the role of language in learning, the transformative power of disciplinary knowledge, curriculum design, assessment and evaluation and fostering conceptual understanding. Dr. Pitterson is dedicated to bridging the gap between theoretical concepts and practical applications. Through her research and teaching, she aims to guide students in developing a critical understanding of core engineering concepts that goes beyond rote memorization so that they can adapt to the changing demands of a global workforce.

Rose Pringle is an Associate Professor in Science Education in the School of Teaching and Learning in the College of Education. Her research agenda extends into two parallel, yet related areas along the continuum of science teacher education. In one line, she focuses on the development of science teachers' disciplinary content knowledge and their response to professional development. In the other line of research, she investigates pedagogical content knowledge as a framework for shifting practices and heightening teachers' stance toward issues of social justice and their roles in positioning learners who are traditionally underrepresented in science. She therefore operates at the nexus between what knowledge teachers need and how it becomes translated into effective and culturally sustaining practices that challenge assumptions and the status quo and lead to increased participation of all groups of learners in science.

Daniel L. Reinholz, Ph.D., is a Professor of Mathematics and Statistics at San Diego State University. Dr. Reinholz engages in transformative research on classroom equity in postsecondary mathematics. This work has been organized around the development of the EQUIP tool and the equity analytics approach, which focuses on generating actionable data to illuminate the subtle and sometimes invisible patterns that play out in classroom participation (by race, gender, disability, etc.). These data can be incorporated into robust professional learning opportunities through equity learning communities that support instructors to transform their teaching. Beyond the classroom, Dr. Reinholz serves as Working Group Leader in the Accelerating Systemic Change Network, which aims to catalyze sustainable and scalable changes to STEM higher education. This work involves developing new models grounded in organizational change and helping STEM departments build their own capacity for change.

Aireale J. Rodgers is an Assistant Professor of Higher Education and Anna Julia Cooper Fellow in the Educational Leadership & Policy Analysis Department at the University of Wisconsin-Madison. Dr. Rodgers is a learning scientist of higher education whose research agenda illuminates how people's everyday (mis)understandings about race and racism shape learning across various higher education ecologies. Most recently, she has used qualitative techniques to study the racialized dimensions of faculty development programs, design and implementation of digital courseware products, student socialization processes, and classroom teaching and learning to better understand how educators can facilitate learning that advances critical race consciousness for faculty and students in postsecondary institutions. Dr. Rodgers earned a Ph.D. in Urban Education Policy with a concentration in Higher Education from the University of Southern California, as well as a master's degree in Learning Sciences and a bachelor's degree in Education and Social Policy from Northwestern University.

Miguel Rodriguez is currently Assistant Professor at California State University Dominguez Hills. Prior to being at CSUDH, Dr. Rodriguez graduated with his PhD in physics in 2020 from Florida International University. In his dissertation, he explored the cognitive and affective benefits of small group learning in introductory physics. Then he worked as a postdoctoral scholar for the University of Utah where he conducted STEM education research for 2 years, until starting his current position in 2022. Dr. Rodriguez's research expertise is in critical studies using CRT, FST, and intersectionality, and he has recent publications in the *Physical Review Physics Education Research* Journal and *Nature Physics* related to equity in physics. He has recently expanded his research agenda to include an investigation of the impact of Large Language Models such as ChatGPT in STEM education.

Vanessa Rosa, Ph.D., is a chemist dedicated to connecting scientific research with societal impact. As the founder of *Cuvette Collective*, she assists researchers and research center directors in translating scientific advancements into meaningful real-world applications.

Her innovative approach extends to *Cuvette Empowered*, a nonprofit organization that supports under-resourced researchers, and "*for CURIOUS Brains*," a YouTube channel that makes scientific research accessible to lifelong learners. Dr. Rosa is committed to advancing science education reform by broadening participation and implementing transformative research communication strategies.

Tatiane Russo-Tait is an Assistant Professor in the Cellular Biology Department at the University of Georgia, where she leads the Advancing Critical Consciousness, Equity, and Social Justice in STEM (ACCESS) lab. Her lab engages in interdisciplinary and collaborative research projects employing various methodological approaches and critical frameworks to examine equity and justice issues in postsecondary STEM education. Her current scholarship explores STEM faculty's equity beliefs and practices; how to best support the critical consciousness development of current and future scientists and educators for collective action and change efforts; the experiences and perspectives of students from minoritized backgrounds in STEM learning environments; and science and social justice curriculum and instruction. Dr. Russo-Tait earned a Ph.D. in STEM Education from the University of Texas at Austin, a master's degree in Stem Cell Biology from San Francisco State University, and a bachelor's degree in Cell Biology from the University of Hawai'i at Hilo.

Diana Sachmpazidi is an Assistant Professor of Physics at the Rochester Institute of Technology and an expert in Science and Physics Education. Her primary research focus is on systemic change within physics programs at both undergraduate and graduate levels. She explores team-based collaboration for instructional change and cultural change of program assessment within physics departments. Sachmpazidi is an expert in qualitative, quantitative, and instrument development methods.

Hadjara Sidibé is a Ph.D. student in the Department of Neurosciences at the Université de Montreal where they have expert knowledge in neuroscience, biochemistry, cellular biology, molecular biology and immunology, with specialization in RNA metabolism and neurodegenerative disorders.

Paulo Tan Paulo Tan's research focuses on advancing intersectional disability justice in and through mathematics education. He is currently an Assistant Professor of Elementary Mathematics Education at the University of California, Santa Cruz. Prior to his academic career, he served as a public school middle-secondary mathematics teacher for 10 years in culturally and linguistically diverse settings. His lived experiences with his son led Paulo to pursue a doctorate in special education with an emphasis in mathematics education and has been reckoning and challenging educational inequities and injustices ever since. His current and future scholarship includes building theories for abolitionist mathematics educational research and practices that center the brilliances of disabled Students of Color. Paulo identifies as Chinese-American, cisgender male, and non-disabled and is a parent of three.

Lisette E. Torres (she/her/ella) is a trained scientist and disabled scholar-activist who is Senior Researcher at *TERC*, a nonprofit made up of teams of math and science education and research experts. She is also the Director of Operations and Communication for the national NSF AISL equity resource center called the *Reimagining Equity and Values in Informal STEM Education (REVISE) Center*. Dr. Torres has a doctorate with a Certificate in Social Justice from the School of Education at Iowa State University and an M.S. in Zoology with a Certificate in

Ecology from Miami University. Her academic research focuses on addressing racialized gender justice and disability in science and higher education. She is a co-founder of *Sines of Disability* and the *National Coalition for Latinxs with Disabilities* (CNLD). Dr. Torres is also an advisory board member of the Women in Engineering Pro-Active Network's (WEPAN) *ADVANCE Resource and Coordination (ARC) Network*.

Paulette Vincent-Ruz (she/ella) is an Assistant Professor in Chemistry Education Research at New Mexico State University. She holds a B.S. in Chemical Engineering from the National Autonomous University of Mexico (UNAM) and a Ph.D. in Learning Sciences and Policy from the University of Pittsburgh. As a Queer Latinx cis-woman of color raised on the rightful lands of the Mexica tribes, her identity deeply informs her research. Dr. Vincent-Ruz investigates systemic inequities in chemistry education, pioneering QuantCrit methodologies and developing the REACT Framework to challenge deficit narratives and incorporate diverse cultural resources into STEM learning. Her work focuses on fostering science identity, belonging, and academic achievement, particularly at Hispanic-Serving Institutions (HSIs). Her unique blend of expertise in learning sciences and chemistry earned her recognition as the first chemistry education researcher to receive the American Chemical Society's Future Leader in Chemistry Award.

Cathery Yeh (she/her/hers) is an Assistant Professor in STEM Education and a core faculty member in the Center for Asian American Studies. Her research examines the intersections of race, language, and disability to provide a more nuanced analysis of the constructions of ability in mathematics classrooms and education systems. Funded by the National Science Foundation, Mathematics Education Fund, National Endowment for the Humanities, and other agencies, her scholarship is collaborative, building research partnerships with school districts and communities to position student and family community and cultural wealth as central sources of knowledge to create change in mathematics classrooms and schools. Dr. Yeh is the recipient of the 2023 American Educational Research Association Special Interest Group for Research in Mathematics Education (SIG-RME) Early Career Publication Award and the 2022 Association of Mathematics Teacher Educators (AMTE) Early Career Award.

Chapter 1
Introduction: The Imperative of Equity-Oriented, Discipline-Based STEM Education Research

Jennifer D. Adams and Geraldine L. Cochran

Abstract The science disciplines have played a critical role in shaping the experiences of humans and more-than-humans. However, the STEM fields maintain social inequities that limit the participation and perspectives of racialized people. The long-standing issue of inequity in STEM is due to colonial logics based on racial hierarchy; these have not only shaped the disciplines but also the ways that non-normalized bodies are positioned in relation to STEM education and practice. Addressing these issues in ways that disrupt and rebuild equitable STEM requires research and practices that center the experiences and perspectives of the most marginalized groups. Centering diverse marginalized identities and critical perspectives, the chapters in this volume create openings for valuing diverse bodies and minds in STEM as equitable learners and participants. Together these chapters forward Critical Discipline-Based Education Research (Crit-DBER) as a transdisciplinary, desettling, humanizing, and creative framework to approach research and pedagogy in postsecondary STEM.

Keywords Equity · DBER · Critical theory · STEM education · Diverse learners

The science disciplines have played a critical role in shaping our experiences as humans and influencing the lives of all beings as well as the diverse ecosystems and environments we inhabit. Science educators, scholars, advocates, and activists are increasingly recognizing the ways that the science disciplines expanded alongside

J. D. Adams (✉)
Department of Chemistry, Werklund School of Education, University of Calgary, Calgary, AB, Canada
e-mail: Jennifer.Adams1@ucalgary.ca

G. L. Cochran
Department of Physics, The Ohio State University, Columbus, OH, USA
e-mail: Cochran.604@osu.edu

G. L. Cochran, J. D. Adams (eds.), *Equity-Oriented, Discipline-Based STEM Education Research*, Critical Studies of Education 19,
https://doi.org/10.1007/978-981-96-9136-4_1

and in the service of colonialism. This has undoubtedly shaped scientific logics, cultures, and priorities. For a long time, the story of science has been told from a European perspective, established as disciplines in which only Europeans were deemed capable of enacting the logic and reason necessary to produce good and valid science (de Sousa Santos, 2007). Up to the present, these logics have persisted and thus produced an enclosure of science and science education around a culture of White supremacy, which Sylvia Wynter (2003) defines as an "overrepresentation," where the White, cis-gendered, heterosexual, able-bodied, able-minded, and affluent man has been at the apex of *homo sapiens*; prototype human. As such, every human is "tested" against the "standards" of this prototype and the ability of racialized and historically minoritized people—particularly women and nonbinary people—to observe and make inferences about phenomenon has been questioned (Prescod-Weinstein, 2020). Any deviation from this standard is considered less than human, less capable of engaging in the scientific enterprise; this applies to both those who are meant to produce scientific knowledge and also to those who are to "benefit" the most from scientific knowledge production. We see this in the example of Henrietta Lacks, where there has been much scientific knowledge produced from her cells—and those who have had the privilege of knowledge production have benefitted financially, medically, and otherwise. The irony is that those who are closer to the prototype human have benefitted most from a line of cells—HeLa—that came from a Black woman who would be the most deviated from the standard. Henrietta Lacks, her immediate family, and the larger demographic of Black-identified people, and especially Black, non/binary women not only have "benefited" the least, but also are most likely to be harmed (i.e., Buskwofie et al., 2020). Le and Matias (2019) note, "If we continue to ignore whiteness and its operation within science, we essentially, regardless of whether or not we intend to do so, dehumanize our Students of Color and choose to uphold a system that maintains White supremacy" (p. 18–19).

In science education, whiteness has been so embedded and normalized that it has remained undetected; thus, emphasis on diversifying STEM has focused on fixing students to fit the system rather than the other way around. As Leyva et al. (2022) note, the pedagogies in STEM education spaces are structured around ideologies of racism and White supremacy. Learners who meet the standards based on whiteness are more likely to be directed toward gifted programs and specialized schools in ways that offer clear pathways into and through STEM. We have seen where renowned specialized high schools have been called out for a lack of inclusion of Black and Latine students, and those who manage to enter are often faced with discrimination and racism (Corcoran & Baker-Smith, 2018). Historically racialized students who are in mainstream schools also face systemic obstacles such as inadequate science learning resources, emphasis on math and literacy, perceptions about STEM, standardized exams, and biases in college guidance counseling (Wade-Jaimes, 2023; Visintainer, 2020). These factors limit the number of students from these groups who enter college as STEM majors with goals of pursuing STEM-related careers. Those who make it to college must navigate the gauntlet of courses designed to filter students out of STEM as well as perceptions about which bodies "belong" in STEM (Nieminen et al., 2024; McCoy et al., 2017). Bodies that feel excluded due to dehumanizing STEM practices and behaviors are pushed out and those who persist do so carrying much trauma and battle wounds from their journeys (El Halwany & Adams, 2025; McGee, 2021; Riegle-Crumb et al., 2019). Those who remain are often forced to develop strategies of persistence in order to counteract and resist negative experiences, perceptions, and interactions (Morton, 2025; White & Morton, 2024). In other words, STEM education is designed for the success of those bodies that have been positioned at the apex of human, as Wynter (2003) posits. As bodies move further from the standard, the challenges and barriers to STEM increase.

In postsecondary schools, gatekeeping or gateway courses serve as one of the barriers in the pathway (Weston et al., 2019; Harper et al., 2019). These courses are entry-level and siloed in the "core" disciplines and serve as prerequisites—courses that students are required to take before they move into more advanced and specialized studies. These courses are supposed to impart fundamental knowledge about the STEM disciplines, and one would think that these courses would be designed with human learners in mind. In contrast, these courses have been traditionally designed to keep certain learning bodies out of STEM while allowing bodies proximal to Whiteness a gentler pathway into advanced STEM studies and careers. As prerequisites, entry-level courses are structured around providing a large quantity of information in a relatively short amount of time through lecturing. Assessments are reduced to a recall of information. Students are not given the time and space to reflect on concepts and apply them to problem-solving activities. This approach reflects the banking model of education as described by Paolo Freire in *Pedagogy of the Oppressed* (2005), "Implicit in the banking concept is the assumption of a dichotomy between human beings and the world: a person is merely in the world, not with the world or with others; the individual is spectator, not re-creator." (p. 75). In science, this also reproduces the dichotomy between feeling and knowing, further disconnecting students from their desired studies. STEM education spaces are an unfortunate fractal of our larger society; the inequities manifested in STEM education are images of those in larger society.

In order to ensure science equity from education to outcomes, we must approach research in postsecondary science teaching and learning in ways that center the experiences of humans who have been historically closed out of science. Our research should aim to delve deeper into our experiences vis-a-vis the historical, colonial structures that have dominated science and breach the enclosure, replacing its content with plurality. This includes shifting teaching and learning practices in ways that would make science learning environments more expansive and welcoming of all identities, with unharmful (i.e., without racist, homophobic, ableist, sexist, xenophobic, elitist) and justice-oriented perspectives (Cochran, Hyater-Adams, Alvarado, Prescod-Weinstein, and Daane, 2021). This would expand the scientific creativity and innovation available to all of us. This is not only important for those students who wish to pursue STEM-related career pathways, but also for non-STEM students to become STEM-literate, critical thinkers as they engage in civic duties and decision-making. This is even more critical as science becomes increasingly politicized and policies are designed to further enclose STEM opportunities to white, able-bodied men, while STEM Scholars of Color are ousted, defunded, and even threatened for engaging in equity and justice-oriented work (Mervis, 2025; Ng et al., 2025).

1.1 Shifting the Fractal

The Fractal Foundation describes fractals as "infinitely complex patterns that are self-similar across different scales. They are created by repeating a simple process over and over in an ongoing feedback loop. Driven by recursion, fractals are images of dynamic systems" (fractalfoundation.org). In relating this to organizational change, organizer adrienne maree brown (2006) notes, "what we practice at the small scale sets patterns for the whole system." Historically marginalized, equity-focused researchers have been unpacking the ways that Whiteness and coloniality structure and constrict participation toward diversifying STEM. These efforts have been shifting the fractal to create more expansive and inclusive learning cultures in STEM. In this book, we have assembled fractal-disrupting DBER authors, most of whom represent identities that have been historically marginalized or excluded from STEM, all of whom provide embedded ways of knowing, navigating, and shifting STEM education contexts to be more welcoming of diverse bodies and worldviews. DBER has been positioned from a place of scientific neutrality, where identities and bodies do not matter in learning the disciplines. This volume reinserts the human in STEM teaching and learning; our collective voices advance research and practices in postsecondary STEM education that desettle normalized practices by creating spaces for identity affirmation, relationship-building and expanding what it means to do science.

Critical Race Theory (CRT) has been foundational to the movement of addressing systemic oppression in STEM education. In his chapter, Rodriguez describes the history of CRT and explores the major tenets of CRT and how those tenets could apply to DBER. Pierre and Pringle also use a CRT tenet of intersectionality to examine the experiences of Black first- and second-generation immigrants in STEM to point out the importance of not essentializing the experiences of racialized populations, but rather seek the nuances that make experiences unique. Boda also utilizes the CRT tenet of intersectionality to unpack the ways that ableist structures restrict STEM learning opportunities for diverse learners while advocating for less emphasis on disciplinary and individual learning toward a more relational praxis. In their chapter, Torres, Yeh, Tan, Reinholz, and Payan bring together DisCrit and Disability Justice as intersectional lenses to analyze a seminal legal case that demonstrates the intersection of abelism and racism in postsecondary education and offer openings for addressing oppressive foundations in STEM education.

Several chapters forward frameworks that are salient to promoting equity through research and pedagogy. Dei-Sharpe, Crandall, Ford, Jreidini, Ozyonum, and Sidibe apply a Decoloniality, Equity, Diversity, Inclusion, and Belonging (DEDIB) framework to examine and dismantle oppressive systems affecting marginalized groups in STEM research and education. Through deconstructing syllabi, they highlight historical and theoretical biases in the STEM fields and present actionable strategies for instructors in the design of STEM courses. Sachmpazidi's chapter reviews equity-oriented research in STEM graduate education, providing a resource for researchers to grasp its scope and identify gaps in the literature. This chapter provides vital information for STEM actors to enhance graduate education while offering ways forward for STEM equity researchers. Vincent-Ruiz, Hosbein, and Rosa present their chapter for DBER researchers interested in critical quantitative principles (QuantCrit) research. They first historicize the development of statistical methods and highlight how these approaches are counter to equity and justice orientations. Importantly, they emphasize using this critical approach to guide decision-making throughout the entire research process. McDermott presents ableism as a normalized manifestation of Whiteness. To address this, they describe Political Disability Identity (PDI) as a salient framework for DBER that centers the lived experiences of disabled students and promotes inclusive practices in postsecondary STEM. Russo-Tait and Rodgers's chapter explores -blind, meritocratic, and equal opportunity ideologies (all manifestations of Whiteness) work to keep STEM faculty from recognizing injustices in STEM for racialized students. The chapter ends with discussion of the promise of educating faculty with a critical consciousness framework in order to forward equitable STEM education opportunities.

Several works in this volume focus on the less considered yet important aspects of DBER. With language being essential to building complex understanding in STEM, Pitterson and Goldschneider examine the role of everyday language in creating equitable and inclusive STEM learning environments. Armstrong, Gonsalves, Danielsson, and Mckay address inequities in postsecondary physics contexts, highlighting promising theoretical and methodological innovations that emphasize the relationship between people, communities, and institutions, aspects not often considered in postsecondary physics education.

Finally, two chapters describe practical projects that produce equitable STEM learning opportunities for historically marginalized students. Lee, Masters, Holloman, Knight, and Lee present a case study of a 3-year project to transform pedagogical practices in an engineering program. The chapter outlines the design, assessment, and lessons learned from this work. Colon and Bayne describe the development and enactment of a field-based, hands-on ecology course designed to promote greater engagement with and enthusiasm for science among traditionally underrepresented biology majors at a community college. This is juxtaposed with their own narratives, as Women of Color, navigating STEM careers.

Bringing these authors together allows us to begin a conversation around the meanings and applications of equity-oriented DBER. In this volume, we extend the disciplinary considerations to focus on the historical, social, and political contexts of postsecondary STEM education. These collective works encompass Critical DBER (Crit-DBER):

- Transdisciplinary—integration of multiple lenses and disciplines to understand the phenomena of diversity, equity, and justice in STEM education.
- Desettling—aims to identify harmful normalized practices toward transforming pedagogical and interactional experiences especially for historically marginalized groups.
- Humanizing—recenteres humans in STEM learning and endeavor; recognizes that identity is an important part of STEM learning and participation.
- Creative—seeks new frameworks, methodologies, and projects to expand meaningful participation in STEM teaching, learning, and practice.

With many pressing issues facing our collective well-being and survival, we view this volume as a beacon of hope that will inspire research and pedagogical practices that will leverage the gifts that diverse learners bring to postsecondary STEM contexts and in their subsequent careers. Furthermore, with contemporary threats to Black, Brown, women, and other groups historically excluded from STEM on many levels and "heightened concerns for social, emotional, psychological, and physical safety" (Morton, 2025), it is important for us to double-down in our pursuit of equity in STEM. As scholar-activist Dr. Terrell Morton (2025) notes, and we extensively quote, "[*these threats and the people perpetuating them] are not stronger than my faith, my determination to live out my purpose, my passion and commitments to racial justice, my love for and joy in being Black, and my dream of having these lands that we currently call the United States and all lands globally be spaces where Black people can just be" (n.p.).* We are encouraged to remember the boldness, courage, and "informed actions" of our ancestors as well as our responsibility to ensure that our future generations not only have rights and a government that serves the people but also that "they will be able to have access to their histories and stories, have social institutions that value, embolden, and advance their needs, desires, stories, and freedom dreams" (Morton, 2025). This is our clarion call as we move forward with the dreams of our ancestors for STEM futures full of creativity and joy.

References

Buskwofie, A., David-West, G., & Clare, C. A. (2020). A review of cervical cancer: Incidence and disparities. *Journal of the National Medical Association, 112*(2), 229–232.

Cochran, G. L., Hyater-Adams, S., Alvarado, C., Prescod-Weinstein, C., & Daane, A. R. (2021). Social justice and physics education. In *Teaching and learning for social justice and equity in higher education* (pp. 125–147). Palgrave Macmillan.

Corcoran, S. P., & Baker-Smith, E. C. (2018). Pathways to an elite education: Application, admission, and matriculation to New York City's specialized high schools. *Education Finance and Policy, 13*(2), 256–279.

de Sousa Santos, B. (2007). Beyond abyssal thinking: From global lines to ecologies of knowledges. *Review (Fernand Braudel Center), 30*, 45–89.

El Halwany, S., & Adams, J. D. (2025). *Affective politics of belonging to STEM: Some conceptual and methodological considerations*. Science Education.

Fractal Foundation. (2006). *Fractal Foundation: Fractals are SMART: Science, Math and Art!* Fractal Foundation. https://fractalfoundation.org

Freire, P. (2005). *Pedagogy of the oppressed (30th Anniversary Edition).* The Continuum International Publishing Group

Harper, R. P., Weston, T. J., & Seymour, E. (2019). Student responses to problematic STEM teaching methods. In *Talking about leaving revisited: Persistence, relocation, and loss in undergraduate STEM education* (pp. 149–195). Springer International Publishing.

Le, P. T., & Matias, C. E. (2019). Towards a truer multicultural science education: How whiteness impacts science education. *Cultural Studies of Science Education, 14*, 15–31.

Leyva, L. A., McNeill, R. T., & Duran, A. (2022). A queer of color challenge to neutrality in undergraduate STEM pedagogy as a White, cisheteropatriarchal space. *Journal of Women and Minorities in Science and Engineering, 28*(2).

McCoy, D. L., Luedke, C. L., & Winkle-Wagner, R. (2017). Encouraged or weeded out: Perspectives of students of color in the STEM disciplines on faculty interactions. *Journal of College Student Development, 58*(5), 657–673.
McGee, E. O. (2021). *Black, brown, bruised: How racialized STEM education stifles innovation.* Harvard Education Press.
Mervis, J. (2025, May 16). NSF's grant cuts fall heaviest on scientists from underrepresented groups. https://www.science.org/content/article/nsf-s-grant-cuts-fall-heaviest-scientists-underrepresented-groups
Morton, T. R. (2025, 2 June). This joy that I have *The Advance Journal*. https://www.advancejournal.org/post/3240
Morton, T. R. (2025). But I just want to be free: A critical race theory-phenomenological variant ecological systems theory analysis of blackness and STEMM in Canada. *Canadian Journal of Science, Mathematics and Technology Education*, 1–15.
Nieminen, J. H., Reinholz, D. L., & Valero, P. (2024). Mathematics is a battle, but I've learned to survive: Becoming a disabled student in university mathematics. *Educational Studies in Mathematics, 116*(1), 5–25.
Ng, E., Fitzsimmons, T., Kulkarni, M., Ozturk, M. B., April, K., Banerjee, R., & Muhr, S. L. (2025). The anti-DEI agenda: Navigating the impact of Trump's second term on diversity, equity and inclusion. *Equality, Diversity and Inclusion: An International Journal, 44*(2), 137–150.
Prescod-Weinstein, C. (2020). Making Black women scientists under white empiricism: The racialization of epistemology in physics. *Signs: Journal of Women in Culture and Society, 45*(2), 421–447.
Visintainer, T. (2020). I think at first glance people would not expect me to be interested in science: Exploring the racialized science experiences of high school students of color. *Journal of Research in Science Teaching, 57*(3), 393–422.
Riegle-Crumb, C., King, B., & Irizarry, Y. (2019). Does STEM stand out? Examining racial/ethnic gaps in persistence across postsecondary fields. *Educational Researcher, 48*(3), 133–144.
Wade-Jaimes, K. (2023). A critical race theory analysis of STEM schools in an urban school district. *Science Education, 107*(5), 1193–1214.
Weston, T. J., Seymour, E., Koch, A. K., & Drake, B. M. (2019). Weed-out classes and their consequences. In *Talking about leaving revisited: Persistence, relocation, and loss in undergraduate STEM education* (pp. 197–243). Springer International Publishing.
White, A. M., & Morton, T. R. (2024). Born into the wrong family: Navigating being a black male in STEM. *Journal of Women and Minorities in Science and Engineering, 30*(1).
Wynter, S. (2003). Unsettling the coloniality of being/power/truth/freedom: Towards the human, after man, its overrepresentation—An argument. *CR: The New Centennial Review, 3*(3), 257–337.

Chapter 2
Liberal Ideologies in Academic STEM: How Faculty Can (Re)produce Racial Inequities

Tatiane Russo-Tait and Aireale J. Rodgers

Abstract This chapter traces how dominant ideologies of meritocracy, equal opportunity, and color-blindness manifest in everyday STEM teaching, learning, and decision-making. With an emphasis on faculty practice, we demonstrate how well-intentioned actions that center these ideologies perpetuate inequity (an unjust and ahistorical distribution of resources and opportunities) that disadvantages racially minoritized students. The chapter starts with an overview of how ideologies become entrenched and operate within higher education. We then outline how hegemonic liberal ideologies inform practices that maintain racial inequities in STEM. Finally, we discuss the concept of critical consciousness and its potential application in faculty development, research endeavors, and disciplinary advancement toward a more equitable STEM education.

Keywords STEM faculty · Liberal ideologies · White supremacy · Critical consciousness

T. Russo-Tait (✉)
University of Georgia, Athens, GA, USA
e-mail: tati@uga.edu

A. J. Rodgers
University of Wisconsin-Madison, Madison, WI, USA
e-mail: arodgers4@wisc.edu

G. L. Cochran, J. D. Adams (eds.), *Equity-Oriented, Discipline-Based STEM Education Research*, Critical Studies of Education 19,
https://doi.org/10.1007/978-981-96-9136-4_2

Dr. A draws on their science communication training to teach. They take care to prepare informative slides and believe they must give lectures that don't drone on. They admit they have seen new approaches to teaching but do not feel the need to change their own teaching.

Dr. A strives to be color-blind and believes that they treat all students as equally and fairly as possible. They make it explicit that if students don't feel that there is an environment of equality in their classroom, they should speak up if somebody is preventing them from having that experience. They worry about explicitly naming identities (e.g., race or gender) because they think it makes students self-conscious. Instead, they share that:

My approach is to try to treat everybody fairly. What I do is I tell the students I want everybody to ask questions... Nobody should be afraid to ask a question in this class. And actually, if you don't ask a question, I may ask you. And also, everybody has to get up and present, and sometimes I get feedback from students that they don't like that. It scares them too much. But the Hispanic students, the Black students, and White students—they all got to do the same thing.

Dr. A is mindful of struggling students and sometimes reaches out and invites them to come to office hours. They report, however, that most students do not take them up on that opportunity. While they are happy to make exceptions to meet with students who must work outside of their office hours, they do not feel comfortable extending deadlines because it would not be fair to the rest of the students. When discussing the underrepresentation of Black, Indigenous, and Latiné students in science, they lament:

[I]n many ways, by the time they get to college we can try our best to make them feel confident and feel that they belong... How much does it do? How well they are prepared to understand my lectures was set many years ago—at least by high school... You really have to solve the problem way back. My stance is that it's not because they feel, let's say they are Black or Hispanic, 'I don't belong, because people will discriminate me.' I think a lot of it has to do with how well they perform, because of past experience.

As such, they tried to support these students by providing out of class study sessions:[F]or the students who were not well prepared, I got a bunch of volunteer undergrads...who agreed to make themselves available to tutor these kids. I make that resource available... I tell students... if you're being offered some options here, it's up to you whether you choose to take them or not. That's your choice. You might say, "This is not important for me. Therefore, I choose not to do it." Then, you own the consequence for that action as well.

At the end of the day, Dr. A says, it's incumbent upon individuals to take advantage of opportunities to the best of their abilities to gain the result that they want.

This vignette may look and feel familiar to many science educators. Some may relate to Dr. A or know colleagues who engage in similar practices. Others may describe this as a regular day in a relatively inclusive classroom led by a professor with progressive ideals. Still, others may feel uneasy with the approaches discussed. We invite readers to take a moment to reflect on how this vignette may or may not align with their own teaching philosophy.

This vignette is an abridged version of a composite narrative about faculty conceptions of equity (Russo-Tait, 2025). A composite narrative is a story that describes research findings, such as themes or patterns within different groups of participants, providing readers with "a picture of the group as a whole" for each theme identified (Willis, 2019, p. 473). This composite narrative was based on results of an exploratory study that examines college science faculty members' conceptions of equity and its relationship to teaching practices—specifically that of twenty-four participants who conceptualized equity as "equality" (Russo-Tait, 2023, 2025). Faculty with this conception of equity shared an assumption that all students, regardless of social identity, had an equal opportunity to succeed in their science, technology, engineering, and math (STEM) classrooms through hard work and merit.

We open with this composite narrative because while it may appear to portray a relatively progressive STEM classroom that is not actively harming students, the practices within them, and the ideologies that inform those practices, are indeed problematic and can work to reproduce inequities for minoritized students. Specifically, the habits of faculty sensemaking described in the narrative represent how liberal ideologies of meritocracy, equal opportunity, and color-blindness creep into and become manifest in everyday, taken-for-granted teaching and learning. Although faculty may be well-intentioned in their pursuit of equality, these practices have been shown to hinder the learning and success of students from marginalized backgrounds, including those who are racially minoritized (NASEM, 2016, 2023). We will explore this phenomenon, whereby well-meaning faculty, drawing from dominant narratives in society and STEM, engage in exclusionary practices in their classrooms and other spheres of control and influence. Because most faculty are socialized in the culture of STEM and are not given the opportunity to critically reflect on the contradictions inherent in the ideological discourses and narratives espoused in that racialized space, they can develop implicit ways of thinking that can lead to harmful practices. These practices range from curricular and instructional choices to mentoring and admissions decisions, all of which work to simultaneously exclude Students of Color and legitimize racial injustices. Further, these practices lead to outcomes that reify narratives that blame people from racially minoritized backgrounds for their exclusion in STEM spaces—while also absolving faculty from the responsibility to ameliorate these racial inequities.

In this chapter, we will review the current literature exploring how meritocratic, equal opportunity, and color-blind ideologies can inform faculty practices. Color-blindness refers to an ideology used to dismiss contemporary forms of systemic racism and rationalize any racial disparities as being due to other factors, unrelated to racial discrimination (Bonilla-Silva, 2006/2018). Equal treatment and opportunity ideals refer to the belief that we now live in a society where everyone has the same opportunity to succeed, and that equal treatment is the only fair approach to address inequality (Secada, 1989). Finally, meritocratic ideals assume a system where everyone can succeed, regardless of background, through individual talent, hard work, and merit. These ideologies can inform STEM faculty thinking about fairness, equity, and justice, and keep them from recognizing—and even allow them to reproduce—the unjust educational experiences and outcomes of students from underrepresented, racially minoritized backgrounds (i.e., Black, Indigenous, and Latiné students). Adherence to these ideologies can also lead to well-intended attempts to "help" racially minoritized students via deficit framings that seek to fix students as opposed to addressing structural barriers. In all, these principles, logics, discourses, and related norms and practices deny the complexities of the systemic barriers faced by racially minoritized students and make it so that faculty members are unable to see their own complicity in perpetuating racial injustice. On the other hand, studies also show that faculty who exhibit a critical consciousness—an understanding and critical analysis of systemic injustice, as well as motivation, agency, and actions to address it—tend to not subscribe to these ideals and see themselves as change agents that have the power to subvert and disrupt racial oppression through specific practices in their spheres of control and influence.

By focusing on faculty thinking and practices, it is not our goal to individualize a systemic problem. Instead, as McNeill and colleagues argue, it is vital to explore faculty perspectives because they "can reveal how ideologies… are resisted or left unchallenged" (2022, p. 3). Science faculty make up a significant group of decision-makers and stakeholders in a system designed to exclude racially minoritized students. By adhering to or failing to notice and disrupt problematic liberal ideologies, they are complicit in perpetuating racial injustice and White supremacy in STEM fields. As such, faculty constitute an important lever for change. The power of ideologies has been implicitly and explicitly explored in academic STEM concerning instruction and class climate, mentoring, admissions, hiring, and leadership. This chapter begins with a brief discussion of how ideologies become hegemonic and how they operate in the context of higher education. We then present a more in-depth exploration of how meritocratic, egalitarian, and color-blind ideologies play a role in maintaining racial inequities in academic STEM, specifically via faculty thinking and practices. The chapter ends with a discussion of critical consciousness and how this concept could be leveraged in faculty development and future research. In examining these ideas, we address some of the central questions framing this volume: "How do we identify and disrupt the teaching, learning, and practice norms in STEM fields in order to make them broader and more inclusive?" and "What can we learn from critical frameworks and research in terms of transforming postsecondary STEM teaching and learning?" We invite DBER scholars and STEM faculty to critically reflect on these important issues, our roles in disrupting them, and engaging in more equitable practices.

2.1 Hegemonic Liberal Ideologies in Higher Education

Ideology has been defined in various ways in the social sciences (Gordon & Noga, 2015). For the purposes of this chapter, ideology refers to a "systematic and integrated set of beliefs whose primary function is explanation." Ideology frames worldviews, values, norms, and practices and helps people to explain and justify the causes of specific behaviors or phenomena. During the process of socialization—when people learn to conform to social norms and become members of a particular group—people are implicitly and explicitly introduced to ideological principles, logics, and discourses that help explain how and why things work the way they do in that context (Gordon & Noga, 2015). These ideologies then become part of people's mental models, shaping their implicit understandings about the world around them. As these ideologies become part of people's individual and collective consciousness, they too are (re)produced as people engage in daily activities and routine cultural practices. For example, it is well documented that meritocracy is a dominant ideology that influences and rationalizes opportunity distribution and resource allocation in higher education (Liu, 2011)—from student admissions through faculty tenure and promotion decisions.

The power of ideology lies heavily in its pervasiveness. As Hall (1982) explains, ideology functions as the presupposed "grammars of culture," functioning as the ways through which our beliefs, practices, and rituals give rise to (or potentially challenge) our understanding of the world. Relying on Gramsci (1971), Hall (1996) emphasizes the disciplining character of ideology. He argues that ideology operates as a shared mental framework that "reconcile[s] and accommodate[s] the mass of the people to their subordinate place in the social formation" (p. 27). That is to say, dominant ideologies become hegemonic and mediate people's understanding of systems of power in society. Antonio Gramsci coined the term "cultural hegemony" to describe how those in power can exercise control in ways that do not involve coercion or violent means. Instead, people in power create and disseminate discourses or narratives that protect their positions by explaining away inequities and normalizing oppression via half-truths or false information (Gordon & Noga, 2015). Much like any ideologies, these ideas are disseminated via socialization processes happening in schooling, media, church, etc., and can be assimilated consciously or unconsciously by members of a society (Kosut, 2012). Ideologies become hegemonic when enough people come to accept these narratives as taken-for-granted realities and common-sense explanations for power differentials and social inequities (Philip, 2011). People and institutions then protect these ideas by claiming them to be inevitable or even natural and by rejecting or ignoring alternative explanations (Kosut, 2012). Ironically, then, ideology both reflects and structures social relations all while tacitly legitimizing them. Thus, hegemonic ideologies are vital to the (re) production of social inequities and oppression (Bonilla-Silva, 2006/2018).

Our previous example of meritocracy as a dominant ideology is a revelatory example of cultural hegemony at work in service of White supremacy. Bonilla-Silva (2006/2018) argues that while liberal ideals have at times supported a progressive political agenda, these ideologies were reshaped in the post-civil rights era to justify contemporary racial inequities and blame racially minoritized people for racial disparities in society (p. 82), including in the U.S. educational system. The narratives engendered by these ideologies claim that all people in the United States have an equal opportunity to succeed in their educational pathways through individual effort and ability (i.e., merit), regardless of their identity (e.g., race). Yet, this ahistorical frame suffers from what Hall (1982) calls the "reality effect" (p. 74), wherein ideological claims are taken-for-granted as intrinsic, concrete truths rather than artifacts of sociohistorical conditions and relations (Makus, 1990). Narratives forwarding meritocracy conceal how discriminatory practices and oppressive forces act as structural barriers for people from racially minoritized backgrounds—and, therefore, work to maintain an illusion of White supremacy in society. For example, Bonilla-Silva (2006/2018) found that White people opposed affirmative action policies by invoking abstract liberal notions of equal opportunity (everyone has a fair chance), meritocracy (the most qualified applicant must win), and individualism (any group-based policies are unfair, and everyone should be judged as individuals). While these claims appear reasonable and even fair to some, they act to individualize a social problem—that is, working from an illusion of meritocracy obscures the relational and structural dynamics that systematically disenfranchise People of Color, such as legacies of exclusion and discrimination in job, housing, and educational systems, to name a few (Bonilla-Silva, 2006/2018, p. 80). In this way, these liberal ideologies and related discourses explicitly dismiss the existence of systemic racism and therefore work to perpetuate a system of racial hierarchy by blaming the individuals positioned at the bottom of the hierarchy for their own oppression.

Debunking the myth of meritocracy—along with other dominant manifestations of cultural hegemony—can do important historicizing work. Dowd and Bensimon (2015) posit that higher education institutions (and their actors) suffer from "historical amnesia" (Taylor, 2009, p. 7). This amnesia absolves the need to grapple with how the development and expansion of the U.S. American higher education system was dependent on the expropriation of Native land, the exploitation of enslaved people's labor, and the continued exclusion and disenfranchisement of Communities of Color across the country (Wilder, 2013). By neglecting to contend with these histories and their afterlives and instead espousing narratives that "propagandize … 'liberty and justice for all,'" Predominantly and Historically White Institutions (PWIs and HWIs) use liberal ideologies and related discourses in ways that appear fair and moral but that instead protect the status quo (Dancy et al., 2018, p. 178).

These taken-for-granted yet highly problematic ideologies in U.S. society and PWIs/HWIs are further amplified in academic STEM, a group of fields that uphold Whiteness and objectivity as standards (Blair-Loy & Cech, 2022; Harding, 2006; Leyva et al., 2022; McGee, 2020b; Posselt, 2020). STEM—as it is often practiced in higher education—is a White supremacist institution; that is, it has a system of beliefs, norms, and practices that privilege and center Whiteness and White culture while dehumanizing non-White people and devaluing non-White cultures and epistemologies (Battey & Leyva, 2016; Callwood et al., 2022; McGee, 2016, 2020b; Morton, 2022). Because scientists, mathematicians, and engineers are socialized to believe that STEM disciplines are objective, neutral, and fair, the taken-for-granted assumptions are that everyone has equal opportunity to succeed in STEM fields through individual effort and merit, regardless of social identity. As discussed, these beliefs dismiss systemic racism and, in the case of STEM, directly undermine the evidence that racially minoritized people have been consistently denied equitable access, opportunities, and outcomes in STEM education and professions (Blair-Loy & Cech, 2022; McGee, 2020a; Morton, 2022; Riegle-Crumb et al., 2019). While the problematic nature of meritocratic, equal opportunity, and color-blind ideologies has been discussed at length at the university level, it was not until recently that scholarship has focused on how these ideologies are present and even reproduced by the work STEM college faculty do in postsecondary institutions (e.g., McNeill et al., 2022; Robertson et al., 2023; Russo-Tait, 2022, 2023, 2025). We argue that STEM faculty can and do, often inadvertently, act as agents that (re)produce White Supremacy and racial injustice in STEM education. By extension, we believe that faculty must learn to recognize and disrupt cultural hegemony in their research, teaching, and service.

To reiterate, this chapter does not aim to put the onus solely on faculty members to solve a systemic problem—but faculty is nevertheless a particularly important group to consider because their impact is far reaching. Most STEM faculty members, regardless of position and rank (i.e., contingent/adjunct, tenured/tenure track), teach and advise hundreds, if not thousands, of students during their careers. Further, many tenured/tenure-track faculty train and mentor graduate students and postdocs (some who will become faculty themselves). They may also vote on admissions, hiring, and promotion/tenure committees. Finally, many faculty members participate in shared governance and make up a large portion of the pool of future university administrators and policymakers. Therefore, faculty members have the collective power to push for change and engage in efforts to transform STEM culture, climate, and policies. Alternatively, and more often the case, they can uphold the status quo via their (in)action. The following sections explore the extant scholarship demonstrating how hegemonic ideologies in STEM can lead faculty to perpetuate racial and other injustices.

2.2 Hegemonic Ideologies in STEM

To study the history of STEM research—and by extension, teaching and learning—in formal schooling environments is to undoubtedly be confronted with the myriad ways histories of racism, sexism, colonialism, and imperialism (among other atrocities) have been deeply intertwined with the development of STEM disciplines (Harding, 2006; Rudolph, 2002; Vakil & Ayers, 2019; Vossoughi & Vakil, 2018). From the legacies of scientific racism to routine and systematic socialization into objectivity and neutrality as prerequisites for "good" science, the disciplinary logics, or system of shared assumptions guiding who is considered legitimate and meritorious in a discipline (Posselt, 2016), reflect the hegemonic ideologies of meritocracy, egalitarianism, color-blindness that help to perpetuate White supremacy in STEM higher education. Although hegemonic ideologies might lead us to believe that "good" science is apolitical, the truth of the matter is that STEM faculty and students are constantly engaging with political sensemaking. Studies of learning in college STEM classrooms have highlighted a number of political and ethical dilemmas in teaching and learning, such as debates about the use of militarized drones (Philip et al., 2018), considerations of the continued use of HeLa cells harvested nonconsensually from Ms. Henrietta Lacks (Virtue et al., 2018), the macroethics of engineering education (Gupta et al., 2019; Morgan et al., 2020), and the imperative of scientists to engage in policy and advocacy work (Gough, 2015). This contradiction—that STEM educators strive to be neutral while simultaneously engaging with the political dimensions of the work—is a noteworthy one because it exposes how these liberal ideologies promote the viewing of successes and failures through lenses that negatively affect faculty's ability to perceive bias or discrimination. Further, these beliefs lead faculty to engage in such bias and discrimination in instruction, advising, mentoring, and admissions and hiring decisions (Burt et al., 2018; Posselt, 2016, 2020; Russo-Tait, 2023). Below, we discuss research concerning how each of the three ideologies play a role in perpetuating racial inequities. It is important to note that these liberal ideologies are inextricably linked and should be understood as collectively contributing to the same neoliberal project. Yet, we divide them in separate sections in this chapter because extant research tends to focus on one or two of them at a time; furthermore, this structure can provide clearer examples of how these related but distinct ideologies manifest in faculty beliefs and practices in ways that perpetuate racial inequities in STEM education.

2.2.1 Meritocratic Ideology

As we have detailed, a dominant ideology that structures social relations in STEM higher education is meritocracy. Scholars have discussed the power of meritocratic narratives in the perpetuation of inequities for men and Women of Color, White women, and LGBTQIA+ people in academic STEM, to name a few (Battey & Leyva, 2016; Blair-Loy & Cech, 2022; Leyva et al., 2022; Posselt, 2020). Meritocratic discourses in STEM are also expressed in conjunction with claims of objectivity and neutrality. These ideals have been heavily critiqued by feminist and critical race scholars in various fields, as related narratives work to protect White supremacy and other systems of oppression via discourses that dismiss historical and sociopolitical power and privilege dynamics in science (Battey & Leyva, 2016; Harding, 2006; McGee, 2020b; Powell et al., 2021).

Meritocratic discourses are particularly problematic for multiply minoritized people in STEM since meritocratic narratives dismiss multiple axes of oppression simultaneously and ignore the compounding injustices people from those backgrounds experience in STEM spaces (McGee, 2020b; Morton et al., 2019; Leyva et al., 2022). Meritocratic logics may appeal to scientists from dominant backgrounds in STEM because these fields are challenging and require concerted effort and dedication to persist. Therefore, faculty members from dominant backgrounds who are not disadvantaged by racialized exclusion and other structural barriers may assume that those who do not make it in STEM failed to do so due to a lack of effort. Moreover, this worldview affirms their own sense of self. If they thought otherwise, it would not only risk dismissing their own efforts, but it would also lead to the recognition that they may have been provided with unfair advantages (the absence of sexist and racist barriers, i.e. White male privilege). This recognition can be disequilibrating. Therefore, meritocratic ideology and related discourses can, at worst, help faculty to protect their privilege by rationalizing inequities and, at best, keep them from being able to identify existing barriers for people from racially minoritized and other marginalized backgrounds.

In the opening narrative, we illuminated how meritocratic ideals play a role in the assumptions well-meaning faculty make about their Students of Color. Faculty may believe that they are engendering a meritocratic classroom environment where all their students have the same opportunity and treatment, regardless of identity, and they all can succeed through their efforts. However, this leads to the assumption that students who are struggling are either lacking in preparation, do not have the ability to learn, or choose not to take advantage of opportunities offered. As a result, these students are not meritorious. These meritocratic logics play out in how STEM faculty make sense or even perceive systemic inequality. Thus, it affects whether they can identify and ameliorate issues they have the power to address.

Cech and colleagues (2018) documented how meritocratic logics can indeed inform STEM faculty's perceptions of inequities. They surveyed over 250 faculty members in STEM departments at a research-intensive PWI. First, they found that faculty who possessed at least one minoritized identity (i.e., LGBT, women, and/or People of Color) reported more personal experiences of marginalization at work and tended to notice more "chilly climates." They then determined if faculty relied on "meritocratic schemas" (discourses emphasizing merit and blaming one's lack of effort, individual motivation, or preparation for unequal outcomes) or schemas regarding structural barriers (acknowledging systemic discrimination and marginalization). After controlling for race/ethnicity, gender, and sexuality, as well as reported personal experiences of marginalization at work, faculty who used meritocratic schemas were significantly less likely to recognize chilly departmental climates and professional cultures than those relying on structural schemas. The authors concluded that faculty who use meritocratic framings to explain inequities were less likely to identify exclusionary issues and could be less disposed to support diversity efforts in the university and their classrooms (Cech et al., 2018). These findings also challenge the notion that all people from minoritized backgrounds automatically subscribe to structural explanations of inequities by virtue of their positionality and lived experiences. It instead highlights the power of meritocratic hegemonic ideology in STEM and how anyone can subscribe to them, regardless of social identity and lived experiences, and thus contribute to the maintenance of unjust systems.

In a case study of the perspectives of STEM faculty members, which surveyed over 250 and interviewed 85 participants, Blair-Loy and Cech (2022) show that taken-for-granted meritocratic ideals were the foundation of two types of hegemonic schemas that (re)produce inequities in STEM fields: work devotion and scientific excellence (both of which have been connected to the culture of Whiteness in STEM by critical scholars). The work devotion schema defines "STEM work as a calling of single-minded devotion, undistracted of other life responsibilities"; while the scientific excellence schema consists of a "constellation of characteristics that serve as a cultural yardstick for measuring up the competence and worthiness of one's professional colleagues and oneself" (Blair-Loy & Cech, 2022. p. 13). They found that the devotion schema (re)produces inequities by allowing others to judge faculty with caregiving responsibilities as less devoted to their work, and therefore be seen as less committed and productive, even when traditional metrics of success (e.g., publications and grant funding) show otherwise. This schema allows others—usually White males without caregiving responsibilities—to be perceived as more devoted and therefore also be seen as more meritorious, even though they are not more productive. The scientific excellence schema reproduces inequities because the characteristics believed to symbolize a competent scientist (i.e., assertiveness, brilliance, creativity, neutrality) are gendered and racialized. That is, heterosexual White men are implicitly perceived to embody these characteristics and are then judged as more meritorious when compared to White women, men and Women of Color, and LGBTQAI+ faculty. When people from these marginalized backgrounds in STEM attempt to embody these characteristics, they are scrutinized and judged harshly by others (e.g., Black and Latiné women are criticized for being assertive). Thus, these schemas work to bias STEM faculty's judgment and lead to the systematic exclusion of current and prospective faculty and students who do not conform to racialized hegemonic heterosexual masculinity (Blair-Loy & Cech, 2022).

Relatedly, Posselt (2016) explored the taken-for-granted inner workings of graduate school admissions. By observing admissions committee meetings and interviewing faculty in 10 highly ranked programs, including some in STEM, she found that even though there are mixed results regarding the ability of the GRE to predict graduate student success, faculty often used GRE scores as an "objective" measure of merit. Further, faculty members often inferred beyond the reasoning and math skills these scores are supposed to measure and presumed intelligence, risk-taking, and belonging from those scores (Posselt, 2016). Lower GRE scores were then used to rationalize and normalize the low number of Black graduate students in the institution as an issue of individual student performance. Moreover, in an ethnographic comparative case study between departments attempting to change their graduate admissions process, Posselt (2020) found that faculty members who held on to traditional measures of merit such as GRE scores and GPA (and other assessment methods wrapped in racialized and gendered meanings) failed to diversify their department's graduate student body.

In summary, meritocratic logics work to maintain STEM as a White heterosexual cis-male space, under the guise of superior ability and effort, taking place in an objective, neutral, and apolitical context. This ethos is powerful in STEM because hard work and dedication are requirements for success in these fields and, coupled with the presumption of objectivity and neutrality, it may be easier for those in privileged positions to assume that failures indicate a lack of ability, motivation, or effort—and therefore merit. Further, socialization into this ideology is so powerful that those who belong to minoritized backgrounds also face the risk of subscribing to these ideals, regardless of their own personal experiences of exclusion (Cech et al., 2018; Doerr et al., 2021; Riegle-Crumb et al., 2022). Ultimately, meritocratic ideals uphold several axes of oppression and social injustices in STEM, including racial inequities, because it frames STEM as a place where anyone can succeed if they try hard enough, regardless of social identity. This masks how structural barriers operate and allow hegemonic beliefs and biases affect the ability of those in power to be truly objective and judge others' merits fairly—which, again, perpetuate inequities.

2.2.2 *Equal Treatment and Opportunity Ideology*

Meritocratic ideologies are presumed to operate in a context where all people are treated equally and have equal opportunities to succeed. Thus, it is important to discuss the taken-for-granted assumption of STEM education as a space with equal opportunity and treatment for all. This ideology is often favored as part of the solution to racial inequities and injustices by most PK-12 educators, administrators, and policymakers (Gutierrez, 2002; Secada, 1989). This ethos is primarily concerned with whether students are given equal access and resources (Gutierrez, 2002; Jordan, 2010; McNair et al., 2020; Secada, 1989), and it does not account for different outcomes, so long as everyone receives the same treatment and opportunities (Jordan, 2010; McNair et al., 2020). As such, this egalitarian logic and discourse ignores that students from different backgrounds do not start their college education on "a level playing field." Therefore, erroneous rationalizations for racial inequities maintain that if all students are treated equally and provided with the same resources and opportunities, disparate outcomes in achievement and attainment must then be due to individual choices and behaviors, such as lack of interest, ability, motivation, or effort. This framing positions students as deficient while dismissing systems of oppression.

The opening composite narrative exemplifies how faculty with "equality" conceptions of equity engage in practices that appear fair but could be disproportionally affecting students from minoritized backgrounds. For example, the assumption that all students have an equal opportunity to attend office hours, or an out-of-class review session is flawed. Some students may be caregivers or must work and thus, their "choice" to not attend has nothing to do with lack of interest or motivation to improve their learning. In addition, research has shown that Students of Color may experience exclusionary interpersonal interactions with peers and teaching assistants. Therefore, their choice to not attend can be informed by that constraint which, once again, does not indicate that a student believes that reviewing material is not important (Russo-Tait, 2023). The related idea of equal treatment, or treating all students the same, is also pervasive in education. What this approach belies is that because students are not starting out in a level playing field, treating them the same way will only lead to the maintenance of disparities. For example, by not being flexible with deadlines due to a concern for treating all students equally, instructors are assuming that all students have the same amount of time to dedicate to the course. This will, in turn, privilege students from higher socioeconomic backgrounds who do not have to work; and those who come from cultures that value individualism (instead of communal values that center family and community responsibilities) and happen to have more time to dedicate to school—values that align with the White supremacist and heteropatriarchal culture of science (Basile & Black, 2019; Miller et al., 2021; Grindstaff & Mascarenhas, 2019; McGee, 2016).

Equal treatment and opportunity discourse and logics allow educators to believe that they are being equitable and just by treating every student in the same way. It also keeps educators from considering that they may be promoting inequitable outcomes themselves, as this type of belief leads one to not recognize their own implicit biases, which position students as intellectually inferior and lead to lowered expectations, increased scrutiny, and treatment that can create a "self-fulfilling prophecy" (Rubie-Davies, 2015). Further, it keeps faculty from recognizing the need to provide differentiated support for students with different needs. For example, faculty with equal treatment beliefs may not think about scaffolding materials to give students with different levels of preparation different entry points and opportunities to engage with the material and build on their prior knowledge (Russo-Tait, 2023). Importantly, research has shown that educators who subscribe to this equal treatment and opportunity ethos also engage in deficit thinking of Students of Color, privilege dominant Western epistemologies, dehumanize the learning environment, and engage in other dynamics that marginalize minoritized students despite equal access and opportunities (Gutierrez, 2002; Hand et al., 2013; Nasir & Hand, 2006; Russo-Tait, 2023; Valencia, 1997).

As mentioned, Russo-Tait (2023) explored how college science faculty conceptualized "equity" and how their conceptions were related to their teaching practices. The most common conception of equity was "equality," defined as providing all students with equal treatment and opportunity. This egalitarian conception of equity was associated with instructor-centered teaching, as well as other practices and beliefs shown to negatively affect student learning, interest, and identification with STEM. Faculty with this conception, for example, reported teaching mostly via lecture, a practice known to be exclusionary (Freeman et al., 2014; Handelsman et al., 2022). These equal treatment and opportunity beliefs kept faculty from attending to class climate and fostering equitable student participation, because they assumed that simply treating all students the same was enough to create a fair system. Further, faculty with equality conceptions tended to use deficit-based explanations when discussing students who were struggling. Because these faculty members assumed that everyone had equal opportunity to succeed in their courses, if some students did not do well, it was because of their own lack of effort, motivation, or preparation.

While this egalitarian ideology appears to be fair and just in theory, in practice, it perpetuates unjust outcomes because it focuses on providing equal treatment and opportunity to people who enter higher education already having experienced inequitable treatment, opportunities, and outcomes. This ethos overlooks the historical inequities that place students from racially minoritized backgrounds at a disadvantage from the start and ignores the multiple contemporary systemic barriers they must navigate within White supremacist higher education settings—on top of the already challenging nature of attaining a STEM degree. Therefore, providing racially marginalized and privileged students with the same treatment and opportunity without accounting for historical injustices that created the contemporary system of racial inequity simply works to keep the racial structure unmarked and unchallenged. That is, it ascribes the "success" of White students as being achieved by merit while also justifying a racial hierarchy and reifying White supremacy. Put simply, equality is not equity, and equality is not always just.

In sum, ideals of equal opportunity and treatment in an unequal system, in conjunction with meritocratic logics allow STEM fields to claim neutrality and objectivity, and therefore fairness, while actively excluding students from racially minoritized backgrounds. Bonilla-Silva (2006/2018) contends that these and other ideologies have coalesced into a new racial ideology, post-civil rights—color-blind racial ideology.

2.2.3 *Color-Blind/Color-Evasive Racial Ideology*

Race is a social construct (i.e., not an innate trait or biological reality). Therefore, its meanings have changed over time and place (Bonilla-Silva, 2006/2018; Omi & Winant, 2015). In the United States, the concept of race was created by colonizers for the purpose of exploitation and power. That is, the social construction of race was necessary for White colonizers to dehumanize Indigenous and African peoples to justify genocide and enslavement. White supremacy-centered Whiteness as the norm, and all non-White people were deemed inferior—and a racial hierarchy was created. In the United States, "from the very inception of the republic to the present moment, race has been a profound determinant of one's political rights, one's location in the labor market, and indeed one's sense of identity" (Omi & Winant, 2015, p. 8). Thus, while race is a social construction, this does not mean that it does not have material (e.g., access to resources, opportunities, civil and human rights) and symbolic (e.g., ideological, psychological) consequences in the lives of *all* people in the racial hierarchy (Bonilla-Silva, 2006/2018).

As time passed and cultural norms and practices changed, the system that upholds the racial structure also had to change. Racial meanings were re-articulated overtime in "a series of steps that culminated in color-blind racial hegemony" (Omi & Winant, 2015). Bonilla-Silva (2001, 2006/2018) proposed the concept of "Color-blind Racism" identifying it as the "new racial structure" in the United States. He demonstrated how racism has evolved to fit in our contemporary pluralistic society, where customs, beliefs, policies, and practices continue to maintain White supremacy without explicitly addressing race or acknowledging systemic racism. This structure is different from the one operating in the pre-civil rights era, i.e., that of overt racist beliefs and discriminatory policies (Bonilla-Silva, 2001). Color-blind racism is an ideology that focuses on ignoring or dismissing race and racism, and, as such, denies the consequences of contemporary racism for People of Color in society, including their exclusionary racialized experiences in educational systems (Bonilla-Silva, 2006/2018). As such, color-blind ideology protects contemporary racism because it affects whether one is able to properly identify the differences in lived experiences between various groups of students imposed by the racial structure; and thus, people and institutions may not recognize their power to perpetuate or disrupt exclusionary policies and practices.

Annamma and colleagues (2017) convincingly argued that researchers should shift from using the term "color-blindness" in favor of a "color-evasiveness" framework that better describes the active processes theorized by Bonilla-Silva and many critical race scholars while also avoiding the use of ableist language. Thus, color-evasiveness will be used to describe this pervasive racial ideology for the remainder of this chapter.

Color-evasiveness has been explored in studies related to STEM college faculty. This work has demonstrated how STEM faculty and other academic staff who adopt a color-evasive stance can disadvantage Students of Color and reify the status quo in admissions decisions, advising, mentoring, curriculum development, teaching, and disciplinary culture (e.g., Aragón et al., 2017; Dancy, 2025; Dowd & Bensimon, 2015; Johnson, 2007; King et al., 2023; McCoy et al., 2015; McNeill et al., 2022; Powell et al., 2021; Robertson et al., 2023; Russo-Tait, 2022). While faculty members may mean well by believing in the importance of being "color-blind" and treating every student in the same way, by not acknowledging race and racism, they dismiss and contribute to the negative experiences their racially minoritized students contend with in STEM spaces. For example, faculty who insist on ignoring race/racism are at risk of allowing implicit bias and deficit-thinking to play a role in their decision-making about racially minoritized students.

Instantiations of color-evasive approaches have been seen in the context of graduate student admissions decisions and have the effect of removing racially minoritized students from the candidate list early in the selection process, as faculty choose to put value on GRE scores (on which racially minoritized students tend to score lower) even though research has shown that higher scores are not always predictive of success in graduate school (Posselt, 2020). Further, due to color-evasive discourses that espouse that racism is a thing of the past and race does not and should not matter, faculty become ambivalent about considering race at all in graduate admission committee discussions. This predicament works to occlude unconscious racist beliefs, as faculty still go on to implicitly consider race, such as when discussing student applicants from China using "model minority" discourses (Posselt, 2020).

Dowd and Bensimon (2015) described instantiations of color-evasion in curriculum development efforts. In a study focusing on racial justice and institutional change, faculty from many disciplines, including STEM, participated in a series of curriculum development workshops. Many participants demonstrated strong color-evasive ideals by refusing to consider race while designing curricula. They believed that doing so would be unfair and that all students should be treated equally. This color-evasive stance was related to an inability to engage in evidence-based equitable practices in curriculum and instruction. Further, faculty members' color-evasive beliefs were often associated with the "deficit model"—beliefs that position racially minoritized students and/or their cultures as deficient or lacking in ways that lead to racial disparities in education (Dowd & Bensimon, 2015; Valencia, 1997). Thus, deficit models place the blame on students for racially unjust outcomes and absolve faculty from taking responsibility in addressing systemic barriers.

Color-evasiveness can also manifest in mentoring relationships. McCoy and colleagues (2015) studied college science faculty members' color-evasive mentoring of undergraduate science Students of Color engaged in research. They found that, while White mentors strived to be seen as "fair and color-blind," they discussed the under-preparation of their individual Students of Color as a group issue, confessed they had lowered standards for those students and did not spend as much time talking with their Students of Color about graduate school because they perceived these students to lack interest. Faculty also ignored structural issues (e.g., racial inequities in the K-12 educational system). Instead, they brought up meritocratic discourses to explain the lower achievement of Students of Color when compared to their White students. The authors argued that this mentoring style could disempower students from sharing their cultural backgrounds and related assets. Relatedly, this erasure of backgrounds could also suggest to students that they are expected to think, behave, and speak like their White mentors, effectively forcing them to assimilate rather than bringing their authentic selves, identities, and lived experiences to STEM (McCoy et al., 2015). Other studies on mentorship have shown that beliefs that align with color-evasiveness can negatively affect students' trust (Johnson, 2007; Reddick & Young, 2012), and conversely, faculty who acknowledge racial differences are better mentors and allies (Dowd & Bensimon, 2015; Reddick & Pritchett, 2015).

Aragón and colleagues (2017) examined the association between faculty endorsement of color-evasive or multicultural ideologies and implementation of inclusive pedagogies. The researchers surveyed 628 STEM faculty, staff, and graduate students who participated in a week-long teaching institute. This institute covered inclusive teaching approaches that have been empirically shown to support students from all backgrounds, including underrepresented, racially minoritized students in STEM. The survey asked faculty to agree or disagree with statements that indicated color-evasive or multicultural perspectives. Color-evasive survey items included "I believe in treating every student equally as a person who has an equal opportunity to be outstanding in science" and "I believe that in an inclusive classroom educators should focus on what makes students the same and not focus on what makes students different (e.g., race and gender)." The multicultural survey items included "I believe that we should make special efforts to be supportive of underrepresented students (i.e., female students and/or minority students)," and "I believe that an inclusive classroom is achieved when we embrace the qualities that make us different." They found that educators who highly endorsed color-evasive ideology were significantly less likely to believe that any given teaching practice was useful and tended to use fewer inclusive teaching practices. On the other hand, those who highly endorsed multicultural beliefs reported high identification with, commitment to, and implementation of inclusive practices. Further, color-evasive beliefs appeared to matter above and beyond the race or gender of the respondent. This study demonstrates how this ideological stance can be held by well-intentioned educators, who self-selected into taking a week of their time to attend a professional development institute, and how those ideals work to uphold the status quo.

More recently, Russo-Tait (2022) investigated how college science faculty members in a HWI made sense of the underrepresentation of Black, Indigenous, and Latiné students in STEM by using color-evasive framings to explain racial inequities. These frames included abstract liberalism, cultural racism, and minimization of racism (Bonilla-Silva, 2006/2018). Abstract liberalism explanations framed racial inequities in STEM as due to a student's individual choices or affect, in a presumed equal opportunity and meritocratic educational system (i.e., students choose not to major in STEM, they lack interest or confidence, or feel discouraged for being the "only one" and opt out). Faculty who used cultural racism frames rationalized racial inequities as the result of deficient family or cultural values and practices that did not align with excelling in STEM education or constrained student success (i.e., Hispanic families not caring about education; Black people's oppositional culture beliefs; and Students of Color not being raised to be curious or to engage in STEM—narratives that have been debunked by many studies). Finally, as demonstrated by Dr. A in the opening vignette, by using the minimization of racism frame, faculty explicitly dismissed discrimination and systemic racism as potential reasons for underrepresentation, and instead focused on presumed poverty and lack of access to rigorous STEM courses in K-12 schools. While racially minoritized students can disproportionately be from lower income backgrounds due to systemic racism in society, this understanding, uncoupled with critiques of systemic racism, allowed faculty to absolve themselves from responsibility in supporting Students of Color, often claiming that there was nothing they could do by the time these students made it to college (Russo-Tait, 2022). However, research has demonstrated the myriad ways students could be supported via instructional changes and other practices (e.g., Theobald et al., 2020; Dewsbury & Brame, 2019; NASEM, 2016, 2023). Further, faculty's assumptions did not account for research demonstrating differential attrition of Black and Latiné students in STEM majors—after controlling for the very predictors that faculty claimed were to blame for racial disparities (i.e., socioeconomic status, parents' education, and prior academic achievement and level of preparation; Riegle-Crumb et al., 2019).

Similarly, Robertson and colleagues (2023) examined how physics faculty used color-evasive frames when describing the discipline, teaching, and learning in physics, and representation issues in the field (e.g., White privilege in physics and the reasons and potential solutions for racial disparities). They found that faculty not only used color-evasive frames but had also adopted physics-specific instantiations of color-evasive frames, which worked to further solidify physics as a supposed race-neutral and objective space due to its focus on rationality. They also documented deficit-thinking in relation to the reasons for the underrepresentation of Students of Color, namely blaming families and local schools (Robertson et al., 2023).

McNeill and colleagues (2022) explored how college mathematics instructors made sense of racialized classroom instructional events using color-evasive racism and racial dysconsciousness (King, 1991) frameworks. They argued that it was important to understand how faculty reasoned through such events because, given how ideologies of neutrality and objectivity frame social issues as unimportant in math, faculty could be engaging in color-evasive racist approaches to teaching (p. 3). Indeed, instructors who used discourses that positioned math as neutral and objective engaged in color-evasive discourses that minimized racism. Further, these perspectives appeared to inform practices related to deficit-perspectives of minoritized students and led to the reported withholding of student support. In these ways, color-evasive discourse seems to lead to instructional practices that work to maintain Whiteness and racial inequities in college math classrooms.

Similarly, King and colleagues (2023) developed narrative cases of college science classroom interactions that depicted common anti-Black events reported in research, such as stereotyping, dismissing, isolating, ignoring, and gaslighting Black students. They asked college STEM instructors to identify problematic practices and interactions in those cases. Data analysis using racial noticing (Shah & Cole, 2020) and color-evasion frameworks showed that many respondents did not notice race/racism in most events, and most responses that did were color-evasive in nature. While some participants acknowledged some events as problematic, they engaged in different rhetorical maneuvers that minimized or dismissed the power of anti-Black racism in the experience of Black STEM students. The authors concluded that if instructors adhere to color-evasive ideologies and cannot notice, identify, and problematize such events, they will at best be unable to address these events as they happen in their classrooms and, at worst, engage in the same problematic behaviors identified in the cases.

It is also important to note that Students of Color recognize the problematic nature of color-evasiveness. For example, Women of Color in science majors reported feeling discouraged by learning environments with supposed "color- and gender-blind cultures" as those invalidated their racialized and gendered experiences of exclusion in those spaces (Johnson, 2007). Science students participating in research experiences who were mentored by faculty with color-evasive inclinations felt discouraged, lacking support, and "weeded out" (McCoy et al., 2017). Finally, a study assessing students' perceptions of STEM faculty with color-evasive philosophies (expressed via syllabus and a video recording) found that faculty who voiced color-evasive philosophies in online courses were perceived as being more biased, and students had lower performance in math and science assessments with those types of instructors (Good et al., 2020).

In summary, recent research using critical frameworks to explore college STEM faculty beliefs and practices show that those who subscribe to color-evasive ideologies are unable to identify and understand racialized systems of power and privilege that constrain Students of Color in multiple ways. Consequently, faculty members engage in racialized exclusionary actions and/or inactions in their teaching, mentoring, and decision-making. Color-evasive ideology also works to conceal unconscious racial biases that lead to deficit models about Students of Color that blame students or their communities for their own oppressive experiences and absolve faculty from responsibility in supporting those students. In all, this ideology limits faculty members' ability to imagine how they could be engaging in equitable practices through their teaching, advising, mentoring, admissions and hiring decisions, and in other spheres of influence—all of which contribute to the maintenance of White supremacy and racial injustice in STEM learning environments.

Ideals of equal opportunity and treatment in conjunction with meritocratic and color-evasive ideologies seamlessly coalesce to maintain White supremacy in STEM, often through the mental models and related practices of faculty members who inadvertently act as gatekeepers. Thus, one important lever for racial justice in STEM is to disrupt faculty members' adherence to these hegemonic liberal ideologies. Next, we discuss one critical framework that could assist DBER scholars in understanding how to best support STEM faculty to renounce these liberal ideologies and adopt justice-centered understandings and practices that can disrupt White supremacy.

2.3 Critical Consciousness: A Framework for Faculty Development and Discipline-Based Education Research

Is it possible to just teach biology? ...Is it possible to discuss, to study the phenomenon of life without discussing exploitation, domination, freedom, democracy, and so on? I think that it's impossible.—Paulo Freire.

Prolific educator and philosopher, Paulo Freire (1968), argued that education is not apolitical or neutral and that, by taking such a stance, an educator is complicit in maintaining oppression and injustice. Freire critiqued the traditional model of teaching (i.e., solely lecturing) by calling it the "banking system." In this system, the teacher and student hold strict, predefined roles. Students are considered "empty receptacles" with no prior knowledge and a teacher then "deposits" knowledge into these passive students who memorize and later repeat what they were told, with no real learning taking place (Freire, 1968). He argued that this type of pedagogy only works to perpetuate inequities, as it does not support learners to think critically about the world and instead indoctrinates them with hegemonic ideologies that justify current world order.

Freire proposed a "problem-posing" approach to teaching whereby the teacher enters in dialogue with students and facilitates their learning of topics by connecting them to what is relevant to students' lives. In this approach, students are not considered objects and blank slates—they are *subjects* of their own learning, who come to educational spaces with prior knowledge and lived experiences that should be leveraged by educators. The instructor must know their students to develop lessons, questions, and pose problems that will resonate with students' lives and expose important contradictions about society. Through problem posing, instructors help students to undergo the process of "conscientização"—critical consciousness raising. Possessing a critical consciousness involves having an understanding and analysis of social injustices, the agency and self-efficacy to challenge the status quo, and engaging in actions in ways that challenge and address unjustice. Students learn to "read and write the world," that is, they can both understand the world and transform it. This approach to teaching and philosophy about the emancipatory power of education was coined *Critical Pedagogy* by Henry Giroux (1983). Critical pedagogy has often been thought of as solely an approach to teaching, but critical pedagogy goes beyond the moves teachers make in the classroom. Embracing critical pedagogy demands adopting and embodying a mindset of education as a liberatory practice wherein the role of the educator is a change agent who acts in solidarity with and alongside their pupils—with the goal being the development of critical consciousness and collective action against injustice. While we discuss Freire's approaches in this chapter, we recognize that these ideas and practices were not exclusive to him. In fact, there is a rich history of Black and Indigenous intellectuals and activists practicing and theorizing liberatory education as a necessary corrective to decades of educational disenfranchisement and exclusion (Givens, 2022; Morales-Doyle, 2023). We honor and respect these legacies. We focus our chapter on Freire because his pedagogical frameworks have been extensively discussed and studied in science education (Basile & Azevedo, 2022; Dewsbury, 2020; Kokka, 2018; Morales-Doyle, 2024; Salinas et al., 2023; Torres Olave et al., 2023).

Freire explained that having a critical consciousness involves both thought and action, theory and practice (i.e., praxis). Thus, developing an understanding of the sociopolitical contradictions, inequities, and systems of oppression of the world (critical reflection) and the individual self-efficacy and motivation for change (critical agency), and individual and collective action to transform the world (critical action) is necessary (Watts et al., 2011). While Freire's work was done in the context of adult literacy in Brazil, he also argued for the role of science instructors in helping students to "read and write the world" in their specific contexts and disciplines. As implied in the quote at the start of this section, Freire believed that students can develop a critical consciousness by becoming aware of contradictions and injustices in STEM fields and learn to use their STEM knowledge as tools to understand and address injustices in that space—and that STEM educators have the responsibility of engaging in such discussions. Research in K-12 science education has shown that critical pedagogy can be a powerful tool for learning STEM content and critical consciousness raising (Calabrese Barton & Tan, 2020; Gutstein, 2012; Morales-Doyle, 2017; Vakil, 2018). On the other hand, STEM students who are not supported in developing a critical consciousness, even those from minoritized populations in STEM, can be socialized in ideologies that justify their own oppression (Doerr et al., 2021; Riegle-Crumb et al., 2022).

For STEM educators to engage in critical pedagogy or any other emancipatory practices in their classrooms and other spheres of influence, they must first possess a critical consciousness themselves (Sengupta-Irving & de Royston, 2020). K-12 researchers have discussed the need for educators to develop a critical consciousness and shown examples of the justice-centered approaches that critically conscious educators use (Gay & Kirkland, 2003; Gutstein, 2012; Ladson-Billings & Tate IV, 1995; Madkins & McKinney de Royston, 2019; Morales-Doyle, 2017). Some have also developed consciousness raising trainings to help K-12 science educators (Crabtree & Stephan, 2023; Register et al., 2022). As Philip (2011) suggests, there is a need for instructors to develop broader, critically historicizing frames to interrogate the routine ways "structure, ideology, and cognition" rearticulate and reproduce problematic yet commonplace beliefs about minoritized groups. In this vein, critical consciousness can be seen as a remedy against the exclusionary hegemonic ideologies discussed in this chapter.

While K-12 STEM education scholars have explored the benefits critical consciousness as well as the complexities in supporting its development, there is little research on how to support postsecondary faculty to develop a critical consciousness (Altranice & Mitchell, 2022; Landreman et al., 2007; Peña, 2012; Tran & Gupta, 2021), and even fewer in STEM fields (Castaneda & Mejia, 2018). Unlike K-12 STEM educators, STEM college faculty often receive no formal training in pedagogical practices, let alone on issues of racial justice. Therefore, supporting faculty in developing the ability to critically reflect on their own social positionalities and the structural causes of racial inequities—including the contradictions inherent in the liberal ideologies discussed throughout this chapter—could be a powerful lever for change. While there are only few studies that center STEM college faculty's development of a critical consciousness, DBER scholars have called for this (e.g., Castaneda & Mejia, 2018; Dewsbury, 2020; King et al., 2023; Russo-Tait, 2022, 2023).

The few extant studies in higher education have shown the potential for exploring critical consciousness of systemic racism in STEM faculty. For example, Haynes (2017) found that faculty with a "racial consciousness"—a deeper understanding of power and privilege dynamics surrounding race and racism—tended to use more inclusive approaches in their classrooms which foster equitable outcomes, as opposed to those who focused on treating students the same. Faculty with this critical consciousness of racial injustice tended to think about equity issues in more expansive ways and engage in equitable instructional practices. Similarly, Russo-Tait (2022) found that faculty who explained the underrepresentation of racially minoritized students in STEM using systemic racism framings disavowed color-evasive rationalizations and recognized that they had a responsibility to work as change agents in their classrooms, departments, and professional societies.

In the study of STEM faculty's conceptions of equity, a small group of science professors conceptualized equity as "justice" (Russo-Tait, 2023). Faculty with this conception of equity possessed a critical consciousness. For example, they discussed systemic racism from historical and sociopolitical perspectives and centered power and privilege dynamics as the underlying contemporary source of inequities for students with minoritized identities in STEM (Russo-Tait, 2023). In addition to active learning and inclusive practices, they also used an emerging critical pedagogy whereby they engaged students in discussions about science and social justice. For example, one professor helped students to explore the racist foundations of their discipline, while also teaching the scientific content and techniques, and others helped students learn how to use STEM as a tool to explore social justice issues such as sexism in sports or environmental racism. These faculty members also used their agency and power to advocate for justice beyond the classroom. This included using identity-conscious mentorship approaches, advocating for policy changes that lead to more equitable hiring, admissions, and evaluation practices, and working to make their professional organizations more justice-centered and inclusive.

2.4 Implications for Discipline-Based Education Research

Research has demonstrated that various individual, organizational, and institutional factors influence faculty decisions about their pedagogy (Hora & Ferrare, 2013; Kane et al., 2002; Kember & Gow, 1994; Posselt, 2020). A core part of faculty's work is to "learn what is appropriate, expected, and accepted in terms of their behavior in the field, including how to teach and how to learn" (Laird et al., 2008, p. 472). Thus, the distinct histories, ideologies, and cultures of the discipline are influential for faculty practice. Indeed, the disciplines often discipline—they shape us as scholars and pedagogues. Because the histories of STEM disciplines and higher education writ large are riddled with racism (e.g., Steinberg, 1998; Wilder, 2013), STEM faculty end up teaching more than just benign subject matter. Students learn what their discipline has been, what it currently is, and what it can (and should) become through their engagement with faculty pedagogy.

Postsecondary STEM education research, combined with insights from extant social sciences and K-12 education literature, demonstrate the potential for leveraging critical consciousness as a framework in future interventions and research that explores faculty beliefs and practices. Many scholars rightly critique individualized solutions to systemic problems. For example, when discussing racism in science, Harding (2006) states "The self-improvement of individuals is never an adequate substitute for collective political action against the White supremacist interests, policies, and practices of dominant social institutions" (p. 21). This is why a critical consciousness framework is particularly powerful to explore in the STEM education context—it combines supporting faculty to both develop a systemic analysis of oppression *and* to recognize and enact their agency to collectively disrupt such injustices. That is, possessing a critical consciousness is not merely about developing an awareness of inequities, it is about recognizing injustice and using one's power to act against it. Faculty members certainly have the power to take individual action in their teaching and mentoring and collective action as a department in shared governance to disrupt the structural barriers affecting their racially minoritized students—but they first need to understand the nature of those barriers and their own roles as gatekeepers. As critical race theorists convincingly argue, the liberal ideologies that people hold, including those that inform faculty's beliefs about the reasons for racial disparities in STEM, must be challenged so that current norms and practices can be transformed by all stakeholders to center racial justice (Morton, 2022).

A critical consciousness framework can be leveraged to design research projects that explore questions like: How can critical pedagogy be used to support current and future STEM faculty in developing a critical consciousness? What are the most effective ways for supporting faculty to engage in sustained critical reflection and critical action? What types of knowledge about racial inequities are necessary and sufficient to promote critical reflection and critical action? What factors mediate or moderate critical consciousness development in adults? How do we support faculty to use critical pedagogy in their own STEM courses? How does this pedagogy support STEM undergraduate and graduate students' consciousness raising, and impact learning outcomes, interest, sense of belonging, communal goal congruity, STEM identity, and advocacy? How do we measure the impact of critically conscious faculty in STEM departments? These types of studies can lead DBER scholars to eventually develop interventions that support the "inside-out change" that needs to occur in STEM faculty, in addition to the top-down and bottom-up approaches necessary for systemic change (Posselt, 2020).

2.5 Conclusion

Education researchers have long argued that faculty and other higher education practitioners do not understand how White supremacy operates in higher education and affect racially minoritized students' educational opportunities and outcomes (Bensimon, 2007). This lack of understanding is at least in part due to powerful hegemonic liberal ideologies that work to maintain White supremacy by allowing people and institutions to rationalize away racial inequities and blame individual Students of Color and their communities for their own oppression, while appearing fair and just. Recent studies in postsecondary STEM disciplines have shown that faculty members who endorse color-evasive, meritocratic, and equal treatment and opportunity ideologies can create and maintain racial injustice. DBER scholars and practitioners must then recognize and contest the ideological and political underpinnings of STEM departments and acknowledge its role as a racialized organization (Ray, 2019).

DBER scholars should consider building on this existing interdisciplinary body of work and conducting research that helps to understand how to effectively challenge and disrupt these problematic liberal ideologies and their related logics, norms, practices, and processes. In terms of supporting faculty in challenging the racial order, DBER scholars could study how to support current and future STEM faculty in developing a critical consciousness (Dowd & Bensimon, 2015; Joseph et al., 2016; Powell et al., 2021; Russo-Tait, 2022, 2023). Critically conscious faculty would be able to critically analyze systems of oppression and recognize their individual agency and collective power to engage in efforts that disrupt White supremacy and racial injustices in their spheres of control and influence in and beyond the academy. The development of a critical mass of critically conscious STEM faculty and students could lead to the collective political struggle necessary to transform postsecondary STEM education into the racially just space it should be.

References

Altranice, K., & Mitchell, B. (2022). Catalysts of conscientization among the professorate: A descriptive phenomenological study. *Journal of Social Work Education*, 1–14.

Annamma, S. A., Jackson, D. D., & Morrison, D. (2017). Conceptualizing color-evasiveness: Using dis/ability critical race theory to expand a color-blind racial ideology in education and society. *Race Ethnicity and Education, 20*(2), 147–162.

Aragón, O. R., Dovidio, J. F., & Graham, M. J. (2017). Colorblind and multicultural ideologies are associated with faculty adoption of inclusive teaching practices. *Journal of Diversity in Higher Education, 10*(3), 201.

Basile, V., & Azevedo, F. S. (2022). Ideology in the mirror: A loving (self) critique of our equity and social justice efforts in STEM education. *Science Education, 106*(5), 1084–1096.

Basile, V., & Black, R. (2019). They hated me till I was one of the "good ones": Toward understanding and disrupting the differential racialization of undergraduate African American STEM majors. *The Journal of Negro Education, 88*(3), 379–390.

Battey, D., & Leyva, L. A. (2016). A framework for understanding whiteness in mathematics education. *Journal of Urban Mathematics Education, 9*(2).

Bensimon, E. M. (2007). The underestimated significance of practitioner knowledge in the scholarship on student success. *The Review of Higher Education, 30*(4), 441–469.

Blair-Loy, M., & Cech, E. A. (2022). *Misconceiving merit: Paradoxes of excellence and devotion in academic science and engineering*. University of Chicago Press.

Bonilla-Silva, E. (2001). *White supremacy and racism in the post-civil rights era*. Lynne Rienner Publishers.

Bonilla-Silva, E. (2006/2018). Racism without racists: Color-blind racism and the persistence of racial inequality in the United States (5th Edition). Rowman & Littlefield Publishers.

Burt, B. A., Williams, K. L., & Smith, W. A. (2018). Into the storm: Ecological and sociological impediments to Black males' persistence in engineering graduate programs. *American Educational Research Journal, 55*(5), 965–1006.

Calabrese Barton, A., & Tan, E. (2020). Beyond equity as inclusion: A framework of "rightful presence" for guiding justice-oriented studies in teaching and learning. *Educational Researcher, 49*(6), 433–440.

Callwood, K. A., Weiss, M., Hendricks, R., & Taylor, T. G. (2022). Acknowledging and supplanting White supremacy culture in science communication and STEM: the role of science communication trainers. *Frontiers in Communication, 35*.

Castaneda, D. I., & Mejia, J. A. (2018). Culturally relevant pedagogy: An approach to foster critical consciousness in civil engineering. *Journal of Professional Issues in Engineering Education and Practice, 144*(2).

Cech, E. A., Blair-Loy, M., & Rogers, L. E. (2018). Recognizing chilliness: How schemas of inequality shape views of culture and climate in work environments. *American Journal of Cultural Sociology, 6*(1), 125–160.

Crabtree, L. M., & Stephan, M. (2023). That exists today: An analysis of emerging critical consciousness in a professional development setting. *Journal of Science Teacher Education, 34*(2), 105–131.

Dancy, M. (2025). Majority group physics instructors struggle to notice, name, and disrupt racist and sexist student interactions. *Physical Review Physics Education Research, 21*(1), 010142.

Dancy, T. E., Edwards, K. T., & Earl Davis, J. (2018). Historically White universities and plantation politics: Anti-Blackness and higher education in the Black Lives Matter era. *Urban Education, 53*(2), 176–195.

Dewsbury, B. M. (2020). Deep teaching in a college STEM classroom. *Cultural Studies of Science Education, 15*(1), 169–191.

Dewsbury, B., & Brame, C. J. (2019). Inclusive teaching. CBE—Life Sciences. *Education, 18*(2), fe2.

Doerr, K., Riegle-Crumb, C., Russo-Tait, T., Takasaki, K., Sassler, S., & Levitte, Y. (2021). Making merit work at the entrance to the engineering workforce: Examining women's experiences and variations by race/ethnicity. *Sex Roles, 85*, 422–439.

Dowd, A. C., & Bensimon, E. M. (2015). *Engaging the "race question": Accountability and equity in US higher education*. Teachers College Press.

Freeman, S., Eddy, S. L., McDonough, M., Smith, M. K., Okoroafor, N., Jordt, H., & Wenderoth, M. P. (2014). Active learning increases student performance in science, engineering, and mathematics. *Proceedings of the National Academy of Sciences, 111*(23), 8410–8415.

Freire, P. (1968). *Pedagogy of the oppressed*. Continuum.

Gay, G., & Kirkland, K. (2003). Developing cultural critical consciousness and self-reflection in preservice teacher education. *Theory Into Practice, 42*(3), 181–187.

Giroux, H. (1983). Theories of reproduction and resistance in the new sociology of education: A critical analysis. *Harvard Educational Review, 53*(3), 257–293.

Givens, J. R. (2022). Fugitive pedagogy. In *Fugitive pedagogy*. Harvard University Press.

Good, J. J., Bourne, K. A., & Drake, R. G. (2020). The impact of classroom diversity philosophies on the STEM performance of undergraduate students of color. *Journal of Experimental Social Psychology, 91*, 104026.

Gordon, B., & Noga, G. (2015). *Ideology: Structuring identities in contemporary life*. University of Toronto Press.

Gough, A. (2015). STEM policy and science education: Scientistic curriculum and sociopolitical silences. *Cultural Studies of Science Education, 10*, 445–458.

Gramsci, A. (1971). *Prison notebooks*. International Publishers.

Grindstaff, K., & Mascarenhas, M. (2019). "No one wants to believe it": Manifestations of White privilege in a STEM-focused college. *Multicultural Perspectives, 21*(2), 102–111.

Gupta, A., Turpen, C., Philip, T., & Elby, A. (2019). Narrative co-construction of stances towards engineers' work in socio-technical contexts. In *Critical, transdisciplinary and embodied approaches in STEM education* (pp. 251–272).

Gutiérrez, R. (2002). Enabling the practice of mathematics teachers in context: Toward a new equity research agenda. *Mathematical Thinking and Learning, 4*(2-3), 145–187.

Gutstein, E. (2012). *Reading and writing the world with mathematics: Toward a pedagogy for social justice*. Routledge.

Hall, S. (1982). The rediscovery of 'Ideology': Return of the repressed in media studies. In M. Gurevitch et al. (Eds.), *Culture, society and the media* (pp. 56–90). Methuen.

Hall, S. (1996). The problem of ideology: Marxism without guarantees. In D. Morley & K. Chen (Eds.), *Stuart Hall: Critical dialogues in cultural studies* (pp. 25–46). Routledge.

Handelsman, J., Elgin, S., Estrada, M., Hays, S., Johnson, T., Miller, S., & Williams, J. (2022). Achieving STEM diversity: Fix the classrooms. *Science, 376*(6597), 1057–1059.

Hand, V., Penuel, W. R., & Gutiérrez, K. D. (2013). (Re) framing educational possibility: Attending to power and equity in shaping access to and within learning opportunities. *Human Development, 55*(5-6), 250–268.

Harding, S. (2006). *Science and social inequality: Feminist and postcolonial issues*. University of Illinois Press.

Haynes, C. (2017). Dismantling the White supremacy embedded in our classrooms: White faculty in pursuit of more equitable educational outcomes for racially minoritized students. *International Journal of Teaching and Learning in Higher Education, 29*(1), 87–107.

Hora, M. T., & Ferrare, J. J. (2013). Instructional systems of practice: A multidimensional analysis of math and science undergraduate course planning and classroom teaching. *Journal of the Learning Sciences, 22*, 212–257. https://doi.org/10.1080/10508406.2012.729767

Johnson, A. C. (2007). Unintended consequences: How science professors discourage women of color. *Science Education, 91*(5), 805–821.

Jordan, W. J. (2010). Defining equity: Multiple perspectives to analyzing the performance of diverse learners. *Review of Research in Education, 34*(1), 142–178.

Joseph, N., Haynes, C., & Cobb, F. (2016). *Interrogating Whiteness and relinquishing power: White faculty's commitment to racial consciousness in STEM classrooms*. Peter Lang.

Kane, R., Sandretto, S., & Heath, C. (2002). Telling half the story: A critical review of research on the teaching beliefs and practices of university academics. *Review of Educational Research, 72*(2), 177–228.

Kember, D., & Gow, L. (1994). Orientations to teaching and their effect on the quality of student learning. *The Journal of Higher Education, 65*(1), 58–74.

King, J. E. (1991). Dysconscious racism: Ideology, identity, and the miseducation of teachers. *The Journal of Negro Education, 60*(2), 133–146.

King, G. P., Russo-Tait, T., & Andrews, T. C. (2023). Evading race: STEM faculty struggle to acknowledge racialized classroom events. *CBE—Life Sciences Education, 22*(1), ar14.

Kokka, K. (2018). Radical STEM teacher activism: Collaborative organizing to sustain social justice pedagogy in STEM fields. *Educational Foundations, 31*, 86–113.

Kosut, M. (2012). Hegemony. In *Encyclopedia of gender in media* (Vol. 1, pp. 160–161). SAGE Publications, Inc.. https://doi.org/10.4135/9781452218540.n69

Ladson-Billings, G., & Tate, W. F., IV. (1995). Toward a critical race theory of education. *The Teachers College Record, 97*(1), 47–68.

Laird, N. T. F., Shoup, R., Kuh, G. D., & Schwarz, M. J. (2008). The effects of discipline on deep approaches to student learning and college outcomes. *Research in Higher Education, 49*(6), 469–494.

Landreman, L. M., Rasmussen, C. J., King, P. M., & Jiang, C. X. (2007). A phenomenological study of the development of university educators' critical consciousness. *Journal of College Student Development, 48*(3), 275–296.

Leyva, L. A., McNeill, R. T., & Duran, A. (2022). A queer of color challenge to neutrality in undergraduate STEM pedagogy as a White, cisheteropatriarchal space. *Journal of Women and Minorities in Science and Engineering, 28*(2).

Liu, A. (2011). Unraveling the myth of meritocracy within the context of US higher education. *Higher Education, 62*, 383–397.

Madkins, T. C., & McKinney de Royston, M. (2019). Illuminating political clarity in culturally relevant science instruction. *Science Education, 103*(6), 1319–1346.

Makus, A. (1990). Stuart Hall's theory of ideology: A frame for rhetorical criticism. *Western Journal of Speech Communication, 54*(4), 495–514.

McCoy, D. L., Winkle-Wagner, R., & Luedke, C. L. (2015). Colorblind mentoring? Exploring White faculty mentoring of students of color. *Journal of Diversity in Higher Education, 8*(4), 225.

McCoy, D. L., Luedke, C. L., & Winkle-Wagner, R. (2017). Encouraged or weeded out: Perspectives of students of color in the STEM disciplines on faculty interactions. *Journal of College Student Development, 58*(5), 657–673. https://doi.org/10.1353/csd.2017.0052

McGee, E. O. (2016). Devalued Black and Latino racial identities: A by-product of STEM college culture? *American Educational Research Journal, 53*(6), 1626–1662.

McGee, E. O. (2020a). *Black, brown, bruised: How racialized STEM education stifles innovation*. Harvard Education Press.

McGee, E. O. (2020b). Interrogating structural racism in STEM higher education. *Educational Researcher, 49*(9), 633–644. https://doi.org/10.3102/0013189X20972718

McNair, T. B., Bensimon, E. M., & Malcom-Piqueux, L. (2020). *From equity talk to equity walk: Expanding practitioner knowledge for racial justice in higher education*. Josey-Bass.

McNeill, R. T., Leyva, L. A., & Marshall, B. (2022). "They're just students. There's no clear distinction": a critical discourse analysis of color-evasive, gender-neutral faculty discourses in undergraduate calculus instruction. *Journal of the Learning Sciences*, 1–43.

Miller, R. A., Vaccaro, A., Kimball, E. W., & Forester, R. (2021). "It's dude culture": Students with minoritized identities of sexuality and/or gender navigating STEM majors. *Journal of Diversity in Higher Education, 14*(3), 340.

Morales-Doyle, D. (2017). Justice-centered science pedagogy: A catalyst for academic achievement and social transformation. *Science Education, 101*(6), 1034–1060.

Morales-Doyle, D. (2023). Putting science education in its place: The science question in social justice education. *Cultural Studies of Science Education, 18*(1), 81–94.

Morales-Doyle, D. (2024). *Transformative science teaching: A catalyst for justice and sustainability*. Harvard Education Press.

Morgan, D. L., Davis, K. B., & López, N. (2020). Engineering political fluency: Identifying tensions in the political identity development of engineering majors. *Journal of Engineering Education, 109*(1), 107–124.

Morton, T. R. (2022). Critical race theory and STEM education. In *Oxford research encyclopedia of education*.

Morton, T. R., Gee, D. S., & Woodson, A. N. (2019). Being vs. becoming: Transcending STEM identity development through Afropessimism, moving toward a Black X consciousness in STEM. *The Journal of Negro Education, 88*(3), 327–342.

Nasir, N. I. S., & Hand, V. M. (2006). Exploring sociocultural perspectives on race, culture, and learning. *Review of Educational Research, 76*(4), 449–475.

National Academies of Sciences, Engineering, and Medicine. (2016). *Barriers and opportunities for 2-year and 4-year STEM degrees: Systemic change to support students' diverse pathways*. The National Academies Press.

National Academies of Sciences, Engineering, and Medicine. (2023). *Advancing antiracism, diversity, equity, and inclusion in stemm organizations: Beyond broadening participation*. The National Academies Press. https://doi.org/10.17226/26803

Omi, M., & Winant, H. (2015). *Racial formation in the United States*. Routledge.

Peña, E. V. (2012). Inquiry methods for critical consciousness and self-change in faculty. *The Review of Higher Education, 36*(1), 69–92.

Philip, T. M. (2011). An "ideology in pieces" approach to studying change in teachers' sensemaking about race, racism, and racial justice. *Cognition and Instruction, 29*(3), 297–329.

Philip, T. M., Gupta, A., Elby, A., & Turpen, C. (2018). Why ideology matters for learning: A case of ideological convergence in an engineering ethics classroom discussion on drone warfare. *Journal of the Learning Sciences, 27*(2), 183–223.

Posselt, J. R. (2016). *Inside graduate admissions: Merit, diversity, and faculty gatekeeping*. Harvard University Press.

Posselt, J. R. (2020). *Equity in science: Representation, culture, and the dynamics of change in graduate education*. Stanford University Press.

Powell, C., Demetriou, C., Morton, T. R., & Ellis, J. M. (2021). A CRT-informed model to enhance experiences and outcomes of racially minoritized students. *Journal of Student Affairs Research and Practice, 58*(3), 241–253.

Ray, V. (2019). A theory of racialized organizations. *American Sociological Review, 84*(1), 26–53.

Reddick, R. J., & Pritchett, K. O. (2015). "I don't want to work in a world of Whiteness": White faculty and their mentoring relationships with Black students. *Journal of the Professoriate, 8*(1), 54–84.

Reddick, R. J., & Young, M. D. (2012). Mentoring graduate students of color. In S. Fletcher & C. Mullen (Eds.), *The SAGE handbook of mentoring and coaching for education* (pp. 412–429). SAGE.

Register, J., Fernandes, A., & Pugalee, D. (2022). Supporting preservice mathematics teachers' culturally responsive teaching: A focus on teaching for social justice. *Mathematics, 10*(6), 896.

Riegle-Crumb, C., King, B., & Irizarry, Y. (2019). Does STEM stand out? Examining racial/ethnic gaps in persistence across postsecondary fields. *Educational Researcher, 48*, 133–144.

Riegle-Crumb, C., Russo-Tait, T., Doerr, K. T., & Nguyen, U. (2022). Critical consciousness of gender inequality in STEM: Considering the viewpoints of racially diverse high school girls with engineering aspirations. *Sociological Perspectives*.

Robertson, A. D., Vélez, V., Hairston, W. T., & Bonilla-Silva, E. (2023). Race-evasive frames in physics and physics education: Results from an interview study. *Physical Review Physics Education Research, 19*(1), 010115.

Rubie-Davies, C. (2015). Teachers' instructional beliefs and the classroom climate. In *International handbook of research on teachers' beliefs* (pp. 266–283). Routledge.

Rudolph, J. L. (2002). *Scientists in the classroom: The Cold War reconstruction of American science education*. Palgrave.

Russo-Tait, T. (2022). Color-blind or racially conscious? How college science faculty make sense of racial/ethnic underrepresentation in STEM. *Journal of Research in Science Teaching*, 1–31. https://doi.org/10.1002/tea.21775

Russo-Tait, T. (2023). Science faculty conceptions of equity and their association to teaching practices. *Science Education, 107*(2), 427–458. https://doi.org/10.1002/sce.21781

Russo-Tait, T. (2025). *Using composite narratives to reflect on equity beliefs and practices* (p. 12). CourseSource. https://doi.org/10.24918/cs.2025.11

Salinas, I., Fernández, M. B., Johnson, D., & Bastías, N. (2023). Freire's hope in radically changing times: A dialogue for curriculum integration from science education to face the climate crisis. *Cultural Studies of Science Education, 18*(1), 21–39.

Secada, W. G. (1989). Educational equity versus equality of education: An alternative conception. In *Equity in education* (pp. 68–88).

Sengupta-Irving, T., & de Royston, M. M. (Eds.). (2020). *STEM and the social good: Forwarding political and ethical perspectives in the learning sciences*. Routledge.

Shah, N., & Coles, J. A. (2020). Preparing teachers to notice race in classrooms: Contextualizing the competencies of preservice teachers with antiracist inclinations. *Journal of Teacher Education, 71*(5), 584–599.

Steinberg, S. (1998). The role of social science in the legitimation of racial hierarchy. *Race and Society, 1*(1), 5–14. https://doi.org/10.1016/S1090-9524(99)80183-1

Taylor, E. (2009). Foundations of critical race theory in education: An introduction. In E. Taylor, D. Gillborn, & G. Ladson-Billings (Eds.), *Foundations of critical race theory in education*. Routledge.

Theobald, E. J., Hilla, M. J., Trana, E., Agrawalb, S., Arroyoc, E. N., Behlingd, S., Chambwee, N., Cintróna, D. N., Coopera, J. D., Dunstera, G., Grummera, J. A., Laceya, M. E., Littlefieldd, C. E., Lowea, A., et al. (2020). Active learning narrows achievement gaps for underrepresented students in undergraduate science, technology, engineering, and math. *Proceedings of the National Academy of Sciences, 117*(12), 6476–6483.

Torres Olave, B., Tolbert, S., & Frausto Aceves, A. (2023). Reflecting on Freire: A praxis of radical love and critical hope for science education. *Cultural Studies of Science Education, 18*(1), 1–20.

Tran, L. U., & Gupta, P. (2021). Rebuilding our teams to be critically conscious in our educational work. *Journal of Science and Environmental Learning, 1*(1.6), 41–62.

Vakil, S. (2018). Ethics, identity, and political vision: Toward a justice-centered approach to equity in computer science education. *Harvard Educational Review, 88*(1), 26–52.

Vakil, S., & Ayers, R. (2019). The racial politics of STEM education in the USA: Interrogations and explorations. *Race Ethnicity and Education, 22*(4), 449–458.

Valencia, R. R. (Ed.). (1997). *The evolution of deficit thinking: Educational thought and practice*. Routledge.
Virtue, E., Wells, G., MacKusick, C., Murphy-Nugen, A., & Rose, A. (2018). The immortal life of Henrietta Lacks: Using a common read to transform a learning community. *Learning Communities: Research & Practice, 6*(1), 4.
Vossoughi, S., & Vakil, S. (2018). Toward what ends? A critical analysis of militarism, equity, and STEM education. In *Education at war* (pp. 117–140). Fordham University Press.
Watts, R. J., Diemer, M. A., & Voight, A. M. (2011). Critical consciousness: Current status and future directions. *New Directions for Child and Adolescent Development, 2011*(134), 43–57.
Wilder, C. S. (2013). *Ebony and ivy: Race, slavery, and the troubled history of America's universities*. Bloomsbury Publishing.
Willis, R. (2019). The use of composite narratives to present interview findings. *Qualitative Research, 19*(4), 471–480.

Chapter 3
Critical Race Theory with Applications to DBER

Miguel Rodriguez

Abstract Critical Race Theory (CRT) is designed to identify and improve the dynamics between race, racism, and power in US society. Since its foundation in legal studies in1989, it has expanded into many fields including DBER. This chapter will begin by providing some brief historical context that led to the foundation of CRT. Then I will cover the major tenets along with some other major themes of CRT, all while providing context relating back to STEM. Next, some direct applications of CRT to qualitative and quantitative research methods will be discussed briefly, with references for the curious reader. Finally, I will articulate how CRT has become a politically divisive topic then briefly rebut some common scholarly critiques. This chapter is designed to provide you with an introduction to CRT, while providing resources for those who wish to learn more.

Keywords Critical studies · Critical race theory · Racism · Racial inequity · STEM

3.1 Introduction to Critical Race Theory

Critical Race Theory (CRT) was created to understand and transform the dynamics between race, racism, and power (Delgado & Stefancic, 2017). While much progress toward equality has been made in the past 70 years, racial disparities continue to persist in many aspects of US society. There are racial disparities in the distribution of wealth (Shapiro et al., 2013), how laws are enforced (Hurwitz & Peffley, 2010), access to healthcare (Williams et al., 2003), and quality of education (Elder

M. Rodriguez (✉)
California State University—Dominguez Hills, Carson, CA, USA

G. L. Cochran, J. D. Adams (eds.), *Equity-Oriented, Discipline-Based STEM Education Research*, Critical Studies of Education 19,
https://doi.org/10.1007/978-981-96-9136-4_3

et al., 2021; Magnuson & Waldfogel, 2005). These systemic inequalities are also embedded in higher education and especially in the STEM fields where many People of Color (PoC) continue to be underrepresented in both Academia and Industry. I will discuss the reality of the racial divide, revealing the persistence of systemic racism in the United States, which is the first tenet of CRT. CRT works to detect and analyze systemic barriers PoC may face, to then study the root causes and find potential solutions. This makes CRT a useful research tool for DBER since there are still a lot of systemic barriers that PoC face in many different STEM fields.

This chapter will explore the four tenets of CRT along with other major themes that are also part of CRT. A very brief historical introduction to CRT will be provided, which is important for setting the foundation for the events that led up to CRT's creation. Then the tenets will be presented in general, with some expansion of possible applications in DBER. Next, some other major themes within CRT will be presented to provide further scope of the theory. I will then provide some guidelines for how to apply CRT to DBER. Further, I will address scholarly critiques to CRT and current assault on CRT by the right-wing media. I hope that readers, initially speculative of CRT, may find this chapter to be reasonably factual and well backed by evidence. However, please note that this chapter is only an introduction to the richness of Critical Race Theory, so please go beyond and explore the topics you find interesting further.

3.2 Brief History of Critical Race Theory

Historical context is extremely important for understanding the purpose and function of CRT. The history of race in the United States is too broad to cover in this chapter, or even one textbook, but some key highlights should be discussed. The use of the word race began to appear in the eighteenth century and was used to identify groups of people according to similar physical characteristics (Harper, n.d.). European Colonialism led to the ideology of White supremacy, the idea that White people are superior to non-White people, which would go on to be accepted worldwide over the following centuries (Dunbar-Ortiz, 2021). Race superiority was, and still is, used to justify the cruel treatment of non-White people, including the mass displacement and murder of many Indigenous groups, and horrific treatment of enslaved peoples. Even after the emancipation proclamation was passed, there were still harsh laws passed known as the Jim Crow laws that would limit the rights of non-White people, which would continue to be enforced until the civil rights movement in the 50s and 60s (Tischauser, 2012). Besides laws, there were several violent atrocities that took place against People of Color such as the 1921 Tulsa race massacres that destroyed the most prominent Black neighborhood in the country at the time, and hundreds of other riots and massacres.

The Brown v. Board of Education ruling declared segregation unconstitutional and ushered in a 15-year period of drastic reforms that lasted until the 1970s. This includes the Civil Rights Act of 1954 which declared racial segregation unconstitutional and the Civil Rights Act of 1964 which prohibited discrimination based on race, religion, sex, or national origin. There were also several key supreme court cases that extended these rights such as Bailey v. Patterson (1962), in which the supreme court ruled to prohibit the racial segregation of public transportation and Loving v. Virginia (1967) which invalidated laws prohibiting interracial marriage. A shift of political power, with the election of President Nixon in 1968, would lead to more conservative supreme court judges that viewed civil rights legislation less favorable than the justices they replaced. This led to several key rulings that were seen by many Scholars of Color, as a retrograde to the progress made in the years before. These rulings include *Washington v.* Davis (1976), where laws having "undesired" racist effects were deemed constitutional (Davis, 1976) and *Regents of the University of California v. Bakke* (1978), which made university racial quotas, a system that guaranteed racial equity in the admissions process, unconstitutional (Knowlton, 1978). Due to these legal setbacks, Scholars of Color began to formulate ideas that would go on to form the foundation of CRT, including Derick Bell's Yale Law Journal article "serving Two Masters" (Bell Jr, 1975) and Harvard Law Review article "Brown v. Board of Education and the interest-Convergence Dilemma" (Bell Jr, 1979), along with Alan Freeman's critical legal reviews of court rulings that posed setbacks to the civil rights progress made decades prior.

The official formulation of CRT as an intellectual movement began during a retreat involving 24 legal Scholars of Color, organized by Crenshaw, entitled "New Developments in Critical Race Theory" at the University of Wisconsin, Madison in 1989. The term Critical Race Theory was used to signify an "intersection of critical theory and race, racism and the law" (Crenshaw et al., 1995). This movement came from the need for a grounded theoretical framework that could provide a foundation to defend PoC against a legal system that upheld systematic and structural racism. Since its inception, there have since been hundreds of legal review articles written, CRT has been successfully used in court to protect PoC, and CRT has also expanded into other fields including education (Taylor et al., 2009), health care (Ford & Airhihenbuwa, 2010), sociology, and others.

Due to coming from legal roots, CRT in education began to be used as a way to to identify education policies that posed systemic barriers for Students of Color, and while there is improvement there is a long way to reaching the goal of racial equity in higher education (Gillborn, 2005). More recently, the concepts of CRT have moved toward pedagogy and practice, with movements to decolonize our education content and redefine the histories we tell and who can be included in them. In STEM, CRT may be critical for the following: (a) changing the majoritarian stories, or master narratives (Stanley, 2007) of who can do science, (b) assessing systemic barriers that negatively impact Students of Color, and (c) finding potential solutions to address current racial divides in representation.

3.3 Tenets of Critical Race Theory

The tenets of Critical Race Theory are fluid and may vary slightly according to scholarship, but I chose to present the four tenets, as described by Delgado and Stefancic (2017). Briefly, these tenets are (1) Racism is Ordinary, (2) Interest Convergence, (3) Race is a social Construct, and (4) Uniqueness of Marginalized Voice (Delgado & Stefancic, 2017). The first tenet will be explored in more detail than the other tenets, to emphasize the need for social justice and the importance of CRT in combatting systemic racial barriers. These tenets will each be presented with current and historical evidence to build the foundation of CRT. For some tenets, applications to Discipline-Based STEM Education Research will be discussed briefly but the overall application of CRT to STEM Education Research will be saved mostly for a later section in the chapter.

3.3.1 First Tenet

Racism is ordinary in US society. This may be the most important tenet in CRT since historically race has been a legal factor in determining who could migrate to the United States, who could be free, and the general quality of life one could have in the United States. The concept of race was purposely constructed by affluent White elites to discriminate and target People of Color; hence race has historically been embedded in much of US laws and policies. While there has been some progress in changing old policies and adding new ones, racial inequalities continue to persist in much of US society. It may take several more generations of activists and scholars pushing toward a racially equitable world, to achieve that dream. As it stands today, there are policies that sustain, and in some cases increase, racial disparities in: the justice system (Delgado & Stefancic, 2017), wealth (Shapiro et al., 2013), health (Williams et al., 2003), and education (Elder et al., 2021; Magnuson & Waldfogel, 2005). Issues regarding justice and wealth are directly linked to education and who can be or who has access to STEM resources. These issues need to be discussed further to educate the readers who may be unaware of these inequalities. Additionally, it is critical as this tenant forms a concrete foundation for other tenets and themes that have originated from CRT. Books can be written for each of these topics, so this will only serve as a starting point, and I encourage the reader to explore each topic further.

3.3.1.1 Justice System

There is systemic racism embedded in all levels of the "justice" system for juveniles and Adults of Color, with African Americans being treated the most inequitable out of all the races in the United States. African Americans have historically and presently been subject to the most violence from the police than any other racial group (Hurwitz & Peffley, 2010), which has become more visible than ever due to technological advances in video recording and sharing media. Even worse, we are currently living in the New Jim Crow Era where a third of male African Americans are incarcerated at some point of their lives (Alexander, 2011). As many as half of African Americans have encounters with police, often receiving longer sentences for the same criminal sentence than White Americans (Rehavi & Starr, 2014). So much more can be said here, from high black populations in prison (Pettit & Western, 2004) to being given higher bail costs (Arnold et al., 2022).

3.3.1.2 Wealth

The racial divergence in wealth has actually increased in the past 30 years (Shapiro et al., 2013) with Black people being the least wealthy of all racial groups. White people own 6 times more per capita than what Black people own (Derenoncourt et al., 2022). What is probably most daunting about the wealth gap are the myths surrounding the causes of wealth inequality, such as myths like *education can help Black people close the wealth gap* or that *homeownerships the primary cause of the wealth gap* to name a few (Darity Jr et al., 2018). These majoritarian stories have been proven false since there are still wealth gaps when accounting for education (Hamilton et al., 2015; Meschede et al., 2017) and the value of the same assets owned by Blacks is often valued less than their White counterparts (Harris, 1999). The effects of inflation on the wealth gap are unclear at the time of writing but may also lead to a further racial divide in wealth.

3.3.1.3 Healthcare

The racial disparities in health is also evident, with White Americans having the better health overall than most PoC (Williams, 2012), with the greatest disparity being between historically marginalized groups such as Native Americans and African Americans (Nazroo & Williams, 2006). PoC may also be subject to having access to lower quality health care overall (Richardson & Norris, 2010). While it has improved, the life expectancy of Black Americans is 3.7 years less than White Americans, as of 2016 (National Center for Health Statistics (US), 2016). More recently, the rates of death and hospitalization were significantly higher for PoC than for White people during the pandemic (CDC, 2021).

3.3.1.4 Education

Closer to the scope of this book, there continues to be a vast racial inequality in education and education remains highly segregated in most of the country. In the US public school system, about 70% of Black kids attend non-White majority schools compared to 13% of White kids (Rothstein, 2013). The quality of education in the United States typically depends on the wealth of the neighborhood where the school is located (Clapp et al., 2008). This means that the education children receive is a function of their family's economic status, since schools with higher budgets can buy more intellectual copyrights to offer more options and bring in and support teachers. Considering the wealth gap, the current system is currently rigged toward those who have more wealth, which in the United States has always and continues to mean mostly White families. The result of this is that it's better to be rich than smart, with kindergartners with high socioeconomic status (SES) with test scores in the bottom half have a 7 in 10 chance of reaching high SES among his or her peers as a young adult (Carnevale et al., 2019). The "achievement gap" should instead be referred to as an "education debt" which is a result of historical, economic, sociopolitical, and moral debt, due to the systemic factors discussed in this chapter, has accumulated in the same way that the national debt has (Ladson-Billings, 2006). Focusing on the debts that Students of Color must overcome as opposed to the "gaps" they have shifts the focus from a deficit to anti-deficit dialogue that acknowledges the extra work that students from underrepresented groups have to do to succeed in higher education, including STEM fields (Gutierrez, 2023).

All these inequities discussed so far act as barriers to PoC, which also limits their access and success in higher education. Racial disparities continue to be prevalent in higher education and are increasing across many disciplines (Garibaldi, 2014). In STEM, the racial disparities are even worse, with Students of Color having less representation, retention, and lower graduation rates than their White counterparts. Examining this further, the racial demographics are different for each field, with the largest disparity in representation being in geoscience. The racial disparities are present in both undergraduate and graduate STEM programs and even go beyond higher education and continue for postdoctoral scholars (Eaton et al., 2020). This includes STEM Researchers of Color, who attain less grants and see less papers published, despite having equivalent qualifications of White peers (Ginther et al., 2011; Taffe & Gilpin, 2020). For Professors of Color in academia, racial inequalities in hiring and promotion continue to persist (Fryer et al., 2013). While researchers can do little to prevent the systemic barriers that students face before attending or during university, a lot can be done to reduce the barriers Students of Color face in universities. Researchers should work toward studying these barriers, to learn what ways the university can better support students from minoritized groups.

There is currently a huge push to reduce racial divide, but there is still so much progress left to be made. However, none of this should be surprising since our Higher Education lives within a US culture in which racism is ordinary.

3.3.2 Second Tenet

Civil Rights only advances when they align with the interests of the elites, specifically affluent White Americans. This idea is known as Interest Convergence, a term coined by Derrick Bell, and it sheds light on how power structures of race are influenced by the elites. There are many historical examples of Interest Convergence, but

I will focus on two specifically: the end of slavery and the civil rights movement (Bell, 2005). While there were genuine abolitionists fighting to end slavery, the reality is that the north was advancing industrially much greater than the south due to the south's reliance of enslaved people as workers. In other words, the north recognized that having slaves meant having slower industrial development. This economic truth is often omitted in the history books, but it clearly highlights Interest Convergence. The second example is the civil rights movement, which shifted US society toward more equitable treatment to People of Color. Again, while there were people that genuinely wanted to see this change, it was necessary for uniting the country during the Cold War against the Soviet Union. These are just a couple historical examples out of the many that have occurred.

Besides historical interpretations, Interest Convergence exists in micro-environments such as schools and universities. Interest Convergence has been used in STEM to see how PoC's interests line up with the dominant group's, at predominantly white institutions (Baber, 2015). Even the goal of increasing representation and retention in STEM is an example of interest convergence since the motivations are fundamentally rooted in capitalist desires and maintaining a technological edge.

3.3.3 Third Tenet

Race is a social construct established through the social relations between different groups of people. Etymologically speaking, the term race was first used to describe people descended from a common ancestor. It wasn't until later that the definition transformed to relate to one's appearance. The modern five race theory originates from Johann Friedrich Blumenbach's categorization of humans, where he claimed there were five varieties of humans: Caucasian, Mongolian, Malayan, Ethiopian, and American. Johann used the term Caucasian and argued that they were the ideal forms of human beings. The original meaning of Caucasian was used regarding skeletal anatomy, without consideration of skin color, but the word would later be transformed to a meaning synonymous to being white. These eugenics ideologies worked to justify colonization overall and a system of power that was centered around race in the United States. These racial categories have no biological significance and only served to perpetuate the systems of power that existed. These racial categories make no sense from a scientific standpoint, which is more evident now, with recent advancements in genetics research showing that the construct of race do not explain any of the measurable genetic differences between people (Tishkoff & Kidd, 2004).

The justification for many wrongs and injustices to Communities of Color was that Whites are superior, consequentially these ideologies continue today in the form of global White supremacy. Global White supremacy takes many forms; a plain example is the billion dollar global industry of skin whitening, which reflects the pandemic nature of White supremacy since this worldwide industry is driven by the opinion that the lighter you are the better you are (Blay, 2011). The brutal history of the exploitation of Black and Indigenous people has had cascading effects that continue to impact us today. Race is not a biological nor scientific term, it is a term to categorize a humans' appearance that continues to play a significant factor in the lived experiences of people. Race was created because of racism and not the other way around. This is perfectly summarized in the quote, "Race is the child of racism, not the father" (Coates, 2015).

This tenet has a direct application to STEM research, since we report and collect data on race to infer statistics about certain groups. As mentioned already, researchers should focus on anti-deficit and systemic approaches to conducting research, from design to publication. Further, much of the statistics used in DBER was founded for eugenics ideologies (Zuberi, 2001), so special attention to how we report on race statistics is warranted. Be sure to avoid doing work that promotes deficit-focused narratives of PoC in STEM, and instead focus on the anti-deficit and systemic critical narratives.

3.3.4 Fourth Tenet

The final tenet expresses the importance of giving a voice to people who are marginalized and exploring their experiences through their lens. To understand racism and discrimination, the people who most feel it must be heard and considered the most. Given that in the United States, race is so tied to one's own experiences, our own experiences are highly subjective and contextually dependent. For example, I am an Afro Hispanic Puerto Rican who lived in South Florida, a highly diverse part of the United States, for most of my life. Hence, I can't compare my racialized experiences with someone of a different race or ethnicity, or even someone of the same race and ethnicity but that has lived in a predominantly white part of the United States. Although this chapter focuses specifically on CRT, it's worth noting that this tenet may also be true for understanding the marginalization of people with other identities such as gender, sexual orientation, religious affiliation, among others.

Since there are racial disparities in STEM, it is critical to hear the voices of marginalized PoC, to learn about their experiences to identify systemic barriers and work toward reducing them in STEM. This will also help scholars to strategize ways to better support Students of Color. Many studies on the experiences of marginalized students tend to use deficit-oriented theoretical frameworks. This includes work focusing on the struggles marginalized students face or comparing students of different races or ethnicities. This wrongly places the blame of these disparities upon students which ignores the systemic barriers they must navigate to achieve success. Instead of deficit-oriented work, I suggest new critical DBER research takes an anti-deficit approach (Mejia et al., 2018) to study how students overcome despite having "deficits." Further research using a systemic critical approach to understand how structures such as school and university systems negatively impact marginalized students is also warranted.

3.4 Other Major Themes of CRT

There are other themes that are also fundamental to CRT which should also be mentioned. Some of these include differential racialization, Black White binary, White privilege, and majoritarian stories. These themes are critical to CRT, with some forming tenets for other scholarly interpretations of CRT. These themes not only form part of CRT but also stand on their own as important concepts.

Differential racialization is a term used to describe how racism changes over time, shifting due to changes in geopolitical issues. Differential racialization is dynamic and is driven by the media coverage of impactful events and public opinion. Being that the United States has been involved in some sort of war for 97% of its history, there have always been groups of people that have been targeted for one reason or another. A few examples include the discrimination and displacement of Indigenous people with the motive to steal their land, placing Japanese Americans in internment camps during WWII when the United States was at war with Japan (Tateishi, 2012), or the discrimination of Muslim people after the attack on the World Trade Center in 2001 (Alsultany, 2013). The most recent example of this is the spike in discrimination and hate crimes against Asian Americans due to the blaming of China for the Covid pandemic (Tessler et al., 2020). There are many examples of racial differentiation throughout US history involving different groups of people. Differential racialization also exists in academia with many STEM disciplines exhibiting their own version of racializing power structure and the mechanisms to obtain privilege (Basile, 2019; Martin et al., 2010).

The Black White binary refers to how racialized experiences are often categorized as White or Black, leaving little room for other racialized experiences or even multiracial experiences. This is due to the historical demographics of the United States, which in 1950 was 89.9% White and 10% Black (Hobbs & Stoops, 2002). This began to change after immigration reform in 1952, finally making it legally possible for non-White people to migrate to the United States (Bennett, 1966). Since then, many diverse non-White people have migrated to the United States, with Hispanics now forming the largest minority group in the United States. As of 2010, many ethnic Hispanics are forced to claim either White or Black as their race, with many claiming White thus expanding the definition of what being White means in the United States (Lopez, 2005). Asians are often lumped together with White Americans in terms of "privilege," but Asian people are physically distinguishable from the dominant group and hence are also subject to discrimination. Finally, the experiences of multiracial people are subject to the same binary, where someone may look more White or Black, but this fails to acknowledge the unique experiences those people have. Hopefully, as the United States becomes more diverse, the use of the construct of race gets thrown out and replaced with how people identify culturally, in DBER and beyond.

White privilege refers to the privileges (46 privileges outlines by McIntosh) that White people have due to being part of the dominant group of the United States (McIntosh, 2007). A few examples include how White people are given educational materials that affirm the dominance of their race, are more likely to have positive experiences with police, or are treated better by school authorities (McIntosh, 2007). If these privileges ever ceased to exist, then White people would lose a huge advantage they have. Hence, there are many White people that would "like to see things change" but without sacrificing their privileges, which really means they want to pretend things are getting better while hoping they stay the same. Sociologically speaking, White privilege may be a natural consequence of Social Identity Theory, since people tend to favor their in-groups over their out-groups (Hogg, 2016). While Social Identity Theory is out of the scope of this chapter, it provides an explanation for understanding how and why discrimination exists. To combat the forces of discrimination and racism, experiences with diverse people and learning about diverse perspectives are necessary and should be a key part of education. In STEM, White privilege plays out in many ways, including who may hold positions of power, who has access to mentorship, and being surrounded by other folks from a similar background. DBER needs to work toward identifying and dismantling these power structures, to make sure opportunities are shared equitably.

Majoritarian stories are the ways that ideologies of race are maintained and established using popular culture. Majoritarian stories are "bundles of presuppositions, perceived wisdoms, and shared cultural understandings persons in the dominant race bring to the discussion of race" (p. 462) that happen over generations (Delgado & Stefancic, 2017). This includes media and stories privileging White heterosexual men while portraying PoC in a way that upholds certain stereotypes. An example is how Black people have been portrayed in movies historically, which is only recently starting to change (Dixon et al., 2003). Majoritarian stories may include how being "different" is a deficit, or how meritocracy should govern how resources are allocated (Mitchell, 2013). Since majoritarian stories are so common in society, PoC and other discriminated groups often propagate these stories without being aware of it. This makes counter stories to those majoritarian stories extremely essential to changing the master narrative of majoritarian stories, which is an important goal of CRT. Majoritarian stories are affluent in STEM; there are majoritarian stories of who can pursue science and what kind of science. For example, physics is often perceived to be a masculine subject (Francis et al., 2017; Gonsalves, 2014), which is a result of the majoritarian stories of who can do physics.

Intersectionality is also an essential theme of CRT. Beyond race, it is also important to acknowledge how everyone has unique experiences, which are also highly contextual, and depend on many factors such as one's class, gender, sexual orientation, socioeconomic background, disabilities, and nationality among others. The combination of oppressed identities is known as intersectionality, a term coined by Crenshaw (1991), although the ideas can be traced back as far as Sojourner Truth's Speech at Women's Convention in 1851 "Ain't I a woman?" (Truth & Kennedy, 1992). The experiences of Heterosexual Black men may vary from the experiences of LBGT Black men, since the way that gender and sexual orientation interact is different for both men. For DBER using critical theoretical frameworks, this means considering all the possible oppressed identities of the participants included in the study, by providing space for people to share those identities. For quantitative research this will involve collecting data about participant identities in modern and inclusive ways (an example is when asking about gender on a survey, make sure there is a choice for everyone, meaning 100% of people). For qualitative research, this involves making time to ask participants about their identities, and how those identities may impact the research being conducted.

3.5 Applying CRT to STEM Education Research

All research, including quantitative studies, are dependent on and interpreted using theoretical frameworks. Much of the modern statistics used in today's research was created by eugenicists and White supremacists such as Galton, Pearson, and Fisher (Zuberi et al., 2008). This fact makes it important to be critical in how quantitative methods are being used as well as how they are interpreted. As such, it is important to form well thought-out research questions while grounding methods to the appropriate theoretical framework needed. I will provide some tenets or elements that should be considered, when doing qualitative and quantitative research in STEM Education, summarized in Fig. 3.1.

Qualitative studies are often conducted with more awareness of the theoretical frameworks they use, than quantitative methods. Qualitative CRT research has been formulated for much longer than Quantitative CRT research, and spans the majority of work done using CRT, thus far. There are five critical elements that should be followed when conducting qualitative CRT education research, which are (Solórzano & Yosso, 2002):

1. Inter-centricity race and racism with other forms of subordination
2. Challenging the dominant ideology
3. Committing to social justice
4. The centrality of experiential knowledge and lived experiences
5. Having transdisciplinary perspective

Applying CRT to STEM Education Research

Qualitative Methods	***Quantitative Methods***
1. Inter-centricity race and racism with other forms of subordination.	1. Racism is difficult to quantify.
2. Challenging the dominant ideology.	2. Numbers are not neutral, and should be evaluated critically.
3. Committing to social justice.	3. Forms of analysis must also be evaluated critically.
4. The centrality of experiential knowledge and lived experiences.	4. Analysis should always be informed by experiences of PoC.
5. Having a transdisciplinary perspective.	5. Statistical analysis has no inherent value but can help social justice.

Fig. 3.1 Applications of CRT to qualitative and quantitative research

The first element is similar to CRT's first tenet and asks that the research be centered upon the ordinary nature of racism and how that connects to other experiences PoC may have. The second element asks that researchers challenge the dominant ideologies, or majoritarian stories with counter stories. The third element asks that researchers commit to social justice by using their research to reduce discrimination and empower research participants so that participants may also benefit, and not just the researchers. The fourth element lines up with the fourth tenet of CRT, which expressed the importance of centering research around the experiences of PoC. Finally, the fifth element expresses how CRT is transdisciplinary, which means a lot can be drawn from outside education and vice versa.

There are five tenets to using quantitative CRT methods that *Gillborn et al.* argued should be adhered to in order to stay close to the fundamental ideas of CRT (Gillborn et al., 2018):

1. Racism is complex and deeply rooted in society, making it difficult to quantify
2. Numbers are not neutral, and the interpretations made should be evaluated critically
3. Categories are also not neutral, so the forms of analysis must also be evaluated critically.
4. Critical analysis should always be informed by the experiences of marginalized group
5. Statistical analysis has no inherent value but can play a role in achieving social justice.

The first tenet for quantitative CRT includes the first tenet of CRT and adds how racism is difficult to quantify. This complexity calls for more methods that consider individualized experiences of every research subject (McCall, 2005). The second tenet calls for an in-depth inspection of how numbers are being used, to make sure that they are not being used to promote White supremacy and other majoritarian stories. The third tenet argues that the units and instruments used in the analysis must also be evaluated for propagating majoritarian stories. The fourth tenet, like the fourth tenet of CRT, stresses the importance of having the voice and insights of people from marginalized groups. And finally, the fifth tenet acknowledges that statistics by itself may not change anything but may be used to fight for social justice.

3.6 Critiques and Challenges to CRT

When Barrack Obama was elected for president, many people thought it was the beginning of "post-racial" society in the United States. This could not be further from the truth since racism is still ordinary and racial inequality continues to persist, which has been articulated throughout this chapter. Even worse, there is a large right-wing movement to hide the atrocious role of race throughout US history in public education, and instead tell the majoritarian version of history by only highlighting the good deeds of White people while hiding the ugly history of conquering, enslaving, and stealing Indigenous lands. Included in this movement is an attack on theoretical frameworks that highlight the ugly truth of racial inequity, such as CRT. The successful use of CRT to legally defend PoC has made it unpopular with politicians and people who want to preserve the status quo of persistent racial inequality in the United States.

CRT has recently become a political buzzword and has been weaponized, as it is used for misinformation campaigns, and has since resulted in at least 35 states introducing legislature to ban teaching about race, diversity, and systemic racism. Further, in January 2025, the federal government moved to eliminate diversity, equity, inclusion, and accessibility (DEIA) initiatives across grant funding, executive agencies, and federal contracting and procurement, principally through Executive Orders 14151 and 14173. Public universities that rely on federal grants are also under attack by the federal government, for their DEI programs and teaching of "woke" ideologies such as CRT, with the most notable being Harvard University. This political attack may in fact be a reaction to the waves that CRT is creating in shifting the dynamics of power, race, and racism. Unfortunately, this attack is also fueled by the spreading of false news that many people who follow right-wing media believe, such as: CRT is taught in elementary schools, CRT is the same as racial inequity so neither should be taught, or that CRT attacks White people, are among many false claims propagated by the media. It is a fact that CRT is not taught in K-12 spaces, as it is a legal and scholarly theoretical framework used in much higher levels of education. CRT talks about racial inequity, but racial inequity is much more general and should absolutely be taught in schools. History is always bias (His-story) but not teaching racial inequality in US history may lead to a feedback loop that feeds greater inequality. CRT does not attack White people, it only highlights the historical truths of racial dynamics, where White people have and continue to be more privileged on average.

There are also several scholarly criticisms against CRT that have been around for a while. Some of these critiques include (1) Critical theorists having personal agendas and only involving Participants of Color, (2) questioning validity of CRT due to success of many Asians and Jewish people, and (3) accusing CRT of being racists in itself. Critical theorists would say that (1) victims of racism must be heard and given a voice since the "truth" is often aligned with the interests of the dominant group, (2) the success of specific marginalized groups, such as Asians and Jewish people, happened despite an unfair system, and (3) CRT only points out racial inequities that exist and only work to dismantle those injustices.

3.7 Conclusion

This chapter is only the tip of the iceberg, serving only to introduce CRT as a theoretical framework. I attempted to relate DBER as much as I could, but there is surely more work out there to explore that utilizes specific aspects of CRT than what has been discussed in this chapter. Some potential direct applications to STEM education research were also noted, for those looking to use CRT for research. Many existing publications in STEM were omitted, and this chapter instead focused on explaining the foundations of CRT. There are already many CRT studies in STEM Education, most of which are qualitative. I strongly encourage the readers to read more on each topic discussed to gain more details and a different perspective of CRT. Like many accepted theoretical frameworks, there may be small differences in scholarships, but the fundamental ideas are the same. Overall, I hope this chapter provided a solid introduction to the fundamental ideas of CRT, especially to those new to the subject, while addressing any misunderstandings regarding CRT.

References

Alexander, M. (2011). The new Jim crow. *Ohio State Journal of Criminal Law, 9*, 7.

Alsultany, E. (2013). Arabs and Muslims in the media after 9/11: Representational strategies for a "postrace" era. *American Quarterly, 65*(1), 161–169.

Arnold, D., Dobbie, W., & Hull, P. (2022). Measuring racial discrimination in bail decisions. *American Economic Review, 112*(9), 2992–3038.

Baber, L. D. B. (2015). Considering the interest-convergence dilemma in STEM education. *The Review of Higher Education, 38*(2), 251–270. https://doi.org/10.1353/rhe.2015.0004

Bailey v. Patterson. (1962). 369 U.S. 31.

Basile, V. (2019). They hated me till I was one of the "good ones". *The Journal of Negro Education, 88*(3).

Bell, D. (2005). *The Derrick Bell reader* (Vol. 75). NYU Press.

Bell, D. A., Jr. (1975). Serving two masters: Integration ideals and client interests in school desegregation litigation. *Yale Law Journal, 85*, 470.

Bell, D. A., Jr. (1979). Brown v. Board of Education and the interest-convergence dilemma. *Harvard Law Review, 93*, 518.

Bennett, M. T. (1966). The immigration and nationality (McCarran-Walter) Act of 1952, as Amended to 1965. *The Annals of the American Academy of Political and Social Science, 367*(1), 127–136.

Blay, Y. A. (2011). Skin bleaching and global white supremacy: By way of introduction. *The Journal of Pan African Studies, 4*(4), 4–46.

Carnevale, A. P., Fasules, M., Quinn, M., & Peltier Campbell, K. (2019). *Born to win, schooled to lose.*

CDC. (2021). *Hospitalization and death by race/ethnicity.*

Clapp, J. M., Nanda, A., & Ross, S. L. (2008). Which school attributes matter? The influence of school district performance and demographic composition on property values. *Journal of Urban Economics, 63*(2), 451–466.

Coates, T.-N. (2015). *Between the world and me*. Text Publishing.

Crenshaw, K. (1991). Mapping the margins: Intersectionality, identity politics, and violence against women of color. *Stanford Law Review, 43*(6), 1241. https://doi.org/10.2307/1229039

Crenshaw, K., Gotanda, N., Peller, G., & Thomas, K. (1995). Critical race theory. In *The key writings that formed the movement* (pp. 276–291).

Darity, W., Jr., Hamilton, D., Paul, M., Aja, A., Price, A., Moore, A., & Chiopris, C. (2018). What we get wrong about closing the racial wealth gap. *Samuel DuBois Cook Center on Social Equity and Insight Center for Community Economic Development, 1*(1), 1–67.

Davis, J. D. (1976). The Supreme Court and Washington v. Davis—A rationale for racism. *North Carolina Central Law Review, 8*, 87.

Delgado, R., & Stefancic, J. (2017). *Critical race theory: An introduction* (Vol. 20). New York University Press.

Derenoncourt, E., Kim, C. H., Kuhn, M., & Schularick, M. (2022). *Wealth of two nations: The US racial wealth gap, 1860–2020.*

Dixon, T. L., Azocar, C. L., & Casas, M. (2003). The portrayal of race and crime on television network news. *Journal of Broadcasting & Electronic Media, 47*(4), 498–523.

Dunbar-Ortiz, R. (2021). *Not "A Nation of Immigrants": Settler colonialism, White supremacy, and a history of erasure and exclusion*. Beacon Press.

Eaton, A. A., Saunders, J. F., Jacobson, R. K., & West, K. (2020). How gender and race stereotypes impact the advancement of scholars in STEM: Professors' biased evaluations of physics and biology post-doctoral candidates. *Sex Roles, 82*(3), 127–141.

Elder, T. E., Figlio, D. N., Imberman, S. A., & Persico, C. L. (2021). School segregation and racial gaps in special education identification. *Journal of Labor Economics, 39*(S1), S151–S197.

Ending Radical and Wasteful Government DEI Programs and Preferencing. (2025, January 29). Exec. Order No. 14151, 90 Fed. Reg. 8339.

Ending Illegal Discrimination and Restoring Merit-Based Opportunity. (2025, January 31). Exec. Order No. 14173, 90 Fed. Reg. 8633.

Ford, C. L., & Airhihenbuwa, C. O. (2010). Critical race theory, race equity, and public health: Toward antiracism praxis. *American Journal of Public Health, 100*(S1), S30–S35.

Francis, B., Archer, L., Moote, J., DeWitt, J., MacLeod, E., & Yeomans, L. (2017). The construction of physics as a quintessentially masculine subject: Young people's perceptions of gender issues in access to physics. *Sex Roles, 76*(3–4), 156–174. https://doi.org/10.1007/s11199-016-0669-z

Fryer, R. G., Pager, D., & Spenkuch, J. L. (2013). Racial disparities in job finding and offered wages. *The Journal of Law and Economics, 56*(3), 633–689.

Garibaldi, A. M. (2014). The expanding gender and racial gap in American higher education. *The Journal of Negro Education, 83*(3), 371. https://doi.org/10.7709/jnegroeducation.83.3.0371

Gillborn, D. (2005). Education policy as an act of white supremacy: Whiteness, critical race theory and education reform. *Journal of Education Policy, 20*(4), 485–505.

Gillborn, D., Warmington, P., & Demack, S. (2018). QuantCrit: Education, policy, 'Big Data' and principles for a critical race theory of statistics. *Race Ethnicity and Education, 21*(2), 158–179.

Ginther, D. K., Schaffer, W. T., Schnell, J., Masimore, B., Liu, F., Haak, L. L., & Kington, R. (2011). Race, ethnicity, and NIH research awards. *Science, 333*(6045), 1015–1019.

Gonsalves, A. J. (2014). "Physics and the girly girl—There is a contradiction somewhere": Doctoral students' positioning around discourses of gender and competence in physics. *Cultural Studies of Science Education, 9*(2), 503–521.

Gutierrez, R. (2023). *A "gap-gazing" fetish in mathematics education? Problematizing research on the achievement gap.*

Hamilton, D., Darity, W., Jr., Price, A. E., Sridharan, V., & Tippett, R. (2015). Umbrellas don't make it rain: Why studying and working hard isn't enough for Black Americans. *New York: The New School, 779*, 780–781.

Harper, D. (n.d.). Etymology of race. In *Online Etymology Dictionary*. https://www.etymonline.com/word/race

Harris, D. R. (1999). "Property values drop when blacks move in, because...": Racial and socio-economic determinants of neighborhood desirability. *American Sociological Review*, 461–479.

Hobbs, F., & Stoops, N. (2002). *Demographic trends in the 20th century* (Vol. 4). US Census Bureau.

Hogg, M. A. (2016). Social identity theory. In *Understanding peace and conflict through social identity theory* (pp. 3–17). Springer.

Hurwitz, J., & Peffley, M. (2010). And justice for some: Race, crime, and punishment in the US criminal justice system. *Canadian Journal of Political Science/Revue Canadienne de Science Politique, 43*(2), 457–479.

Knowlton, R. E. (1978). Regents of the University of California v. Bakke. *Arkansas Law Review, 32*, 499.

Ladson-Billings, G. (2006). From the achievement gap to the education debt: Understanding achievement in U.S. schools. *Educational Researcher, 35*(7), 3–12. https://doi.org/10.3102/0013189X035007003

Lopez, I. H. (2005). Race on the 2010 census: Hispanics & the shrinking white majority. *Daedalus, 134*(1), 42–52.

Loving v. Virginia. (1967). 388 U.S. 1.

Magnuson, K. A., & Waldfogel, J. (2005). Early childhood care and education: Effects on ethnic and racial gaps in school readiness. *The Future of Children*, 169–196.
Martin, D. B., Gholson, M. L., & Leonard, J. (2010). Mathematics as gatekeeper: Power and privilege in the production of knowledge. *Journal of Urban Mathematics Education, 3*(2). https://doi.org/10.21423/jume-v3i2a95
McCall, L. (2005). The complexity of intersectionality. *Signs: Journal of Women in Culture and Society, 30*(3), 1771–1800.
McIntosh, P. (2007). White privilege and male privilege. In *Race, ethnicity and gender: Selected readings* (pp. 377–385).
Mejia, J. A., Revelo, R. A., Villanueva, I., & Mejia, J. (2018). Critical theoretical frameworks in engineering education: An anti-deficit and liberative approach. *Education Sciences, 8*(4), 158.
Meschede, T., Taylor, J., Mann, A., & Shapiro, T. M. (2017). "Family achievements?": How a college degree accumulates wealth for Whites and not for Blacks. *Review—Federal Reserve Bank of St. Louis, 9*, 121–137.
Mitchell, K. (2013). Race, difference, meritocracy, and English: Majoritarian stories in the education of secondary multilingual learners. *Race Ethnicity and Education, 16*(3), 339–364.
National Center for Health Statistics (US). (2016). *Health, United States, 2015: With special feature on racial and ethnic health disparities*.
Nazroo, J. Y., & Williams, D. R. (2006). The social determination of ethnic/racial inequalities in health. *Social Determinants of Health, 2*, 238–266.
Pettit, B., & Western, B. (2004). Mass imprisonment and the life course: Race and class inequality in US incarceration. *American Sociological Review, 69*(2), 151–169.
Regents of the University of California v. Bakke. (1978). 438 U.S. 265.
Rehavi, M. M., & Starr, S. B. (2014). Racial disparity in federal criminal sentences. *Journal of Political Economy, 122*(6), 1320–1354.
Richardson, L. D., & Norris, M. (2010). Access to health and health care: How race and ethnicity matter. *Mount Sinai Journal of Medicine: A Journal of Translational and Personalized Medicine, 77*(2), 166–177.
Rothstein, R. (2013). *For public schools, segregation then, segregation since*. Economic Policy Institute.
Shapiro, T., Meschede, T., & Osoro, S. (2013). *The roots of the widening racial wealth gap: Explaining the black-white economic divide*.
Solórzano, D. G., & Yosso, T. J. (2002). Critical race methodology: Counter-storytelling as an analytical framework for education research. *Qualitative Inquiry, 8*(1), 23–44. https://doi.org/10.1177/107780040200800103
Stanley, C. A. (2007). When counter narratives meet master narratives in the journal editorial-review process. *Educational Researcher, 36*(1), 14–24. https://doi.org/10.3102/0013189X06298008
Taffe, M., & Gilpin, N. W. (2020). *Racial inequity in federal grant funding*.
Tateishi, J. (2012). *And justice for all: An oral history of the Japanese American detention camps*. University of Washington Press.
Taylor, E., Gillborn, D., & Ladson-Billings, G. (2009). *Foundations of critical race theory in education*.
Tessler, H., Choi, M., & Kao, G. (2020). The anxiety of being Asian American: Hate crimes and negative biases during the COVID-19 pandemic. *American Journal of Criminal Justice, 45*(4), 636–646.
Tischauser, L. V. (2012). *Jim crow laws*. ABC-CLIO.
Tishkoff, S. A., & Kidd, K. K. (2004). Implications of biogeography of human populations for 'race' and medicine. *Nature Genetics, 36*(Suppl 11), S21–S27.
Truth, S., & Kennedy, A. P. (1992). *Ain't I a Woman?* AP Kennedy, Jr.
Washington v. Davis. (1976). 426 U.S. 229.
Williams, D. R. (2012). Miles to go before we sleep: Racial inequities in health. *Journal of Health and Social Behavior, 53*(3), 279–295.
Williams, D. R., Neighbors, H. W., & Jackson, J. S. (2003). Racial/ethnic discrimination and health: Findings from community studies. *American Journal of Public Health, 93*(2), 200–208.
Zuberi, T. (2001). *Thicker than blood: How racial statistics lie*. U of Minnesota Press.
Zuberi, T., Bonilla-Silva, E., et al. (2008). *White logic, white methods: Racism and methodology*. Rowman & Littlefield Publishers.

Chapter 4
QuantCrit as a Methodological Framework for Equity-Centered Discipline-Based Education Research

Paulette Vincent-Ruz, Kathryn N. Hosbein, and Vanessa Rosa

In trying to become 'objective,' Western culture made 'objects' of things and people when it distanced itself from them, thereby losing 'touch' with them.
—Gloria E. Anzaldúa

Abstract If Discipline-Based Education Research (DBER) aims to address educational inequity, our research requires research lenses rooted on solidarity and liberation. That is, research lenses aligned with subverting harmful narratives and emphasizing patterns of marginalization and minoritization. The goal should be to build an educational system capable of setting up students for success and so their identities are reflected on the practice of science. The main objective of this chapter is to provide a quantitative methodological framework utilizing a critical lens for Discipline-Based Education Researchers focused on QuantCrit. It uses a Critical lens to guide DBER researchers in applying critical quantitative principles and evaluating their decisions throughout the research process. QuantCrit involves constant engagement in reflexivity, improvement, and adaptation, as research is a nonlinear process. The framework serves as a blueprint to guide DBER researchers in establishing research questions, choosing theoretical frameworks, experimental design, and supporting analytical decisions. The chapter illustrates the utility of QuantCrit in three sections. Firstly, the chapter provides historical context around the develop-

P. Vincent-Ruz (✉)
Chemistry and Biochemistry Department, New Mexico State University, Las Cruces, NM, USA
e-mail: pvr@nmsu.edu

K. N. Hosbein
Department of Chemistry, Middle Tennessee State University, Murfreesboro, TN, USA
e-mail: katy.hosbein@mtsu.edu

V. Rosa
Division of Science Education Research, The Cuvette, Wauwatosa, WI, USA
e-mail: vanessarosa@thecuvette.edu

G. L. Cochran, J. D. Adams (eds.), *Equity-Oriented, Discipline-Based STEM Education Research*, Critical Studies of Education 19,
https://doi.org/10.1007/978-981-96-9136-4_4

ment of statistical methods used today and how they are at odds with equity and justice goals. Secondly, it discusses QuantCrit as a methodological framework and its theoretical tenets. Finally, the chapter illustrates how to apply QuantCrit tenets throughout the research process. By adopting QuantCrit, DBER can better promote educational equity and justice for marginalized students.

Keywords QuantCrit · Quantitative methods · Methodological framework · Critical quantitative methods

4.1 Why Do We Need Quantitative Methods Through a Critical Lens?

While there have been recent calls to address educational injustice and inequity in STEM (Asai, 2020; Burrows et al., 2020; Gosztyla et al., 2021; Herbert et al., 2022; Mensah & Sadler, 2021; Mervis, 2022; Reisman et al., 2020; Riegle-Crumb et al., 2019; Rollins, 2021; Sanford, 2020; Thompson et al., 2020; Wilson-Kennedy et al., 2020), access to opportunities, experiences, and broader success of marginalized student groups in STEM classrooms within reforms remains largely unchanged (Gonzales et al., 2002; O'Brien et al., 2020; Prince, 2004; Riegle-Crumb et al., 2019). Historically, educational research has failed to reform systemic issues due to framing inequities as inherent to marginalized individuals' culture, background, family life, or identity rather than deconstructing the contexts and systems in which these individuals are embedded.

Without a theoretical framework, scholars are biased in interpreting data based on their personal experiences, perpetuating narratives of biological determinism and failing to separate what is biological from what is societal. This perpetuates narratives of biological determinism and fails to separate what is biological from what is societal (Shen & Feldman, 2022). For example, meritocracy (or the narrative that success is earned through hard work, even in an unjust society) fails to acknowledge the societal systems that prevent equity; instead, centering failure as a trait inherent to those who share socially constructed categorizations such as race, ethnicity, class, gender, disabilities, etc. (Au, 2016; Blair-Loy & Cech, 2022; Cabrera et al., 2017; Guinier, 2015; Lemann, 2000; Liu, 2011; McNamee & Miller, 2009).

Integrating a critical lens into quantitative methods of equity-centered education research would enhance our nuanced understanding of the issues at hand. While quantitative methods have been built for statistical analysis and generalizability, they often fail to capture the complexity of social phenomena and the contextual factors that shape them. By incorporating a critical lens, researchers can examine the power structures, cultural norms, and historical legacies contributing to educational disparities. This approach can lead to more meaningful and actionable findings that address the root causes of inequities rather than just the symptoms. Additionally, integrating a critical lens can promote inclusivity and center the voices and experiences of marginalized groups in the research process, leading to a more just and equitable education system.

Quantitative methods in equity-centered education research can benefit from a critical lens because it allows for a deeper understanding of the complex and interconnected systems of oppression that affect marginalized communities. By adopting a critical lens, researchers can analyze how power and privilege influence educational outcomes and identify areas where change is needed to promote equity and justice. By incorporating a critical lens into quantitative research, researchers can move beyond a deficit education model and begin to address the root causes of educational disparities (Gillborn & Ladson-Billings, 2017; Ladson-Billings, 2006). This approach can lead to more meaningful and impactful interventions that promote equity and justice in education. A critical research approach allows for a more nuanced understanding of how racism operates in educational settings. Researchers must be willing to challenge dominant narratives and examine how power and privilege shape educational outcomes.

A paradigm can be thought of as a *lens* through which we view the world (Lincoln & Guba, 1985). Researchers have a specific lens they use when conducting research, even if it is not explicitly described. To address educational inequity, it is necessary to use a lens capable of accounting for social systems of imbalanced power. Equity researchers in education should use a critically informed framework to avoid interjecting personal biases from their experiences living in an unjust system. This means emphasizing anti-deficit and systemic perspectives over assigning deficits to individuals subject to the context of an unjust system. QuantCrit, the methodological framework discussed in this chapter, can help researchers use quantitative methods with a critical lens to advance equity and justice in education. Critical frameworks allow researchers to refocus their efforts from reforming marginalized individuals to rebuilding an educational system capable of broadening participation and success for all students, which brings us to the main objective of this chapter: to provide a methodological framework for using quantitative methods with a critical lens for Discipline-Based Education Researchers. We will do this by (1) providing the historical context surrounding the development of the statistical methods we use today and how it is at odds with equity and justice goals; (2) discussing QuantCrit as a methodological framework; and (3) illustrating reflexive questions to consider when applying QuantCrit tenets throughout aspects of the research process—from establishing a research question and choosing a theoretical framework, to experimental design and supporting analytical decisions. Ultimately, we hope to provide an accessible bridge between our field of DBER and the decades of critical scholarship outside of DBER, enriched by philosophies that could help educators and researchers advance equity and justice.

4.2 Historical Context of Quantitative Methods Through Competing Lenses

4.2.1 A Post-Positivist Lens Grounded in Objectivity

Methodologies are inherently neither oppressive nor liberatory (Cokley & Awad, 2013; Zuberi & Bonilla-Silva, 2008). However, quantitative methods are often rooted in a post-positivist lens and therefore framed through an "objective" lens. This lens removes the ideologies of isms (e.g., racism, sexism, etc.) (Hughes, 1995; Zuberi, 2001) and can be, and have been, used to cement oppression. When using a post-positivist lens, theoretical frameworks and practices aiming to reveal and challenge power structures are not applied. Consequently, this frames inequities as a shortcoming of marginalized groups (Garcia et al., 2018; Zuberi, 2001). For example, consider how categories and variables comprised of social constructs (e.g., race, sex, course grades, etc.) are defined. Sex is often conflated with gender and presented in the binary (i.e., male or female), actively excluding those with more fluid gender identities (Nicolazzo, 2018; Suen et al., 2020). This presents the construct of gender as a biological (more objective) truth rather than a social construct researchers need to apply throughout the research process. This means that subjectivity is heavily introduced in the social construction of a key variable; from defining research questions, establishing parameters of data collection, determining rules for data cleaning, conducting analysis, and interpreting results. There is a tremendous amount of subjectivity and opportunity for bias in quantitative work hidden behind the guise of post-positivism.

The dangers of using a post-positivist lens can be found on a larger scale in the foundation of modern statistics and its application in social sciences, and applications in human genetics, genomics, and epigenetics research as well. In the eyes of eugenicists, *biology* is *destiny* (Zuberi, 2001; Zuberi & Bonilla-Silva, 2008); thus, statistics was established as part of the eugenics project aimed to "prove scientifically" the superiority of the White race over the Black race, colonizers over Indigenous nations, and slave-owners over slaves. Scientific history is riddled with examples of quantitative methods used to support pseudoscientific theories that justify oppressive power structures like racism (Rollins, 2021; Wingfield & Chavez, 2020), economic segregation (Pager & Shepherd, 2008), and bio-determinism (Anand, 1999; Byrd & Hughey, 2015; Garcia & Mayorga, 2018; Pearson & Moul, 1927). The previous examples illustrate substantive challenges for fields applying quantitative methods to advance equity research, including DBER. There is a need to delineate liberatory methodological practices and guidelines that can empower researchers to rise to the challenge of advancing equity in education by countering post-positivism and notions of "objectivity" by acknowledging the social context, and thereby the subjectivity, of the data we collect and analyze (Pearson et al., 2022).

4.2.2 A Critical Lens Grounded in Justice and Liberation

In general terms, research that follows a critical lens is explanatory, emancipatory, practical, and normative (Covarrubias & Vélez, 2013; Gillborn & Ladson-Billings, 2017; Misiaszek & Torres, 2019; Torres, 2002; White, 1983). Critical quantitative inquiry refers to the application of a critical lens to shift quantitative inquiry from *post-positivist* perspectives to perspectives that *challenge the status-quo* (Hernández, 2015). Critical research seeks to explain systems of oppression and how they operate, identify the actors to change it, and provide clear norms for criticism and achievable practical goals for social transformation (Gottesman, 2016). However, the lens shift from *post-positivism* to *critical* comes with challenges and questions about how to apply critical quantitative inquiry in education, as these two lenses hold conflicting philosophical assumptions (Denzin & Lincoln, 2011; Lincoln et al., 2011). Table 4.1 provides the assumptions that generally underlie quantitative methods through a post-positivist lens and questions to encourage the shift to a critical lens.

Table 4.1 Post-positivist assumptions and critical questions for the purpose of shifting quantitative methods from a post-positivist to critical lens

Post-positivist assumption	Critical questions
Quantitative methods strive for generalization, defaulting to majoritarian perspectives rather than marginalized perspectives (Hernández, 2015)	How can quantitative methods be used to look at patterns of marginalization and elevate communities that have been traditionally ignored because of where they lie in a "normal distribution?"
Systems of oppression are not amenable to quantification, and this has often resulted in research that attributes causation to social constructions of identity (Sablan, 2019)	How do we account for systems of oppression in educational processes and research?
Quantitative methodologies through a post-positivist lens have engaged in methodological violence by projecting institutionalized definitions of communities and reifying power structures and hierarchies (Nicolazzo, 2018)	How do we subvert these institutional definitions from within the system while also engaging in reconciliation with communities harmed by these practices?
A crucial underpinning of post-positivism is to let the data "speak for itself" (Covarrubias & Vélez, 2013); however, a critical lens asks us to reflect on the values and positionality we bring as researchers	How will this practice of acknowledging positionality be presented and framed in terms of methodological decisions?
The field of DBER emerged from many STEM disciplines that operate under a post-positivist lens by default (Hernández, 2015; Lincoln et al., 2011). Therefore, oppressive narratives and assumptions are encoded into our training and socialization as researchers within DBER, depending on the training and socialization of those who trained us (Pérez Huber et al., 2018)	What guidelines can support researchers interested in engaging with this lens shift?

In conclusion, we argue that our field of DBER could improve its progress toward educational equity by transitioning from a majoritarian-centered, post-positivist lens to lenses specifically designed to reveal and dismantle oppressive barriers. In the following section, we address the challenges listed above by proposing ways for DBER to incorporate a critical lens to research using quantitative methods.

4.3 QuantCrit Tenets Transformed into an Action-Oriented Methodological Framework for DBER

4.3.1 Tenets of QuantCrit

First, it is important to distinguish between a *methodological framework* and *specific methods*. *Methods* are the ways in which we collect and analyze data (e.g., survey, linear regression). A *methodological framework* provides practical guidance structured along a particular theory to guide the user through a process aligned with their research lens (McMeekin et al., 2020). QuantCrit is a specific methodological framework that takes Critical Race Theory (CRT) tenets and adapts them to our understanding of numbers and data (Garcia et al., 2018; Gillborn et al., 2018; Pérez Huber et al., 2018). We have aligned the QuantCrit tenets with characteristics of various critical theories (e.g., LatCrit, Indigenous, Crip, and Queer Theory) to expand the original conception of QuantCrit to an action-oriented methodological framework. Additionally, we have developed strategies to support DBER researchers seeking to engage with critical quantitative inquiry throughout the research process.

The tenets of QuantCrit are presented in Table 4.2 (Original QuantCrit Tenets column). We have paired each QuantCrit tenet to one of the five characteristics of critical theories (Table 4.2, Critical Characteristic column). These critical characteristics were combined by looking at common themes across social critical theorists or theories that were the product of specific social movements that have identified varied forms of oppression of human beings in modern societies (Gottesman, 2016; Levinson et al., 2015; White, 1983). The five characteristics of a critical study are:

1. **Explanatory**—Explain what is wrong with the status quo or current social reality.
2. **Practical**—Identify stakeholders and actors that can change the system.
3. **Reflexivity**—Examine one's beliefs, judgments, and biases.
4. **Counterstories**—Challenge master narratives and elevate the voices of those who are marginalized.
5. **Normative**—Questioning established "norms" that are presented as objective or scientific but actually serve to maintain oppression. For example, challenging the belief that quantitative methods are inherently more "objective" or "valuable" than qualitative methodologies.

Table 4.2 QuantCrit tenets transformed into an action-oriented methodological framework for DBER

Original QuantCrit tenets	Critical characteristic	Reflection	Actionable tenet
The centrality of racism as a complex and deeply rooted aspect of society that is not readily amenable to quantification	Explanatory	What is wrong with the current social reality?	(1) Contextualize Systems of Oppression
Numbers are not neutral and should be interrogated for their role in promoting deficit analyses that serve White racial interests	Practical	Who are the actors that can make changes?	(2) Reorient from Student to System Level
Categories are neither "natural" nor given and so the units and forms of analysis must be critically evaluated	Reflexivity	What beliefs, values, and practices am I bringing into the project? And what is the sociohistorical context of the place?	(3) Grounded in Sociohistorical Context
Voice and insight are vital: critical analyses should be informed by the experiential knowledge of marginalized groups	Counterstories	How is this analysis amplifying the experiences, narratives, and truths of marginalized communities?	(4) Elevate and Honor Marginalized Voices
The understanding that statistical analyses have no inherent value, but they can play a role in struggles for social justice	Normative	What is the limit of quantitative methods within your study design? (The limits will be highly dependent on a particular study's' context and design)	(5) Acknowledge Limitations of Quantitative Methods

Combining both the original tenets and the critical characteristics, we pose questions to reflect on the tenet's purpose through a critical lens (Table 4.2, Reflection). These reflection questions, and those throughout the rest of the chapter, have the purpose of engaging in reflexivity as a field, rather than us acting as gatekeepers of the "truth" or "correct ways of knowing." That is, it would be contrary to the spirit of critical theory for us to claim this chapter encompasses *the only correct way* to engage in QuantCrit. Finally, these characteristics helped us reorient the original QuantCrit tenets in action-oriented form (Table 4.2, Actionable Tenet). We hope action-oriented tenets empower researchers in applying QuantCrit and offer more detailed suggestions for reflection during research implementation throughout the chapter.

4.3.2 *Applying QuantCrit to DBER*

Research is a non-linear process requiring constant engagement in reflexivity, improvement, and adaptation (Gear et al., 2022). Albeit simplified, in this chapter, the research process will be illustrated in four stages: (I) Defining a Research Problem, (II) Building a Critical Theoretical Framework, (III) Research Design, and (IV) Interpretation & Reporting of Conclusions. Through the rest of this chapter, we walk through each of the four stages, describing how a researcher could engage with applied QuantCrit tenets. Throughout each research stage, we highlight a *research milestone.* These milestones are paired with reflexive questions to support coherence of the QuantCrit tenets throughout the research process. We want to emphasize that this chapter is not a step-by-step guide in how to apply QuantCrit tenets. Rather, it provides an inquiry-guided structure for the application of QuantCrit tenets through reflexive questions that should be addressed during every step of the research process.

4.4 Stage I: Defining a Research Problem

DBER focuses on understanding how students develop and learn concepts, practices, and ways of thinking in science and engineering (National Research Council, 2012). However, measuring "success" in a learning environment should not be solely assigned to individual qualities as it amplifies deficit perspectives and narratives of marginalized student failure (Harper, 2010). Deficit-centered framing is often referred to as the "achievement-gap," using White, cis-male, able-bodied students as the benchmark for all student progress and behavior (Gutiérrez, 2008; Gutiérrez & Dixon-Román, 2011; Rodriguez, 2001). DBER researchers must avoid perpetuating or subverting these narratives of supremacy and meritocracy when framing their research questions.

Questions about measurement of knowledge, skills, and dispositions are often used by Discipline-Based Education Researchers to frame their work. However, this framing could lead to work that either perpetuates or subverts narratives of supremacy and meritocracy. When applying a QuantCrit methodological framework to DBER, researchers need to reflect on: (1) Our positionality(ies), (2) Our responsibility toward the marginalized groups we study, and (3) The language used to describe marginalized groups. In other words, reflexivity about one's perspective, the perspectives of the participants, and the broader discourse surrounding these systems can help researchers identify key relations for study and, ultimately, reform.

> The milestone at the end of Stage I is to have a specific research question or questions, a clear understanding of our relationships with the communities we want to study, and who in the research team has experiential knowledge that is important to the endeavor.

4.4.1 Why Do We Need Positionality in Quantitative Research?

The adage "data speak for themselves" assumes data are not (1) Reflective of subjective categorical and quantitative reduction, (2) Subject to sociohistorical context, and (3) Interpreted through the experiences of the reader and researcher (Covarrubias & Vélez, 2013; D'ignazio & Klein, 2020). As a socially constructed system, data pertaining to education is inherently subjective. Whether quantitative or qualitative, acknowledging our role in the systems we influence (positionality) and describing the lens through which data are interpreted (reflexivity) are key to engaging in equity-centered research (Bourke, 2014; Dery, 2020; Lu & Hodge, 2019; Milner IV, 2007; Reich, 2021). Education researchers can enhance their research's transparency, accountability, and credibility by providing positionally statements.

Therefore, *reflecting on your positionality as a researcher* (Table 4.3, *Tenet 3*) is key to defining a research problem using QuantCrit. We acknowledge this is not a common practice in quantitative methods; however, we call on scholars to engage and publish this exercise alongside quantitative studies to make this practice more normative. Positionality statements help readers understand the context from which the research is generated and the potential impact of the researcher's position on the research process and the findings. Positionality statements should include information about the education researcher's research training (i.e., the frameworks they work with, the literature bases they read), their positionality regarding the power dynamics between themselves and their participants, and other stakeholders in the research. When writing positionality statements, authors may struggle with disclosing certain identities or traumatic experiences to their readers. Sharing one's positionality is important because it acknowledges the author's relationship with power, and also can help readers see themselves reflected on the research (Ahmed, 2012). However, it is crucial that authors do not disclose information that may put them in danger or make them feel uncomfortable. Sharing positionality should not be about "bleeding ourselves dry" to gain credibility, but rather about providing readers with a clear understanding of the author's perspective and potential biases (Corlett & Mavin, 2018; Holmes, 2020). Ultimately, positionality statements should be written thoughtfully and with the intention of creating transparency and trust between the author and the reader.

Reflecting on positionality is particularly valuable in STEM education wherein scholars are disproportionately representative of majority-aligned social categories (Lu & Hodge, 2019; Milner IV, 2007; Reich, 2021; Roegman, 2018) and the sociohistorical context of Western science. Western science is fundamentally heteropatriarchal, white supremacist, transphobic, and ableist. For example, any modern scientific endeavor is built over the foundations of scientific colonialism (Anand, 1999; Clarke & Walsh, 2009). Accepted practices and knowledge are designed for those who conform to Eurocentric ways of knowing even though much of what we consider foundational to science today was initially thought, developed, and appropriated from non-European cultures (Bang et al., 2007; Laituri, 2011; Prescod-Weinstein, 2020; Zuberi & Bonilla-Silva, 2008). In addition, the centering of white

Table 4.3 QuantCrit principles applied to Stage I: Defining a research problem and reflexivity questions

Actionable tenet	Applied tenet	Reflexive questions
(1) Contextualize Systems of Oppression	Research problem must explicitly state the oppressive system/s to be studied (e.g., racism, sexism, ableism, etc.)	What intersectional relations are important for your problem of study? • Does your research question center "comparisons" as a key aspect of the research problem? If so, how can you ensure the framing doesn't center individual identities as the source of the differential outcomes?
(2) Reorient from Student to System Level	Research problem needs to center a part of the system that affects students' experiences (e.g., practice, policy, curriculum, or norm)	Who are the possible stakeholders/actors involved in the problem of study (e.g., teachers, administrators, department)? How do the actions of people in power affect student's knowledge, skills, and dispositions? What are the systemic mechanisms?
(3) Grounded in Sociohistorical Context	Reflect on your positionality as a researcher	How do your identities (privileged and/or marginalized) relate to the problem of study? • Is there a history of members of your community committing violence or harm against the community you want to study? • How has your privileged identity allowed you access to STEM in ways other communities have been gatekept? • How can you build trust with the people you want to study? • What lived experiences have shaped your research decisions, practices, approaches, epistemologies, and agendas?
(4) Elevate and Honor Marginalized Voices	Reflect on whether you have the cultural knowledge necessary to understand the community of study or need collaborators or mentors with experiential knowledge	How do you negotiate and balance your own identity and research agenda with the identities and needs of the research participants in your study? • What have you done to educate yourself about the marginalized groups in your study? • How much have you personally engaged with the communities of the marginalized groups in your study? • How are you contributing to injustice, inequity, or oppression by benefiting from the status quo? • What does the community have to gain from your research? • Do any researchers on your team belong to the marginalized community of study and do they have agency to direct the course of the research project? Does your research agenda include ways to listen and include the voices of the community you want to study?

(continued)

Table 4.3 (continued)

Actionable tenet	Applied tenet	Reflexive questions
(5) Acknowledge Limitations of Quantitative Methods	Consider the role of comparisons in advancing progress	Does your research question center "comparisons" as a key aspect of the research problem? • If so, how does the answer to this comparison inform change that must happen in the system? Is this comparison using a privileged group as a benchmark of "ideal" scientific success and behavior?

cis-women in STEM within quantitative studies due to sample sizes omits the intersectional experiences of BIWOC (Black, Indigenous, and Women of Color) and erases Trans and Non-Binary identities (Bauer et al., 2021; Nicolazzo, 2018).

Positionality and *reflexivity* allow scholars to challenge the status quo by acknowledging our individual and collective roles in perpetuating or subverting these systems and resources, as well as the scholarship or experiences on which our perspectives in equity-centered research are grounded. The purpose of engaging in positionality is for the researcher to understand the relations of power, knowledge construction, and social context between themselves, their research question and the community studied (Milner IV, 2007). Reflecting on positionality is especially crucial for researchers who share mostly majoritarian identities e.g. White, cis, straight, able-bodied, neurotypical men. When studying marginalized communities, researchers who share mostly majoritarian identities need to be reflexive about the inclusion of those who have marginalized perspectives pertaining to the study at each stage of research design and implementation. This work is a prerequisite to meeting our responsibilities as stewards of the data and reforms impacting marginalized communities. Moreover, positionality should be specifically tied to the research problem and community we want to engage in. That is, it goes beyond a list of social identities and experiences we have. Rather, it is relational and contextually situated (Sultana, 2007) within the interactions we have with marginalized communities while conducting a study (Spanò et al., 2016). In Table 4.3, *Tenet 3* includes reflexive questions to help you *reflect on your positionality as a researcher.*

4.4.2 Our Responsibility to the Communities We Study

Given the relationship of power between the researchers and the people studied, it is important to state our collective responsibility toward the communities we study. That is, when studying marginalized groups, we must understand that it is not only about the benefits we, as researchers, get from the project (e.g., publications, grants, tenure). When engaging with marginalized groups, it is our responsibility to be intentional about the immediate and sustaining benefits they get from our work. To benefit marginalized groups more effectively, it is crucial to *reflect on whether you have the cultural knowledge necessary to understand the community of study or need collaborators or mentors with experiential knowledge* (Table 4.3, *Tenet 4*).

Furthermore, educational research has a history of researchers from privileged identities studying marginalized groups they have no experiential knowledge with. This has often led to exploitation, community damage, symbolic violence, and misrepresentations (Marshall & Batten, 2004). In no way does this suggest that researchers can only study their communities of origin and avoid participating in equity and justice work if they come from a privileged background (Milner IV, 2007; Roegman, 2018; Stewart & Goddard, 2023; Tillman, 2002). However, it does call for building on one's positionality to get ahead of what Milner IV (2007) called "Dangers Seen, Unseen, and Unforeseen." In other words, how does one know what they do not know? Tillman (2002) argues that to avoid these dangers, it is important to ensure both "researcher 's and research participant's voices are represented in the question of study." This can be achieved in different ways from educating oneself as a researcher deeply on the cultural knowledge necessary to understand the community of study (Roegman, 2018), to ensure the presence of stakeholders as part of the research development and interpretation (e.g., participatory action research) (Bang et al., 2007); or creating a research team that includes people with experiential knowledge. Table 4.3 includes reflexive questions to help you *reflect on whether you have the cultural knowledge necessary to understand the community of study or need collaborators or mentors with experiential knowledge.*

Another key topic regarding responsibility and reciprocity is regarding data ownership. At this stage, it is important to start having conversations and a plan surrounding collecting data from communities. Researchers should engage in dialogue with communities about data ownership and their participation in the research process to avoid falling into extractive colonial practices (Walter et al., 2021). For example, Indigenous peoples have long struggled with the theft and misappropriation of their data, highlighting the need for researchers to work with Indigenous communities to ensure that data ownership remains with the community (Tsosie et al., 2021). This requires a deep understanding of Indigenous sovereignty and the role that data collection can play in either supporting or undermining that sovereignty (Foxworth & Ellenwood, 2022; Kukutai & Taylor, 2016). Researchers must also be aware of the potential power dynamics at play when working with marginalized communities, and work to ensure that these communities are able to participate in the research process in meaningful ways (Wilson, 2001). Ultimately, it is only by engaging in ongoing dialogue with communities about data ownership and participation that researchers can avoid perpetuating harmful and extractive practices.

4.4.3 The Language We Use to Describe Education Problems

We construct research through language, influencing the public dialogue toward a given topic (Wetherell & Edley, 1999). The language we use to discuss educational problems reflects our philosophical assumptions and shapes the perceptions of others as to the role of education in addressing inequity (Ladson-Billings, 2006). For example, consider the quote: "the persistent achievement gaps between various demographic groups" (Granovskiy, 2018); this phrase conveys a perception of equity decontextualized from the education system in several ways. The phrase implies "*persistent achievement gaps*" are an inherent property of "*various demographic groups*," using "*achievement*" to suggest some "*demographic groups*" have accomplished something because of effort or skill, and "gap" to imply mere differences rather than consequences of systemic injustice. This language minimizes the moral urgency of the problem and attributes deficits to students of marginalized, socially constructed identities. To counter this minimization, our research studies need to *explicitly state the oppressive system/s to be studied (e.g., racism, sexism, ableism, etc.)* and *center part of the system that affects student experiences (e.g., practice, policy, curriculum, or norm)* and *consider the role of comparisons in advancing progress* (Table 4.3, *Tenets 1, 2, and 5*).

More than semantics, the phrase reflects a deficit-oriented epistemology. It is likely to inform methodologies seeking resolution by reforming individuals to be more persistent, resilient, or otherwise overcome their societal oppression (Bekki et al., 2013; Duckworth et al., 2007; Harris et al., 2020; Knaggs et al., 2015), rather than improving the societies that produce persistent inequities across individuals. We contribute to the propagation of deficit-oriented language by attributing risk to students (e.g., "at-risk" students with low math "aptitude" scores, underrepresented minorities), conflating sex with gender, and conveying deterministic predictions via gendered behaviors (Mertens, 2019). There is nothing neutral, passive, or objective about the words we use to label objects, constructs, measures, identities, or phenomena because the language we use conveys the meaning we derive from our reality (Frowe, 2001; Scotland, 2012) and (as researchers and educators) this has the power to influence the existence of others. Critical scholars have proposed different conceptualizations to look and name educational outcomes in a way that centers marginalized populations and long-term solutions. We suggest DBER uses terms like "Opportunity Gap" (Akiba et al., 2007), "Educational Debt" (Ladson-Billings, 2006), or "Community Cultural Wealth" (Yosso, 2005) that center the assets and wealth of knowledge of the marginalized groups of study (Harper, 2010). Furthermore, this forces us to really think about *the role of comparisons in advancing progress* (Table 4.3, *Tenet 5*). By doing so, we provide both context to the systems of oppression (Table 4.3, *Tenet 1*) and the systemic context (Table 4.3, *Tenet* 2) for the groups we study.

4.5 Stage II: Building a Critical Theoretical Framework

A theoretical framework represents the formal theory or theories that explain or connect what we know about learning to the empirical work in question. Because the theoretical framework explains the relations between variables, it affects both the outcomes of interest and the interpretation of the results. Therefore, it's crucial that your chosen theoretical framework has theoretical underpinnings suited to answer your specific research question.

Many commonly used theoretical frameworks in DBER are not centered around systems of oppression or marginalized groups' experiences (Vincent-Ruz, 2020). As a result, these status-quo frameworks are oftentimes not adequate to design research pertaining to the experiences of marginalized groups. For example, Santos-Díaz and Towns (2020) used the theory of communities of practice (CoP) as a framework to situate the experiences of underrepresented communities in chemistry outreach. Through this study, the authors investigated factors related to diversity and inclusion that contribute to facilitators' experiences in outreach events. While CoP helped the authors describe new factors, they acknowledged that the CoP framework was a limitation, as it did not explicitly address diversity in its theoretical underpinnings. As a result, the authors encouraged the use of other frameworks when studying diversity in chemistry outreach (Santos-Díaz & Towns, 2020).

To address inequities more fully, we need to either expand these status-quo theories through a critical lens or use existing critical theoretical frameworks that focus on problems and solutions specific to marginalized groups. We are, by far, not the first scholars to have called for the use of critical frameworks in DBER (Metcalf, 2016) and scholars have already begun incorporating critical frameworks within DBER (Cochran et al., 2020; Hyater-Adams et al., 2018; Nardo, 2021; Robertson et al., 2023; Rodriguez et al., 2022; Smith et al., 2022). Examples of existing critical theoretical frameworks that address systems of oppression are Opportunity Gaps (Akiba et al., 2007), Educational Debt (Ladson-Billings, 2006), Community Cultural Wealth (Yosso, 2005), Ethics of Care (Noddings, 1988), Funds of Knowledge (González et al., 2006), and Border Crossing (Anzaldúa, 1987; Bang et al., 2007). Using critical theoretical frameworks can help to move DBER away from using deficit-based frameworks and harmful practices such as gap gazing (Gutiérrez & Dixon-Román, 2011; Shukla et al., 2022).

> The milestone at the end of this stage is to have a literature review that can help you to build a critical theoretical framework to situate your study. By (1) Contextualizing the Research Topic Within Background Literature and (2) Deconstructing Prior Quantitative Work, you will have a clearer understanding of the context in which the topic of your research has been explored and the pieces of your research question or questions that need to be supported through critical theories.

4.6 Contextualizing the Research Topic Within Background Literature

After identifying your possible research question or questions, ask yourself how you can most carefully and accurately describe the problems without the majoritarian (i.e., the group in power) perspective on this issue. A QuantCrit researcher must explicitly reject the oppression-neutrality of a setting that distorts findings toward the deficit perspective, and instead *identifying the oppressive systems* (Table 4.4, *Tenet 1*) that exist historically and politically, within the study context. One must remain consistently grounded in the experiences of the marginalized groups they are working with and must be intimately familiar with the history and past ramifications of policymaking and research findings in their chosen setting. Presenting literature and informed commentary on this operationalized oppression will help define the social fabric upon which the data is situated.

The synthesis of background literature can be thought of as establishing a playing field, which will explicitly reject oppression-neutral settings. Calling these forces out, directly and by name in one's introduction is a strategy to disrupt majoritarian assumptions. In Pérez Huber et al. (2018) study, the authors explicitly "challenge the dominant discourse that frames degree 'value' from a purely economic perspective and provide space for People of Color to define their own meaning of value through their stories." Weatherton and Schussler (2021) have called for researchers to reexamine what "success" means for students. These examples reframe supposedly neutral terms, value and success, to the context of oppression (Pérez Huber et al., 2018; Weatherton & Schussler, 2021). By doing this, the authors explicitly eject the neutrality of their spaces and instead move forward in redefining terms and phenomena applicable to their specific context. The redefining exposes deeper meanings and engages the readers in conversation concerning majoritarian norms that might be construed as neutral, normal, or without fault. By explicitly rejecting the oppression-neutrality of a setting, the conversation is reoriented to center the realities of these "norms" as different systematizations of oppression. Table 4.4 includes reflexive questions to reflect on as you *identify systems of oppression*.

Identifying the systems of oppression will lead to *critically evaluating the prior research* (Table 4.4, *Tenet 2*). By challenging the majoritarian perspective, this process will have already begun. A theoretical framework that assumes that the individual innate characteristics are the strongest predictor of their choices and outcomes will inherently lead to deficit-thinking and analysis when marginalized groups are compared to privileged ones. Whether you are starting a new study or working with previously collected data, it is important to critically assess the way prior literature has either reinforced or challenged deficit-thinking approaches or stereotypes about the marginalized groups you are working with. For example, Van Dusen and Nissen (2020) question how the literature has traditionally focused on reporting the higher DFW rates by racially marginalized groups in physics. To further contextualize this data, they reported prior research on how faculty and advisors discourage BIPOC

Table 4.4 QuantCrit principles applied to Stage II: Building a critical theoretical framework

Actionable tenet	Applied tenet	Reflexive questions
(1) Contextualize Systems of Oppression	Identify systems of oppression	How have systems of oppression been tied to your research topic in prior works? • What norms are present in the extant literature base surrounding your study? • How might neutrality in your research topic have affected the framing of data in the past? • Have authors previously contextualized gaps that were found, or has the presence of marginalization been erroneously termed differentiation? • What information about contextualized oppression is missing? • Can different outcomes be connected to social, political, historical, or institutional actions, or inaction? • How is this acknowledged in prior literature, or how is this argument missing?
(2) Reorient from Student to System Level	Critically evaluate prior research	Which variables (e.g., practices, policies, outcomes) have been previously explored? • Are studies concerning different outcomes generally framed in the light of systemic oppression, or do authors take a more societally neutral stance that blames individual students, groups, or actions for this difference? • What indicators of this oppression have been measured in the past, e.g., scores, stress, etc.? • How have prior results reified the deficit thinking about this group? • How have prior results erased groups related to this system of oppression? • With which purpose and groups in mind were prior measures designed?
(3) Ground on Sociohistorical Context	Situate the marginalized groups within shared experiences	How and why did this problem come to be? • What similar experience, oppression, etc., do these groups of people have to put them under the same category? • Were the categories situated in a similar sociohistorical context as the present study? • Did the authors describe how the categories were chosen and why? • How does the way in which these specific categories were used differ from the way the categories will be defined in this study? • What groups have had their experiences erased by grouping participants in a monolithic way? What assumptions are we making by breaking people into groups?

(continued)

Table 4.4 (continued)

Actionable tenet	Applied tenet	Reflexive questions
(4) Elevate and Honor Marginalized Voices	Utilize past studies that uplift voices from the members of this group	How have minoritized communities described or evaluated this problem? • What have other studies done to express the experiences of the group? What voices still need to be included?
(5) Acknowledge Limitations of Quantitative methods	Critically evaluate prior quantitative work	What are the most common methods used to study this research topic? • What couldn't we know from the methods used to examine this problem? • How has prior work helped us on the road to liberation and positive change for this group? • How has this prior work reified existing oppression?

from pursuing STEM careers, the way peers marginalize Black students, and how an ethos of "filtering out students" further exacerbates the problem. By contextualizing the previous reports of high DFW rates, the issue is reframed from the student deficit to the oppressive system (Van Dusen & Nissen, 2020).

In presenting a study, it is often useful to establish either an additive or corrective focus, as literature concerning systemic oppression is very often deficit framed, or under-cited because of silencing voices who do not resemble or promote majoritarian oppressive norms. An additive focus might address a hole in a literature base previously disconnected from issues of equity and specifically call out this shortcoming. A corrective focus may reject a previous finding in the light of systemic oppression present in this study context and use QuantCrit as a lens through which to reframe these questions in this context. López et al. (2018a, 2018b) argue that "QuantCrit" coupled with intersectional knowledge projects can help us reframe the question from *"what's more important: race, gender, class?"* to *"how can we examine unique constellations of race, gender, class social location as categories of experience in a given sociohistorical and political economic context?"* (López et al., 2018a). Table 4.4 includes reflexive questions to reflect on as you *critically evaluate prior literature.*

4.6.1 Deconstructing Prior Quantitative Work

When study participants are grouped into monolithic categories, it assumes the people within that group have shared experiences. These shared experiences should be expanded upon within the background literature. Studies that highlight the category of interest in your research study should be highlighted. Not only should these studies be mentioned, but they should be critically analyzed for the way the categories were used to avoid the propagation of further marginalization. For example, Irizarry (2015) points out that the previous categorization of demographic groups across educational scholarship on race needed to be reconsidered. They proposed a way to use multidimensional measures of race to capture diverse racialized experiences (Irizarry, 2015). This can be done by *situating marginalized groups within shared experiences* (Table 4.4, *Tenet 3*). For example, Black Americans and Black Immigrants generally face different oppressive structures. If study participants are categorized in a group labeled "Black" and contain both Black Americans and Black Immigrants, the shared experiences of these groups in reference to the purpose of the research study needs to be carefully considered (Griffith et al., 2011). Table 4.4 includes reflexive questions to reflect on as you work to *situate marginalized groups within shared experiences* (López et al., 2018b).

Just as many physicists may struggle to critically evaluate the work of a chemist due to the different disciplinary norms, a person who does not share the lived experiences of a population under study can provide limited insight as to the challenges faced or advancement of efficacious solutions. Therefore, we should aim to *utilize past studies that uplift voices from the members of marginalized groups* (Table 4.4, *Tenet 4*). For example, Howe et al. (2022) investigated graduate student women's perception of faculty careers. In their synthesis of the literature, the authors highlight existing literature studying women's perceptions of faculty careers from the perspective of women. They also bring attention to the issue that by only including the perspective of women who have already chosen to pursue a faculty career, the studies are not highlighting voices of women who are marginalized before reaching that stage in their career choices (Howe et al., 2022). Being conscious of the voices that have been included in the research topic you're studying is crucial to understanding the voices that have been left out. Table 4.4 includes reflexive questions to reflect on as you work to *utilize past studies that uplift voices from the members of marginalized groups.*

Finally, it is necessary to *critically evaluate prior quantitative work* (Table 4.4, *Tenet 5*) when synthesizing the background literature surrounding your research topic. Depending on its application and interpretation, statistics can be used to uncover power structures of oppression and how these structures manifest or to reinforce the homogenization of oppressive, majoritarian perspectives (Zuberi & Bonilla-Silva, 2008). One-way scholars can unpack which role statistics plays in a field or topic is to examine the work along a spectrum of methodological formalism (a single approach) and pluralism (the application and appreciation of multiple approaches). By situating prior research in the context of methodological pluralism, scholars can begin to identify considerable gaps in the methodology (and thereby the epistemological stances used) to approach a phenomenon (Moss & Haertel, 2016). Table 4.4 includes reflexive questions to reflect on as you work to *critically evaluate prior quantitative work.*

4.7 Stage III: Methods and Study Design

As we have previously emphasized, the use and interpretation of statistics are part of socially constructed processes reflected by our decisions as researchers. The data we collect is the product of stances we take during the research design process. Quantitative research design requires the linking of theory, measurement, and statistical tools. Strong research design ties these three elements together to build a cohesive whole. A comprehensive guide of each step in the research design process is beyond the scope of this book chapter. However, we want to highlight three key steps: (1) Choosing Critical Outcomes and Measures, (2) Hypothesized Relations Between Variables, and (3) Representing Social Identities. A discussion on choosing a specific statistical technique is intentionally left out. QuantCrit does not aim to change the mathematical models applied to our studies, rather it aims to rethink the assumptions, theories, and inferences surrounding them.

> The milestone at the end of this stage is to have the start of a data collection plan.

4.7.1 *Choosing Critical Outcomes and Measures*

Measuring systemic inequities is an imperfect and complicated process. The oppressive systems are not amenable to quantification in the sense that we can calculate a systemic racism score (Garcia et al., 2018). Part of applying QuantCrit to DBER is stepping away from simplistic outcome measurements that only equate success to final grades in a specific course. Therefore, it is first important to *embrace the complexity of data and the complexity of meaning of outcomes we choose* (Table 4.5, *Tenet 1*). Our choice of measure is just an approximation to the concepts we state in our framework, and we must understand and explicitly acknowledge that whatever measure we collect will be limited and imperfect.

Systems of oppression are simultaneously enmeshed in society through history and structures (Bonilla-Silva, 2021), while being perpetuated by overt and covert individual and institutional actions (Banaji et al., 2021; Boynton-Jarrett et al., 2021; Christian, 2019; Jones, 2000). Therefore, measurement of systemic oppression in education is a current methodological challenge in DBER. Oppression acts through the actions of people, exclusionary policies, barriers to access, language, etc. (Hardeman et al., 2022; Irizarry & Cohen, 2019; Rodriguez-Lonebear, 2021; Thornhill, 2019). These variables are more amenable to quantification. The fields of health and medicine, for example, have studied indices of neighborhood segregation, economic indexes, etc. Other studies have taken this a step further by treating oppression as multifaceted measures through latent variable models where multiple

Table 4.5 QuantCrit principles applied to Stage III: Methods and study design

Actionable tenet	Applied tenet	Reflexive questions
(1) Contextualize Systems of Oppression	Embrace the complexity of data	What specific systems of oppression do you want to measure? • What processes can you measure and collect data from that serve as proxy or transmitters of systemic oppression
(2) Reorient from Student to System Level	Design studies capable of informing systemic change	What systems need to be measured in this study? How useful are these data in describing the system from which data were collected? Who has the power to influence change within the systems that we are measuring?
(3) Ground on Sociohistorical Context	Ground social identities within sociohistorical context	What are the social identities that your participants share? • What specific social identities share the types of oppression in your study? What are the intersectional social identities that could describe different experiences within the social identity categories you decide to use?
(4) Elevate and Honor Marginalized Voices	Ground outcomes and measures within experiences and values of marginalized groups	How has success been defined in your research topic? • Whose voices are included in the ascription of value to your chosen outcomes? • What validity and reliability evidence exists for your measures of interest, specifically with the marginalized groups in your study?
(5) Acknowledge Limitations of Quantitative Methods	Collect descriptive data	Which social identities won't be amenable to quantification in your study? What qualitative method could be used to represent the experiences of social identities with small sample sizes?

proxies are used at once (Brown & Homan, 2022; Chantarat et al., 2021; Dougherty et al., 2020). Regardless of the chosen method, we invite DBER researchers to innovate on this topic. To account for systemic oppression quantitatively, our theories and hypothesized relations need to explicitly name the mechanisms through which oppression operates and use them as proxies for hegemony (Hardeman et al., 2022). Table 4.5 includes reflexive questions to reflect on as you work to *embrace the complexity of data.*

For example, standardized testing has been the key gatekeeper to the educational system in the twentieth century (Berger & Berger, 2012). The origins of standardized testing go back to the eugenics movement in statistics initiated by Francis Galton (Gillham, 2001). The eugenics movement took shape in the United States, where privileged ethnic groups were looking for validation of race and class hierarchies (Helms, 2012). It was this mainstream societal push that allowed for the creation of standardized tests like the SAT (Berger & Berger, 2012). Despite claiming to measure student's aptitude for college, several studies have provided evidence that performance in the SAT is highly correlated with family income, school resources, and racial inequality (Dixon-Román et al., 2013; Lemann, 2000). Therefore, it is important that we question how the use of these types of measures in research perpetuates myths of meritocracy when we use them as measures of success.

As a field, we need to expand our understanding and definition of outcomes and measures by *grounding them within experiences and values of marginalized groups* (Table 4.5, *Tenet 4*). It is important that we expand our definitions of what it means to be successful in STEM (Weatherton & Schussler, 2021). Moreover, we need to understand that certain communities value different outcomes or ascribe different meanings to outcomes like degrees, grades, etc. (Pérez Huber et al., 2018). These differences in priority do not make marginalized groups less interested in education or invested in their own definitions of success (Major & Godwin, 2018; Quadlin & Conwell, 2021). For example, many measures of success center college completion "on time" in terms of 4 years. However, this benchmark is predicated on a higher-education model where students enroll full-time and have no family responsibilities (Callender, 2008; Darolia, 2014). This benchmark uses a privileged group of individuals as "ideal" and comparisons against the group is only setting students, who do not fit this model, up for failure (Levine & Levine, 2014; Nam, 2020). Moving forward, we must conceptualize success more broadly as both a collection of malleable, individual traits (e.g., knowledge, skills, and dispositions) and the dismantling of barriers to access (Vincent-Ruz, 2020).

A discussion of measures used to quantify abstract concepts must happen within broader discussions of validity and reliability. QuantCrit is complementary to these topics and applying its tenets will lead to improved evidence for validity and reliability for data surrounding marginalized groups. Whenever selecting a prior developed measure is paramount to look for validity evidence that this measure has been successfully applied in the community we want to study. Furthermore, measure design must consider the experiences of the people you are studying. When comparing across marginalized groups, we need to take care to provide validity and reliability evidence for each group in the comparison. In addition, there must be evidence that the measures function in the same way for both groups (Rocabado et al., 2020). If we do not provide this evidence, the comparisons will not reflect what we intended to compare. Table 4.5 includes reflexive questions to reflect on as you work to *ground outcomes and measures within experiences and values of marginalized groups.*

4.7.2 Hypothesized Relationship Between Variables

From the very beginning of our hypothesized relations, we must include systemic variables which leads us to design studies capable of informing systemic change (*Tenet* 2, Table 4.5). A theoretical framework guides our decisions about how concepts are related. One of the most important considerations in QuantCrit method development is the decision of how marginalized social identities play a role in these relations and the hypothesized mechanisms behind disparities. Using terms like "opportunity gap" or "education debt" is not useful if the hypothesized relations center social identities as predictors without additional measurement of how the systemic structures are oppressive. While measuring individual level variables is not problematic by itself, centering of individual-level variables as the cause of disparities is problematic, because it perpetuates reification of scientific racism and centers the fault on individuals rather than systemic structures (Conwell, 2021; Sewell, 2016; Stanfield, 1995). Having systemic variables will connect results to the actors that have the power to make change in the oppressive system. Table 4.5 includes reflexive questions to reflect on as you work to *design studies capable of informing systemic change*.

4.7.3 Representing Social Identities

Another important methodological consideration within QuantCrit is the need for indicators that represent social identities that go beyond traditional demographic measures. It is crucial that we *ground social identities in sociohistorical contexts* (Table 4.5, *Tenet 3*) because social identities act as imposed and chosen categorizations that seldom can be described in simplistic categorical variables. For example, Latinx students are often grouped together in statistical analyses due to their belonging to the same ethnicity (Hargrove, 2019; Irizarry, 2015; Lopez, 2003). However, studies seldom discuss the limitations that this label has. Latinx students can have very different educational experiences due to their skin color, racial identity, immigration status, and socioeconomic status (López et al., 2018b). As a second example, the racial category of Asian represents an array of ethnicities and cultures that are often conflated, hiding the difficulties of students who identify as Asian but do not benefit from the status quo (Goyette & Xie, 1999; Ngo & Lee, 2007). People with different social identities experience systemic oppression in intersectional ways and it is important that we develop multidimensional measures that can allow us to capture the complexity of identity and intersecting systems of oppression by grounding these identities within the appropriate sociohistorical contexts (López et al., 2018b). Table 4.5 includes reflexive questions to reflect on as you work to *ground social identities within sociohistorical contexts* (Jenkins, 1994).

Finally, we want to emphasize the importance of *collecting descriptive data* (Table 4.5, *Tenet 5*). Methodological violence is committed when erasing social identities from our studies when they represent a very small sample size in our final sample. We acknowledge quantitative methods have various requirements for sample sizes, but limiting analysis to groups with sufficient sample sizes perpetuates hegemonic systems. For example, when studying gender, oftentimes quantitative measures only report results for cis-women and cis-men. This completely erases trans students from the narrative (Nicolazzo, 2018). A second example can be found in populations of Indigenous heritage, wherein sample sizes tend to be small due to colonization and other violence (Rodriguez-Lonebear, 2021). We contend that in our research design, we must include qualitative ways to collect data on the groups of people we know may not reach sufficient sample size, so we can increase their visibility and experiences.

We have observed (and in some cases even participated in) summarizing these populations as a single category ("less than 5%" of the population). Instead, we recommend listing these populations with the sample sizes reflected and acknowledging the context and limitations of educational systems (and broader social systems) that may not effectively represent or support these populations. There is also a benefit to be gleaned from small-sample quantitative (e.g., single case or AB testing) or qualitative research. Adding small open-ended questions on our quantitative surveys could provide ways to elevate the voices of those who won't be present in our analysis. Table 4.5 includes reflexive questions to reflect on when you *collect descriptive data*.

4.8 Stage IV: Interpretation and Reporting of Conclusions

At its core, this stage is about the negotiation of your perspective as a researcher, your participant's voice, and the prior perspectives and narratives of the community that have been reported in the literature (Milner IV, 2007). This dialogue between narratives and counter-narratives has been discussed in the critical literature in terms of contribution to policy, research, and theory (Ladson-Billings, 1998; Solorzano & Yosso, 2001). Furthermore, explicitly discussing this tension between status-quo and counter-narratives allows for a discussion of how the DBER field has, in the past, perpetuated hegemonic perspectives and helps deconstruct the past to create a way forward as a field. To discuss the application of QuantCrit tenets to this stage, we will center the discussion around (1) tensions while interpreting data and (2) limitations of data.

4.8.1 Tensions When Interpreting Data

When interpreting your results, they need to be *grounded within the oppressive systems described* (Table 4.6, *Tenet 1*). In addition, we need to *emphasize systemic solutions to systemic problems* (Table 4.6, *Tenet 2*). Using the systems of oppression

Table 4.6 QuantCrit principles applied to Stage IV: Interpretation and reporting of conclusions

Actionable tenet	Applied tenet	Reflexive questions
(1) Contextualize Systems of Oppression	Ground implications within the oppressive systems described	How can the systems of oppression explain the results? • What is the mechanism through which these systems operate over individuals? • Through which mechanisms did this system oppress and how could this mechanism be reformed?
(2) Reorient from Student to System Level	Emphasize systemic solutions to systemic problems	Who are the actors that can make change? • Which actor has the most influence over the system of study? • How is systemic impact reflected in the measures and results presented?
(3) Ground on Sociohistorical Context	State limitations to the categories chosen and whose voices were erased	What nuance is lost with the categorical variables you chose to describe social identities? • In which ways could these measures be improved in the future? What could be addressed in future studies with more data, measure development, etc.?
(4) Elevate and Honor Marginalized Voices	Avoid majoritarian, monolithic, and deficit-oriented interpretations	How does this analysis amplify voices of minoritized communities? • How is your analysis telling a counter-story to prior status-quo narratives?
(5) Acknowledge Limitations of Quantitative Methods	Bound implications by the limitations of constructs and assumptions	How can future research address the limitations of this approach? • Which aspects of how the system influences the variables measured cannot be readily quantified or accounted for within these analyses? • How could the perspective of the research team be limiting these interpretations? • How can future research address the limitations of this approach? What variables could be better addressed through a qualitative study?

as context and focusing on systemic solutions will provide the clearest path of where efforts should be made to provide justice for the marginalized groups within our study.

We highlight Tillman's (2002) comment questioning "whether the researcher has the cultural knowledge to accurately interpret and validate the experiences" (p. 4) of others in a study. This comment was made under a qualitative context, but equally applies to quantitative interpretations. When looking at and making sense of patterns of data, we filter our interpretations through the theoretical frameworks and voices we chose to use (Pérez Huber et al., 2018). To *avoid majoritarian, monolithic, and deficit-oriented interpretations* (Table 4.6, *Tenet 4*), we suggest supporting your interpretation with prior qualitative studies that have a connection to your research topic. Furthermore, we suggest comparing this interpretation with prior explanations that have been deficit-oriented to help deconstruct harmful narratives for other researchers in the field. Table 4.6 includes reflexive questions to reflect on as you work to *ground implications within the oppressive systems described, emphasize systemic solutions to systemic problems,* and *avoid majoritarian, monolithic, and deficit-oriented interpretations.*

4.8.2 Limitations of Data

Every study will have its limitations. To minimize harm to marginalized communities, it is key that we recognize the limits of our results and implications. *Bounding implications by the limitations of constructs and assumptions* (Table 4.6, *Tenet 5*) involves reflecting on every application of Tenet 5 during the research process and discussing how that affects our interpretation and generalization of results. Each of the guiding questions presented previously for Tenet 5 should serve well to write limitations in addition to the additional reflexive questions we include for *stating limitations to the categories chosen and whose voices were erased* (Tenet 3) and *bounding implications by the limitations of constructs and assumptions* (Tenet 5) in Table 4.6.

Depending on your study you will be able to speak to implications for research, practice, or policy. We suggest framing your reported conclusions using those categories, to help other researchers in the field build on your results to push systemic change forward.

4.9 Closing Remarks

The QuantCrit methodological framework emphasizes a transdisciplinary approach that draws upon critical theories from various fields to address inequities in education. To apply this framework to DBER, the following key implications should be considered:

1. Characterizing the practices, policies, and norms of a learning environment (and other oppressive, socially constructed systems) allows researchers to interpret data with the recognition that not all reforms will have the same impact on diverse groups of individuals experiencing a spectrum of intersectional privilege and oppression. Researchers must consider how systemic oppression affects the experiences of marginalized groups in STEM education and analyze the ways in which reforms may have differential impacts on these groups.

2. To prevent advancing oppressive narratives despite the original intention of promoting educational equity, it is crucial to provide context for the experiences of marginalized students in STEM education. By contextualizing the challenges faced by underrepresented groups in STEM, researchers can avoid framing these students as deficient and instead recognize the systemic factors that contribute to their marginalization.
3. It is crucial to account for sociohistorical factors when studying education systems and outcomes. By failing to account for the sociohistorical influence on systems of education, we severely biasing studies to the norms and experiences of the majority. Researchers must consider the historical and cultural contexts that shape STEM education and acknowledge the ways in which these factors contribute to inequities in the field.
4. Without elevating the voices of marginalized groups, we seek to support, we run the risk of misinterpreting the meaning of our measures (e.g., what "success" means to and within a specific community), why our reforms were or were not successful, and are actively participating in perpetuating majoritarian narratives/silencing. Researchers should prioritize the voices and experiences of marginalized groups and actively work to center these perspectives in their research.
5. Recognize that subjectivity is inherent to quantitative measures and acknowledge the limitations and boundaries of the data used to develop equity-centered reforms. Researchers must recognize the limitations of quantitative measures in capturing the complexity of equity issues in STEM education and consider how these limitations may affect the development and implementation of equity-centered reforms.

Below, we have expanded on these points, providing more resources for individuals seeking to broaden their understanding of crucial literature from Discipline-Based Education Researchers integrating Critical perspectives.

First, characterizing the practices, policies, and norms of a learning environment (and other oppressive, socially constructed systems) allows researchers to interpret data with the recognition that not all reforms will have the same impact on diverse groups of individuals experiencing a spectrum of intersectional privilege and oppression. Understanding the practices, policies, and norms of a learning environment is important for interpreting data in a way that recognizes the intersectional impact of reforms on diverse groups of individuals. A critical approach to studying education must consider the historical and social context in which learning occurs (Jenks, 2011). Oppressive, socially constructed systems can significantly impact marginalized groups' experiences in learning environments. Intersectionality, or how different aspects of identity intersect to create unique experiences of privilege and oppression, must be taken into account when analyzing data from learning environments (Stewart & Nicolazzo, 2018).

To prevent advancing oppressive narratives despite the original intention of promoting educational equity, it is crucial to provide context for the experiences of marginalized students in STEM education. Adding context to the experiences of marginalized students in STEM education is not only important for preventing oppressive narratives, but also for making invisible systems of oppression tangible and visible for change. By providing historical and societal context, we can understand how systemic inequities have shaped the experiences of marginalized groups in STEM education (Gloria, 2006). This understanding can help us identify and address the root causes of inequities and work toward dismantling oppressive structures. Without context, we risk perpetuating harmful narratives and failing to address the systemic barriers that prevent educational equity. Therefore, a critically informed approach that centers the experiences and perspectives of marginalized groups is essential for advancing educational equity and disrupting oppressive narratives (Harper, 2010; Kolluri & Tichavakunda, 2022; Milner IV, 2012).

It is crucial to account for sociohistorical factors when studying education systems and outcomes. Education is not a neutral entity, and its development and practices are shaped by historical and societal factors. Ignoring these factors can result in biased studies that reinforce normative assumptions and fail to address the root causes of inequality in education. A critical perspective that considers sociohistorical factors can help challenge dominant narratives and promote more equitable outcomes for historically marginalized groups (Au, 2016; Brathwaite, 2017). Researchers need to engage in ongoing critical reflection and inquiry, drawing on diverse perspectives and approaches to account for the complex social and historical contexts in which educational systems operate. Only by doing so can we effectively address issues of bias and inequality in education and work toward achieving educational equity for all.

Without elevating the voices of marginalized groups, we seek to support, we run the risk of misinterpreting the meaning of our measures (e.g., what "success" means to and within a specific community), why our reforms were or were not successful, and are actively participating in perpetuating majoritarian narratives/silencing. Elevating the voices of marginalized groups is essential to accurately understanding the meaning of the success and failure of educational reforms within specific communities. As noted by scholars in critically informed discipline-based education research, marginalized groups have unique perspectives and experiences that can inform and enhance our understanding of educational practices and outcomes (Pérez Huber et al., 2018; Solorzano & Yosso, 2001). Ignoring the voices of marginalized groups can perpetuate majoritarian narratives that silence and marginalize these groups, hindering progress toward equity and inclusivity in education. Inclusive educational practices that elevate the voices of marginalized groups can lead to more equitable and effective educational outcomes. As noted by scholars in culturally responsive teaching, incorporating diverse perspectives and experiences into education can enhance student engagement and achievement (Sanders Johnson, 2022; Smith et al., 2022).

Recognize that subjectivity is inherent to quantitative measures and acknowledge the limitations and boundaries of the data used to develop equity-centered reforms. "Quantitative measures are inherently subjective because they involve human judgment and interpretation" (Strauch, 1976). Therefore, it's essential to recognize that any data derived from such measures will also involve some level of subjectivity. Data used for equity-centered reforms has limitations: quantitative data often fails to capture the complete picture of the issues being addressed and that context-specific factors can significantly influence the interpretation of data. Data can only provide a limited perspective on complex social issues, and it's essential to recognize data limitations to avoid oversimplification or misinterpretation.

The QuantCrit framework offers a transdisciplinary approach that integrates critical theories from diverse fields to tackle inequities in education. Its application to DBER requires researchers to scrutinize the learning environment and other oppressive systems, contextualize the experiences of marginalized students in STEM, consider sociohistorical factors, amplify marginalized voices, and acknowledge the subjectivity inherent in quantitative measures. By adopting a critical perspective, researchers can unearth and confront the root causes of educational inequality and strive toward educational equity for all. This involves engaging in continuous critical reflection and inquiry to challenge dominant narratives that perpetuate systemic oppression and promote equitable outcomes for historically marginalized groups.

DBER research holds immense potential to not only impact the scientific community but also society at large. Its findings can inform the development of evidence-based policies in education and science, leading to advancements that benefit humanity. DBER research can also help develop more effective teaching methods and curricula in STEM fields, enabling educators to better prepare future generations of scientists and researchers. By understanding how people learn and identifying effective teaching practices, we can make significant strides in advancing science and technology for the betterment of society. Therefore, it is crucial for DBER researchers to reconnect with the communities they study and regain their trust, as our work impacts how we can best educate future generations of scientists and researchers. The ideas presented in this chapter offer a pathway toward achieving this goal and unleashing the transformative potential of DBER research.

References

Ahmed, S. (2012). *On being included*. Duke University Press.

Akiba, M., LeTendre, G. K., & Scribner, J. P. (2007). Teacher quality, opportunity gap, and national achievement in 46 countries. *Educational Researcher, 36*(7), 369–387.

Anand, S. S. (1999). Using ethnicity as a classification variable in health research: Perpetuating the myth of biological determinism, serving socio-political agendas, or making valuable contributions to medical sciences? *Ethnicity and Health, 4*(4), 241–244.

Anzaldúa, G. (1987). *Borderlands/la frontera*.

Asai, D. (2020). Excluded. *Journal of Microbiology & Biology Education, 21*(1), 10.

Au, W. (2016). Meritocracy 2.0: High-stakes, standardized testing as a racial project of neoliberal multiculturalism. *Educational Policy, 30*(1), 39–62.

Banaji, M. R., Fiske, S. T., & Massey, D. S. (2021). Systemic racism: Individuals and interactions, institutions and society. *Cognitive Research: Principles and Implications, 6*, 1–21.

Bang, M., Medin, D. L., & Atran, S. (2007). Cultural mosaics and mental models of nature. *Proceedings of the National Academy of Sciences, 104*(35), 13868–13874.

Bauer, G. R., Churchill, S. M., Mahendran, M., Walwyn, C., Lizotte, D., & Villa-Rueda, A. A. (2021). Intersectionality in quantitative research: A systematic review of its emergence and applications of theory and methods. *SSM-Population Health, 14*, 100798.

Bekki, J. M., Smith, M. L., Bernstein, B. L., & Harrison, C. (2013). Effects of an online personal resilience training program for women in STEM doctoral programs. *Journal of Women and Minorities in Science and Engineering, 19*(1).

Berger, S., & Berger, S. J. (2012). The rise and demise of the SAT. *American Educational History Journal, 39*(1), 165–180.

Blair-Loy, M., & Cech, E. A. (2022). *Misconceiving merit: Paradoxes of excellence and devotion in academic science and engineering*. University of Chicago Press.

Bonilla-Silva, E. (2021). What makes systemic racism systemic? *Sociological Inquiry, 91*(3), 513–533.

Bourke, B. (2014). Positionality: Reflecting on the research process. *The Qualitative Report, 19*(33), 1–9.

Boynton-Jarrett, R., Raj, A., & Inwards-Breland, D. J. (2021). Structural integrity: Recognizing, measuring, and addressing systemic racism and its health impacts. *EClinicalMedicine, 36*.

Brathwaite, J. (2017). Neoliberal education reform and the perpetuation of inequality. *Critical Sociology, 43*(3), 429–448. https://doi.org/10.1177/0896920516649418

Brown, T., & Homan, P. (2022). *Structural racism and health stratification in the US: Connecting theory to measurement*.

Burrows, C. J., Huang, J., Wang, S., Kim, H. J., Meyer, G. J., Schanze, K., Lee, T. R., Lutkenhaus, J. L., Kaplan, D., & Jones, C. (2020). Confronting racism in chemistry journals. *ACS Applied Nano Materials, 3*(7), 6131–6133.

Byrd, W. C., & Hughey, M. W. (2015). Biological determinism and racial essentialism: The ideological double helix of racial inequality. *The Annals of the American Academy of Political and Social Science, 661*(1), 8–22. https://doi.org/10.1177/0002716215591476

Cabrera, N. L., Franklin, J. D., & Watson, J. S. (2017). *Whiteness in higher education: The invisible missing link in diversity and racial analyses: ASHE Higher Education Report* (Vol. 42, Number 6). Wiley.

Callender, C. (2008). The impact of term-time employment on higher education students' academic attainment and achievement. *Journal of Education Policy, 23*(4), 359–377.

Chantarat, T., Van Riper, D. C., & Hardeman, R. R. (2021). The intricacy of structural racism measurement: A pilot development of a latent-class multidimensional measure. *EClinicalMedicine, 40*, 101092.

Christian, M. (2019). A global critical race and racism framework: Racial entanglements and deep and malleable whiteness. *Sociology of Race and Ethnicity, 5*(2), 169–185.

Clarke, S., & Walsh, A. (2009). Scientific imperialism and the proper relations between the sciences. *International Studies in the Philosophy of Science, 23*(2), 195–207.

Cochran, G. L., Boveda, M., & Prescod-Weinstein, C. (2020). Intersectionality in STEM education research. In *Handbook of research on STEM education* (pp. 257–266). Routledge.

Cokley, K., & Awad, G. H. (2013). In defense of quantitative methods: Using the "master's tools" to promote social justice. *Journal for Social Action in Counseling & Psychology, 5*(2), 26–41.

Conwell, J. A. (2021). Diverging disparities: Race, parental income, and children's math scores, 1960 to 2009. *Sociology of Education, 94*(2), 124–142.

Corlett, S., & Mavin, S. (2018). Reflexivity and researcher positionality. In *The SAGE handbook of qualitative business and management research methods* (pp. 377–399).

Covarrubias, A., & Vélez, V. (2013). Critical race quantitative intersectionality: An anti-racist research paradigm that refuses to "let the numbers speak for themselves". In *Handbook of critical race theory in education* (pp. 290–306). Routledge.

Darolia, R. (2014). Working (and studying) day and night: Heterogeneous effects of working on the academic performance of full-time and part-time students. *Economics of Education Review, 38*, 38–50.

Denzin, N. K., & Lincoln, Y. S. (2011). *The Sage handbook of qualitative research.* Sage.

Dery, I. (2020). Negotiating positionality, reflexivity and power relations in research on men and masculinities in Ghana. *Gender, Place & Culture, 27*(12), 1766–1784.

D'ignazio, C., & Klein, L. F. (2020). *Data feminism.* MIT Press.

Dixon-Román, E. J., Everson, H. T., & McArdle, J. J. (2013). Race, poverty and SAT scores: Modeling the influences of family income on black and white high school students' SAT performance. *Teachers College Record, 115*(4), 1–33.

Dougherty, G. B., Golden, S. H., Gross, A. L., Colantuoni, E., & Dean, L. T. (2020). Measuring structural racism and its association with BMI. *American Journal of Preventive Medicine, 59*(4), 530–537.

Duckworth, A. L., Peterson, C., Matthews, M. D., & Kelly, D. R. (2007). Grit: Perseverance and passion for long-term goals. *Journal of Personality and Social Psychology, 92*(6), 1087.

Foxworth, R., & Ellenwood, C. (2022). Indigenous peoples and third sector research: Indigenous data sovereignty as a framework to improve research practices. *Voluntas: International Journal of Voluntary and Nonprofit Organizations*, 1–8.

Frowe, I. (2001). Language and educational research. *Journal of Philosophy of Education, 35*(2), 175–186.

Garcia, N. M., & Mayorga, O. J. (2018). The threat of unexamined secondary data: A critical race transformative convergent mixed methods. *Race Ethnicity and Education, 21*(2), 231–252.

Garcia, N. M., López, N., & Vélez, V. N. (2018). QuantCrit: Rectifying quantitative methods through critical race theory. *Race Ethnicity and Education, 21*(2), 149–157.

Gear, C., Eppel, E., & Koziol-McLain, J. (2022). If we can imagine it, we can build it: Developing complexity theory-informed methodologies. *International Journal of Qualitative Methods, 21*, 16094069211070936.

Gillborn, D., & Ladson-Billings, G. (2017). Critical race theory. In *SAGE research methods foundations*. SAGE Publications.

Gillborn, D., Warmington, P., & Demack, S. (2018). QuantCrit: Education, policy, 'Big Data' and principles for a critical race theory of statistics. *Race Ethnicity and Education, 21*(2), 158–179.

Gillham, N. W. (2001). Sir Francis Galton and the birth of eugenics. *Annual Review of Genetics, 35*(1), 83–101.

Gloria, L.-B. (2006). From the achievement gap to the education debt: Understanding achievement in U.S. schools. *Educational Researcher, 35*(7), 3–12. https://doi.org/10.3102/0013189x035007003

Gonzales, P. M., Blanton, H., & Williams, K. J. (2002). The effects of stereotype threat and double-minority status on the test performance of Latino women. *Personality and Social Psychology Bulletin, 28*(5), 659–670.

González, N., Moll, L. C., & Amanti, C. (2006). *Funds of knowledge: Theorizing practices in households, communities, and classrooms*. Routledge.

Gosztyla, M. L., Kwong, L., Murray, N. A., Williams, C. E., Behnke, N., Curry, P., Corbett, K. D., DSouza, K. N., de Pablo, J. G., & Gicobi, J. (2021). Responses to 10 common criticisms of anti-racism action in STEMM. *PLoS Computational Biology, 17*(7), e1009141.

Gottesman, I. (2016). *The critical turn in education: From Marxist critique to poststructuralist feminism to critical theories of race*. Routledge.

Goyette, K., & Xie, Y. (1999). Educational expectations of Asian American youths: Determinants and ethnic differences. *Sociology of Education*, 22–36.

Granovskiy, B. (2018). *Science, Technology, Engineering, and Mathematics (STEM) education: An overview*. CRS Report R45223, Version 4. Updated. Congressional Research Service.

Griffith, D., Metzl, J., & Gunter, K. (2011). Considering intersections of race and gender in interventions that address US men's health disparities. *Public Health, 125*(7), 417–423.

Guinier, L. (2015). *The tyranny of the meritocracy: Democratizing higher education in America*. Beacon Press.

Gutiérrez, R. (2008). A "Gap-Gazing" fetish in mathematics education? Problematizing research on the civil. In *Transnational and borderland studies in mathematics education* (pp. 69–94).

Gutiérrez, R., & Dixon-Román, E. (2011). Beyond gap gazing: How can thinking about education comprehensively help us (re)envision mathematics education? In M. G. B. A. W. Secada & P. Valero (Eds.), *Mapping equity and quality in mathematics education* (pp. 21–34). Springer. https://doi.org/10.1007/978-90-481-9803-0_2

Hardeman, R. R., Homan, P. A., Chantarat, T., Davis, B. A., & Brown, T. H. (2022). Improving the measurement of structural racism to achieve antiracist health policy: Study examines measurement of structural racism to achieve antiracist health policy. *Health Affairs, 41*(2), 179–186.

Hargrove, T. W. (2019). Light privilege? Skin tone stratification in health among African Americans. *Sociology of Race and Ethnicity, 5*(3), 370–387.

Harper, S. R. (2010). An anti-deficit achievement framework for research on students of color in STEM. *New Directions for Institutional Research, 2010*(148), 63–74.

Harris, R., Mack, M., Bryant, J., Theobald, E., & Freeman, S. (2020). Reducing achievement gaps in undergraduate general chemistry could lift underrepresented students into a "hyperpersistent zone". *Science Advances, 6*(24), eaaz5687.

Helms, J. E. (2012). A legacy of eugenics underlies racial-group comparisons in intelligence testing. *Industrial and Organizational Psychology, 5*(2), 176–179.

Herbert, J. M., Head-Gordon, M., Hratchian, H. P., Head-Gordon, T., Amaro, R. E., Aspuru-Guzik, A., Hoffmann, R., Parish, C. A., Payne, C. M., & Van Voorhis, T. (2022). Words matter: On the debate over free speech, inclusivity, and academic excellence. *The Journal of Physical Chemistry Letters, 13*(30), 7100–7104.

Hernández, E. (2015). What is "good" research? Revealing the paradigmatic tensions in quantitative criticalist work. *New Directions for Institutional Research, 2014*(163), 93–101.

Holmes, A. G. D. (2020). Researcher positionality—A consideration of its influence and place in qualitative research—A new researcher guide. *Shanlax International Journal of Education, 8*(4), 1–10.

Howe, M. E., Kim, M. M., & Pazicni, S. (2022). Graduate student women's perceptions of faculty careers: The critical role of departmental values and support in career choice. *JACS Au, 2*(6), 1443–1456.

Hughes, D. M. (1995). Significant differences: The construction of knowledge, objectivity, and dominance. *Women's Studies International Forum, 18*(4), 395–406.

Hyater-Adams, S., Fracchiolla, C., Finkelstein, N., & Hinko, K. (2018). Critical look at physics identity: An operationalized framework for examining race and physics identity. *Physical Review Physics Education Research, 14*(1), 010132.

Irizarry, Y. (2015). Utilizing multidimensional measures of race in education research: The case of teacher perceptions. *Sociology of Race and Ethnicity, 1*(4), 564–583.

Irizarry, Y., & Cohen, E. D. (2019). Of promise and penalties: How student racial-cultural markers shape teacher perceptions. *Race and Social Problems, 11*, 93–111.

Jenkins, R. (1994). Rethinking ethnicity: Identity, categorization and power. *Ethnic and Racial Studies, 17*(2), 197–223.

Jenks, A. C. (2011). From "lists of traits" to "open-mindedness": Emerging issues in cultural competence education. *Culture, Medicine, and Psychiatry, 35*, 209–235.

Jones, C. P. (2000). Levels of racism: A theoretic framework and a gardener's tale. *American Journal of Public Health, 90*(8), 1212.

Knaggs, C. M., Sondergeld, T. A., & Schardt, B. (2015). Overcoming barriers to college enrollment, persistence, and perceptions for urban high school students in a college preparatory program. *Journal of Mixed Methods Research, 9*(1), 7–30.

Kolluri, S., & Tichavakunda, A. A. (2022). The counter-deficit lens in educational research: Interrogating conceptions of structural oppression. *Review of Educational Research*, 00346543221125225.

Kukutai, T., & Taylor, J. (2016). *Indigenous data sovereignty: Toward an agenda*. ANU Press.

Ladson-Billings, G. (1998). Just what is critical race theory and what's it doing in a nice field like education? *International Journal of Qualitative Studies in Education, 11*(1), 7–24.

Ladson-Billings, G. (2006). From the achievement gap to the education debt: Understanding achievement in US schools. *Educational Researcher, 35*(7), 3–12.

Laituri, M. (2011). Indigenous peoples' issues and indigenous uses of GIS. In *The SAGE handbook of GIS and society, 1996* (pp. 202–221).

Lemann, N. (2000). *The big test: The secret history of the American meritocracy*. Macmillan.

Levine, M., & Levine, A. G. (2014). Coming from behind: A historical perspective on Black education and attainment. *American Journal of Orthopsychiatry, 84*(5), 447.

Levinson, B. A., Gross, J. P., Hanks, C., Dadds, J. H., Kumasi, K., & Link, J. (2015). *Beyond critique: Exploring critical social theories and education*. Routledge.

Lincoln, Y. S., & Guba, E. G. (1985). *Naturalistic inquiry*. Sage.

Lincoln, Y. S., Lynham, S. A., & Guba, E. G. (2011). Paradigmatic controversies, contradictions, and emerging confluences, revisited. *The Sage Handbook of Qualitative Research, 4*(2), 97–128.

Liu, A. (2011). Unraveling the myth of meritocracy within the context of US higher education. *Higher Education, 62*, 383–397.

Lopez, I. H. (2003). White Latinos. *The Harvard Latin American Law Review, 6*, 1.

López, N., Erwin, C., Binder, M., & Chavez, M. J. (2018a). Making the invisible visible: Advancing quantitative methods in higher education using critical race theory and intersectionality. *Race Ethnicity and Education, 21*(2), 180–207.

López, N., Vargas, E., Juarez, M., Cacari-Stone, L., & Bettez, S. (2018b). What's your "street race"? Leveraging multidimensional measures of race and intersectionality for examining physical and mental health status among Latinxs. *Sociology of Race and Ethnicity, 4*(1), 49–66.

Lu, H., & Hodge, W. A. (2019). Toward multi-dimensional and developmental notion of researcher positionality. *Qualitative Research Journal, 19*(3), 225–235.

Major, J., & Godwin, A. (2018). *Towards making the invisible engineer visible: A review of low-socioeconomic students' barriers experiencing college STEM education* (pp. 1–9).

Marshall, A., & Batten, S. (2004). Researching across cultures. *Issues of Ethics and Power, 5*(3).

McMeekin, N., Wu, O., Germeni, E., & Briggs, A. (2020). How methodological frameworks are being developed: Evidence from a scoping review. *BMC Medical Research Methodology, 20*(1), 1–9.

McNamee, S. J., & Miller, R. K. (2009). *The meritocracy myth*. Rowman & Littlefield.

Mensah, F. M., & Sadler, T. D. (2021). A vision that focuses on diversity and broader impact. *Journal of Research in Science Teaching, 58*(3), 307–309.

Mertens, D. M. (2019). *Research and evaluation in education and psychology: Integrating diversity with quantitative, qualitative, and mixed methods*. Sage.

Mervis, J. (2022). *Can US physics overcome its record of exclusion*?

Metcalf, H. (2016). Broadening the study of participation in the life sciences: How critical theoretical and mixed-methodological approaches can enhance efforts to broaden participation. *CBE—Life Sciences Education, 15*(3), rm3.

Milner, H. R., IV. (2007). Race, culture, and researcher positionality: Working through dangers seen, unseen, and unforeseen. *Educational Researcher, 36*(7), 388–400.

Milner, H. R., IV. (2012). Beyond a test score: Explaining opportunity gaps in educational practice. *Journal of Black Studies, 43*(6), 693–718.

Misiaszek, G. W., & Torres, C. A. (2019). Ecopedagogy: The missing chapter of pedagogy of the oppressed. In *The Wiley handbook of Paulo Freire* (pp. 463–488).

Moss, P. A., & Haertel, E. H. (2016). Engaging methodological pluralism. *Handbook of research on teaching, 5*, 127–247.

Nam, Y. (2020). Parents' financial assistance for college and Black-White disparities in post-secondary educational attainment. *Children and Youth Services Review, 110*, 104828.

Nardo, J. E. (2021). Case for narrative inquiry: Understanding the experiences of Elena, a historically minoritized graduate student in chemistry. *Journal of Chemical Education, 99*(1), 113–121.

National Research Council. (2012). *Discipline-based education research: Understanding and improving learning in undergraduate science and engineering*. National Academies Press.

Ngo, B., & Lee, S. J. (2007). Complicating the image of model minority success: A review of Southeast Asian American education. *Review of Educational Research, 77*(4), 415–453.

Nicolazzo, Z. (2018). Starting with trans*: The effects of methodological violence through gender categorization in higher education research. In *Starting with gender in international higher education research*. Routledge.

Noddings, N. (1988). An ethic of caring and its implications for instructional arrangements. *American Journal of Education, 96*(2), 215–230.

O'Brien, L. T., Garcia, D. M., Blodorn, A., Adams, G., Hammer, E., & Gravelin, C. (2020). An educational intervention to improve women's academic STEM outcomes: Divergent effects on well-represented vs. underrepresented minority women. *Cultural Diversity and Ethnic Minority Psychology, 26*(2), 163.

Pager, D., & Shepherd, H. (2008). The sociology of discrimination: Racial discrimination in employment, housing, credit, and consumer markets. *Annual Review of Sociology, 34*, 181–209.

Pearson, K., & Moul, M. (1927). The problem of alien immigration into Great Britain, illustrated by an examination of Russian and Polish Jewish children. *Annals of Eugenics, 2*(1–2), 111–244.

Pearson, M. I., Castle, S. D., Matz, R. L., Koester, B. P., & Byrd, W. C. (2022). Integrating critical approaches into quantitative STEM equity work. *CBE—Life Sciences Education, 21*(1), es1.

Pérez Huber, L., Vélez, V. N., & Solórzano, D. (2018). More than 'papelitos:' a QuantCrit counterstory to critique Latina/o degree value and occupational prestige. *Race Ethnicity and Education, 21*(2), 208–230.

Prescod-Weinstein, C. (2020). Making Black women scientists under white empiricism: The racialization of epistemology in physics. *Signs: Journal of Women in Culture and Society, 45*(2), 421–447.

Prince, M. (2004). Does active learning work? A review of the research. *Journal of Engineering Education, 93*(3), 223–231.

Quadlin, N., & Conwell, J. A. (2021). Race, gender, and parental college savings: Assessing economic and academic factors. *Sociology of Education, 94*(1), 20–42.

Reich, J. A. (2021). Power, positionality, and the ethic of care in qualitative research. *Qualitative Sociology*, 1–7.

Reisman, S. E., Sarpong, R., Sigman, M. S., & Yoon, T. P. (2020). Organic chemistry: A call to action for diversity and inclusion. *Organometallics, 39*(16), 2931–2936.

Riegle-Crumb, C., King, B., & Irizarry, Y. (2019). Does STEM stand out? Examining racial/ethnic gaps in persistence across postsecondary fields. *Educational Researcher, 48*(3), 133–144.

Robertson, A. D., Vélez, V., Hairston, W. T., & Bonilla-Silva, E. (2023). Race-evasive frames in physics and physics education: Results from an interview study. *Physical Review Physics Education Research, 19*(1), 010115. https://doi.org/10.1103/PhysRevPhysEducRes.19.010115

Rocabado, G. A., Komperda, R., Lewis, J. E., & Barbera, J. (2020). Addressing diversity and inclusion through group comparisons: A primer on measurement invariance testing. *Chemistry Education Research and Practice, 21*(3), 969–988.

Rodriguez, A. J. (2001). From gap gazing to promising cases: Moving toward equity in urban education reform. *Journal of Research in Science Teaching: The Official Journal of the National Association for Research in Science Teaching, 38*(10), 1115–1129.

Rodriguez, M., Barthelemy, R., & McCormick, M. (2022). Critical race and feminist standpoint theories in physics education research: A historical review and potential applications. *Physical Review Physics Education Research, 18*(1), 013101. https://doi.org/10.1103/PhysRevPhysEducRes.18.013101

Rodriguez-Lonebear, D. (2021). The blood line: Racialized boundary making and citizenship among Native Nations. *Sociology of Race and Ethnicity, 7*(4), 527–542.

Roegman, R. (2018). Seen, unseen, and unforeseen dangers: What a White emerging scholar learned about positionality in research with racially diverse practitioners. *International Journal of Qualitative Studies in Education, 31*(9), 836–850.

Rollins, O. (2021). Towards an antiracist (neuro) science. *Nature Human Behaviour, 5*(5), 540–541.

Sablan, J. R. (2019). Can you really measure that? Combining critical race theory and quantitative methods. *American Educational Research Journal, 56*(1), 178–203.

Sanders Johnson, S. (2022). Embracing culturally relevant pedagogy to engage students in chemistry: Celebrating Black women in the whiskey and spirits industry. *Journal of Chemical Education, 99*(1), 428–434. https://doi.org/10.1021/acs.jchemed.1c00504

Sanford, M. S. (2020). Equity and inclusion in the chemical sciences requires actions not just words. *Journal of the American Chemical Society, 142*(26), 11317–11318.

Santos-Díaz, S., & Towns, M. H. (2020). Chemistry outreach as a community of practice: Investigating the relationship between student-facilitators' experiences and boundary processes in a student-run organization. *Chemistry Education Research and Practice, 21*(4), 1095–1109.

Scotland, J. (2012). Exploring the philosophical underpinnings of research: Relating ontology and epistemology to the methodology and methods of the scientific, interpretive, and critical research paradigms. *English Language Teaching, 5*(9), 9–16.

Sewell, A. A. (2016). The racism-race reification process: A mesolevel political economic framework for understanding racial health disparities. *Sociology of Race and Ethnicity, 2*(4), 402–432.

Shen, H., & Feldman, M. W. (2022). Diversity and its causes: Lewontin on racism, biological determinism and the adaptationist programme. *Philosophical Transactions of the Royal Society B, 377*(1852), 20200417.

Shukla, S. Y., Theobald, E. J., Abraham, J. K., & Price, R. M. (2022). Reframing educational outcomes: Moving beyond achievement gaps. *CBE—Life Sciences Education, 21*(2), es2.

Smith, T., Avraamidou, L., & Adams, J. D. (2022). Culturally relevant/responsive and sustaining pedagogies in science education: Theoretical perspectives and curriculum implications. *Cultural Studies of Science Education, 17*(3), 637–660. https://doi.org/10.1007/s11422-021-10082-4

Solorzano, D. G., & Yosso, T. J. (2001). Critical race and LatCrit theory and method: Counter-storytelling. *International Journal of Qualitative Studies in Education, 14*(4), 471–495.

Spanò, E., Grimaldi, E., & Serpieri, R. (2016). Positionality, symbolic violence and reflexivity: Researching the educational strategies of marginalized groups. In *Researching marginalized groups* (pp. 134–148). Routledge.

Stanfield, J. H. (1995). The myth of race and the human sciences. *Journal of Negro Education*, 218–231.

Stewart, N. D., & Goddard, Y. (2023). Discussion boards as a pedagogical tool engendering critical race conversations: Meeting anti-racist aims to raise consciousness and disrupt whiteness. *Social Sciences & Humanities Open, 7*(1), 100426. https://doi.org/10.1016/j.ssaho.2023.100426

Stewart, D.-L., & Nicolazzo, Z. (2018). High impact of [whiteness] on trans* students in postsecondary education. *Equity & Excellence in Education, 51*(2), 132–145.

Strauch, R. E. (1976). Critical look at quantitative methodology. *Policy Analysis*, 121–144.

Suen, L. W., Lunn, M. R., Katuzny, K., Finn, S., Duncan, L., Sevelius, J., Flentje, A., Capriotti, M. R., Lubensky, M. E., & Hunt, C. (2020). What sexual and gender minority people want researchers to know about sexual orientation and gender identity questions: A qualitative study. *Archives of Sexual Behavior, 49*, 2301–2318.

Sultana, F. (2007). Reflexivity, positionality and participatory ethics: Negotiating fieldwork dilemmas in international research. *ACME: An International Journal for Critical Geographies, 6*(3), 374–385.

Thompson, S. K., Hebert, S., Berk, S., Brunelli, R., Creech, C., Drake, A. G., Fagbodun, S., Garcia-Ojeda, M. E., Hall, C., & Harshman, J. (2020). A call for data-driven networks to address equity in the context of undergraduate biology. *CBE—Life Sciences Education, 19*(4), mr2.

Thornhill, T. (2019). We want black students, just not you: How white admissions counselors screen black prospective students. *Sociology of Race and Ethnicity, 5*(4), 456–470.

Tillman, L. C. (2002). Culturally sensitive research approaches: An African-American perspective. *Educational Researcher, 31*(9), 3–12.

Torres, C. A. (2002). From the pedagogy of the oppressed to a Luta Continua: The political pedagogy of Paulo Freire 1. In *Paulo Freire* (pp. 119–144). Routledge.

Tsosie, K. S., Yracheta, J. M., Kolopenuk, J. A., & Geary, J. (2021). We have "gifted" enough: Indigenous genomic data sovereignty in precision medicine. *The American Journal of Bioethics, 21*(4), 72–75.

Van Dusen, B., & Nissen, J. (2020). Associations between learning assistants, passing introductory physics, and equity: A quantitative critical race theory investigation. *Physical Review Physics Education Research, 16*(1), 010117.

Vincent-Ruz, P. (2020). What does it mean to think like a chemist? In *Integrating professional skills into undergraduate chemistry curricula* (pp. 57–79). ACS Publications.

Walter, M., Kukutai, T., Carroll, S. R., & Rodriguez-Lonebear, D. (2021). *Indigenous data sovereignty and policy*. Taylor & Francis.

Weatherton, M., & Schussler, E. E. (2021). Success for all? A call to re-examine how student success is defined in higher education. *CBE—Life Sciences Education, 20*(1), es3.

Wetherell, M., & Edley, N. (1999). Negotiating hegemonic masculinity: Imaginary positions and psycho-discursive practices. *Feminism & Psychology, 9*(3), 335–356.

White, S. K. (1983). The normative basis of critical theory. *Polity, 16*(1), 150–164.

Wilson, C. (2001). Decolonizing methodologies: Research and indigeneous peoples. *Social Policy Journal of New Zealand*, 214–218.

Wilson-Kennedy, Z. S., Payton-Stewart, F., & Winfield, L. L. (2020). Toward intentional diversity, equity, and respect in chemistry research and practice. *Journal of Chemical Education, 97*(8), 2041–2044.

Wingfield, A. H., & Chavez, K. (2020). Getting in, getting hired, getting sideways looks: Organizational hierarchy and perceptions of racial discrimination. *American Sociological Review, 85*(1), 31–57. https://doi.org/10.1177/0003122419894335

Yosso, T. J. (2005). Whose culture has capital? A critical race theory discussion of community cultural wealth. *Race Ethnicity and Education, 8*(1), 69–91.

Zuberi, T. (2001). *Thicker than blood: How racial statistics lie*. U of Minnesota Press.

Zuberi, T., & Bonilla-Silva, E. (2008). *White logic, white methods: Racism and methodology*. Rowman & Littlefield Publishers.

Chapter 5
Moving Beyond Deficit: DisCrit and Disability Justice in STEM Postsecondary Education

Lisette E. Torres, Cathery Yeh, Paulo Tan, Daniel L. Reinholz, and Roy Payan

Abstract This chapter focuses on DisCrit and Disability Justice as intersectional lenses to understanding oppression and liberation in STEM education. We introduce these two critical frameworks and provide our own synthesis of their theoretical tenets. We then use these frameworks to analyze the case *Payan v Los Angeles Community College District*, which offers a poignant example of the intersections of ableism and racism in higher education. Using this case as an example, we discuss broader takeaways for STEM and STEM education.

Keywords DisCrit · Disability Justice · Mathematics · STEM education · Intersectionality

L. E. Torres (✉)
TERC, Cambridge, MA, USA
e-mail: lisette_torres@terc.edu

C. Yeh
The University of Texas at Austin, Austin, TX, USA
e-mail: cathery.yeh@austin.utexas.edu

P. Tan
University of Missouri, St. Louis, St. Louis, MO, USA
e-mail: paulotan@umsl.edu

D. L. Reinholz
San Diego State University, San Diego, CA, USA
e-mail: daniel.reinholz@sdsu.edu

R. Payan
University of Southern California, Los Angeles, CA, USA

G. L. Cochran, J. D. Adams (eds.), *Equity-Oriented, Discipline-Based STEM Education Research*, Critical Studies of Education 19,
https://doi.org/10.1007/978-981-96-9136-4_5

5.1 Introduction

Equity has recently become a focal area of educational transformation and research in STEM education. For example, many campuses are rolling out workshops on implicit bias and microaggressions with a hope of transforming their campus culture. Despite the appeal of this approach—given its low cost and ease of implementation—the impact of this work is questionable at best (Chang et al., 2019), and problematic at worst, bringing tensions to the fore but not meaningfully addressing them. Similarly, active learning in STEM higher education is often positioned as a panacea to reduce racial disparities (Theobald et al., 2020). Again, this line of reasoning is overly optimistic and fails to recognize that unless active learning is implemented with meaningful attention to equity, it can result in further inequity (Reinholz et al., 2022). These disparities are driven by a variety of mechanisms, including biased teacher expectations (Copur-Gencturk et al., 2019), interpersonal microaggressions between students (Cooper & Brownell, 2016), stereotype threats (Steele, 1997), and inequitable opportunities to participate (e.g., Reinholz & Wilhelm, 2022). These mechanisms play out across a variety of intersectional identities, including race, class, gender, sexuality, disability, etc., and across postsecondary settings, illuminating the complexities of disrupting the status quo to rightfully transform STEM education in ways that provide equitable opportunities and access to all learners. Minimal reformist approaches (e.g., 1 day workshops, diversifying curriculum, etc.) to change are little more than window dressing that will not result in meaningful change.

To move beyond cheap talk toward meaningful change, we draw upon Disability Critical Race Theory (DisCrit) and Disability Justice as guiding frameworks. These frameworks are inherently intersectional, and we argue that such an intersectional approach is required to make meaningful change. In the absence of cross-identity solidarity, minoritized communities are often pitted against one another to compete for minimal resources in the so-called game of Oppression Olympics. To be clear, pitting communities against one another is a textbook example of White supremacy, capitalism, and colonialism in action. Intersectional frameworks like DisCrit and Disability Justice can help us break out from this trap, as we work to build movements together. In this chapter, we analyze the case of *Payan v Los Angeles Community College District (LACCD)* to show how DisCrit and Disability Justice offer analytic and organizing tools that help us move toward meaningful justice and liberation in STEM education.

5.2 Literature Review

Before delving into the case study, we would like to provide the reader with a contextual understanding of disability within STEM and higher education. It is important to note that more research is still needed within this area, especially as it pertains to the lived experiences of disabled Students of Color at the undergraduate and graduate levels. In the following sections, we share what we know about disability in higher education, highlighting predominant models of disability and focusing on STEM disciplines. We then transition to describing two theoretical frameworks—DisCrit and Disability Justice—that can allow us to identify and describe nuanced understandings of the intersection of race/ethnicity and disability in postsecondary educational spaces as well as address inequities. We spend much of our time describing Disability Justice, as (1) it is an important grassroots activist theory, practice, and way of being and knowing that we want to properly honor; and (2) it is relatively new to the academy. Though we offer Disability Justice as an analytical lens, we are highly aware of and concerned about the academy's long history of knowledge co-optation from marginalized communities. If Disability Justice is something that the reader wants to engage in, then they must be and will be held accountable to sick and disabled queer trans People of Color (SDQTPOC) everywhere.

5.2.1 Disability and STEM in Higher Education

Approximately one in five post-secondary students identify as disabled (National Center for Education Statistics [NCES], 2018). Postsecondary institutions in the United States are required by law under the Americans with Disabilities Act (ADA) and Section 504 of the 1973 Rehabilitation Act to provide accommodations for disabled students. When provided in a timely manner, these accommodations have resulted in positive outcomes, such as improved grade point averages (Kim & Lee, 2015) and retention rates (Mamiseishvili & Koch, 2011). However, disabled college students often express hesitation in seeking accommodations due to perceived negative attitudes from instructors or adverse past experiences with instructors when inquiring about accommodations (Hartman-Hall & Haaga, 2002; Marshak et al., 2010). Notice how the burden of seeking accommodations falls on disabled students (Nieminen & Pesonen, 2022). In fact, only 24% of undergraduates who have a documented IEP or 504 plan in high school decide to register with their university's disability services office prior to the start of classes and only 60% of registered students at 4-year colleges received accommodations (Newman et al., 2011). We agree with scholars (e.g., Prema & Dhand, 2019) who have argued that the burden, duty, care, and responsibilities of accommodations should be equitably shared with instructors, administrators, and higher education systems, particularly in the STEM disciplines.

In STEM fields, the stigma around disability is remarkably strong. For example, Rao and Gartin (2003) found evidence that engineering and law faculty members were significantly less willing to provide accommodations compared to other faculty. Jensen et al. (2004) reported similar findings, highlighting how faculty were reluctant to provide accommodations to students with learning disabilities because they were not convinced of the diagnoses and were preoccupied with fairness related to extended time on exams and assignments. More recently, Mirin and Tan (2022) report that in their study involving 117 college students, mathematics instructors often prevented students' access to accommodations. The authors report limited instances of mathematics instructors' compliance with U.S. federal law. STEM fields, such as engineering and mathematics, have enacted their gatekeeper role (Martin et al., 2010) through educational practices, such as ability tracking and preventing marginalized students from accessing mathematics courses beyond those that are remedial (Treisman, 2013). Thus, compliance with federal law may not be enough to shift practices in higher education, in general, and in the STEM fields, in particular. Saia (2019) points to the importance of disability cultural centers on college campuses as sites of system-level transformations that can assist in the training of college instructors in countering the pervasive deficit-oriented models of disability.

Disability in STEM higher education research has been largely framed through medical and psychological lenses (Reinholz & Ridgway, 2021; Saia, 2019). In the medical and psychological lenses, disability is seen as an individual impairment that needs to be fixed by an expert. From this perspective, disabled students are seen as less than their nondisabled or "normal" peers and additional supports and interventions are required to help disabled students achieve "normality" to their nondisabled peers. Ironically, this traditional, deficit-oriented conception of disability leads to low expectations and a loss of student independence, recreating the inequities we hope to address. The medical model dominates higher education STEM education research, policy, and practice, with the majority of work focused on what disabled students lack (and not what they need) and interventions to fix students. The medical model ignores the historical and structural influences and intersections of disability with other difference markers (e.g., race, gender, social class, language). These factors are regarded as negligible—merely background noise that can be statistically controlled. In contrast to the medical model, the social model examines how disability is caused by the way classrooms, educational systems, and society are organized around disabled students. The social model looks at barriers—physical and ideological—that restrict access, agency, and choice for disabled people. Specifically here, we highlight two frameworks that refuse deficit-oriented master narratives and attend to structural inequities and the agency of students at intersecting oppressions.

5.3 Theoretical Frameworks

5.3.1 DisCrit

DisCrit is a theoretical framework that racializes disability while disabling race (Annamma et al., 2013). It confronts binary notions of normality and simplified conceptions of identity as well as the often-hidden ways in which ableism and racism are simultaneously built into institutional policies and practices (Annamma et al., 2013). It also recognizes "the material and psychological impacts of being labeled as raced or dis/abled" (Annamma et al., 2013, p. 11), articulating the following seven tenets:

1. DisCrit focuses on the ways that racism and ableism move interdependently, often in neutralized and invisible ways, to uphold notions of normalcy.
2. DisCrit values multidimensional identities and troubles singular notions of identity such as race or dis/ability or class or gender or sexuality, and so on.
3. DisCrit emphasizes the social constructions of race and ability and yet recognizes the material and psychological impacts of being labeled as raced or dis/abled, which automatically marks one as "other" and on the margins of Western expectations of what it means to be the "norm."
4. DisCrit privileges voices of marginalized populations, traditionally not acknowledged within research.
5. DisCrit considers legal and historical aspects of dis/ability and race and how both have been used separately and together to deny the rights of some citizens.
6. DisCrit recognizes Whiteness and Ability as Property and that gains for people labeled with dis/abilities have largely been made as the result of interest convergence of White, middle-class citizens.
7. DisCrit requires activism and supports all forms of resistance.

These tenets used individually or as a whole offer a unique lens by which to question and contest equity issues related to disability in STEM higher education.

DisCrit as a theoretical grounding draws attention to how STEM education has often ignored the interwoven relationship between racism and ableism. In a critical review of educator and disability research in mathematics education, Tan and colleagues (2022) noted that there were substantial shifts toward humanization in mathematics education for disabled students; however, studies continued to avoid meaningful intersectional considerations of race and disability. The oppression disabled Students of Color experience through regularities of classroom practice, including otherization, conditional participation, and dehumanization, is systemic and enduring, manifesting daily in local practice in higher education STEM classrooms (Mirin & Tan, 2022; Rao, 2003; Saia, 2019). Inequity is persistent, complex, and (re)produced through social structures, including assumptions embedded in models of teaching and learning, assessments, differentiation, and management. These inequities are made invisible through regularities in practice, allowing racism and ableism to circulate interdependently through notions of normalcy. This includes from whose access needs are met in classroom and lab spaces to whose abilities are recognized and valued as well as who has authority to make decisions in classrooms and school systems (Annamma et al., 2013; Calabrese Barton & Tan, 2020; Reinholz et al., 2020). We will use DisCrit in this chapter to examine *Payan v Los Angeles Community College District (LACCD)*.

As noted by its creators, DisCrit "was not meant to be an all-encompassing theory" (Annamma et al., 2018, p. 63). In fact, the hope was that scholars would pick up the theory and interweave it with other critical theories to uncover entrenched inequalities related to disability and race beyond special education. We believe that we can expand and deepen DisCrit by bringing it in conversation with Disability Justice and by trying to reconnect it to its origins in law by examining a recent court case.

5.3.2 Disability Justice

We extend DisCrit by drawing from the Disability Justice framework, which was developed through the grassroots efforts of disabled queer activists and disabled People of Color (Sins Invalid, 2019) and their demand to shift away from treating disability as a single-issue concern. Disability Justice focuses on the relationship between ableism and other interlocking systems of oppression, such as racism, sexism, imperialism, and capitalism. Disabled activists of color—Patty Berne, Mia Mingus, Stacey Park Milbern, Leroy Moore, Aurora Levins Morales, and others—launched the Disability Justice framework as an expansion upon the Disability Rights Movement, which advanced a philosophy of independent living and opened possibilities for people with disabilities. The original movement, however, only considered ableism as a single-axis issue, leaving the lives of sick and disabled queer trans People of Color (SDQTPOC) invisibilized; the voices, perspectives, and contributions of SDQTPOC were and continue to be erased. As with DisCrit, a guiding principle of Disability Justice is to center the voices of the people most impacted by systems of oppression, rather than the voices of "external experts" or perceived authorities. In this way, the development of the Disability Justice perspective has contrasted other lineages of disability-focused work, such as special education, which is largely driven by nondisabled people.

Research shows that a disability designation can be a marginalizing factor in itself, despite intentions to help disabled people (Artiles, 2019). Thus, it is crucial to listen directly to the voices of disabled people, rather than make assumptions about what is needed. Disability Justice aims to reclaim the term "disabled" as a political statement (e.g., #SayTheWord campaign devised by Lawrence Carter-Long from the National Council on Disability), much in the same way that the term "queer" has been reclaimed. Therefore, consistent with a disability studies perspective, it recognizes the role of society in *disabling* people through ableist systems (Oliver, 2013). Paying attention to the ways that ableism is linked to multiple other systems of oppression, Disability Justice centers the needs and voices of "disabled People of Color, immigrants with disabilities, queers with disabilities, trans and gender nonconforming people with disabilities, people with disabilities who are houseless, people with disabilities who are incarcerated, people with disabilities who have had their ancestral lands stolen, amongst others" (Sins Invalid, 2019, p. 5). Where Disability Justice departs from more academic perspectives is its unapologetic commitment to collective liberation and movement building.

To guide activists and researchers in forming broad-based movements for change, the activists and artists of Sins Invalid developed the *10 Principles of Disability Justice*. The principles, briefly summarized, are: (1) intersectionality, (2) leadership of the most impacted, (3) anti-capitalist politics, (4) cross-movement solidarity, (5) recognizing wholeness, (6) sustainability, (7) cross-disability solidarity, (8) interdependence, (9) collective access, and (10) collective liberation. The ten principles each, and as a collective, offer opportunities for us to recognize the intersecting legacies of White supremacy, settler colonialism, and gendered oppression that are interwoven with and (re)produce ableism in STEM higher education.

5.3.3 Social Identities: Intersectionality and Recognizing Wholeness

Disability Justice centers intersectionality and the ways diverse systems of oppression amplify and reinforce one another. An intersectional framework recognizes, affirms, and interrogates and then works to disrupt hegemonic structures of oppression. Crenshaw (2017) defines intersectionality as a vehicle to represent the lived experiences of Women of Color who simultaneously navigate the complexities of gender, race, language, ability, and other identity politics. This recognizes the need to attend to multiple ways of knowing and being, especially from the ways in which ability is socially constructed in STEM education and how whiteness and ableism are embodied by traditional educational designs. STEM education is not above the fray of racial and ableist reality and politics. It has always been imbued with contestation, hierarchy, and ideologies that privilege those who are already in power (Joseph et al., 2016). Structural intersectionality "refers to the connectedness of systems and structures in society and how those systems affect individuals and groups differently" (Few-Demo, 2014, p. 181) and must be considered in STEM.

While a growing number of critical scholars are attending to STEM education from an intersectional lens, disability has been absent from the conversation. In broader society, racial justice efforts have also not focused on disability. When it comes to disability, the dominant narrative has been around the issues faced by White disabled people, which led Vilissa Thompson to start the social media campaign #DisabilityTooWhite. Disability Justice makes visible the lives of disabled People of Color, immigrants with disabilities or immigration status that led to forced disablement, and disability and carcerality. It provides a richer and more holistic perspective on the complexity of disability and attends to how disability is only one component of a fully embodied experience (alongside race, gender, class, etc.). While disability is seen as a central frame of analysis, Disability Justice activists and scholars attend to how disabled identities are in intersections with queer, gendered, and classed identities. For example, how does racism disable people? How does colonialism disable people?

5.3.4 Sustainability and Change Through Leadership of Those Most Impacted

> There is no neutral body from which our bodies deviate. Society has written deep into each strand of tissue of every living person on earth. What it writes into the heart muscles of five star generals is distinct from what it writes in the pancreatic tissue and intestinal tracts of Black single mothers in Detroit, of Mexicana migrants in Fresno, but no body stands outside the consequences of injustice and inequality … What our bodies require in order to thrive, is what the world requires. If there is a map to get there, it can be found in the atlas of our skin and bone and blood, in the tracks of neurotransmitters and antibodies.—Aurora Levins Morales, *Kindling* (2013)

The Disability Rights Movement often seeks to fight back against a system that often speaks and acts for individuals with disabilities without consulting them, a sentiment encapsulated in the phrase "Nothing about us, without us" (Charlton, 2000, p. 3). Disabled folks possess what DuBois calls "double consciousness" in understanding their subordinate social positions and how others perceive them. They are socialized to feel ashamed of and apologize for their disabled bodies (Taylor, 2021). That is, "dominant groups, by definition, set the parameters within which the subordinates operate" (Tatum, 2000, p. 11). In the realm of education, disabled students are often passive agents in their education trajectories because educational specialists, parents, and educators make decisions about their placements, goals, and accommodations (Annamma et al., 2018). Traditional research has failed to attend to students' lives and experiences in ways that critique or counter the disabling forces of the educational environment.

Disability Justice formed outside of academic spaces, with activist roots from the work and struggle of SDQTPOC folx as a movement for collective liberation. As aforementioned, in 2005, disabled activists of color launched the framework of Disability Justice as a reaction to the U.S. Disability Rights Movement. While the Disability Rights Movement did have "many successes in advancing a philosophy of independent living and opening possibilities for people with disabilities" (Sins Invalid, 2019), considering ableism as a single issue invisibilized the lives of peoples who lived at intersecting junctures of oppression. Consistent across Disability Justice is the leadership of SDQTPOC, centering the priorities and approaches of those most impacted.

Disability Justice activists, organizers, and cultural workers understand that able-bodied supremacy has formed in relation to other systems of domination and exploitation. Ableism is a normal feature of society (e.g., designing a building with no ramps or lifts; failing to make information available in an accessible format; intentional or subconscious discrimination against disabled individuals by believing that intelligence and productivity is connected to able-bodiedness and English fluency) and is embedded within systems and institutions, invisible to most because it is endemic and permanent. Disability Justice centers those most impacted, and their experiential knowledge and ways of being are crucial to understanding, analyzing, and challenging ableist subordination. Nondominant students and communities have ways of knowing, doing, and being to navigate through dynamics of power, ableism, and hierarchies every day. Disability Justice recognizes the resilience, ingenuity, and agency they have had all along. An Anishinaabe scholar, Gerald Vizenor (2008) talks about this as "survivance"—not simply surviving in the face of oppression and marginalization but finding ways to grow, thrive, reimagine, and evolve. It is this way of knowing and being that holds the potential for transforming our education systems.

Disability oppression is deeply rooted in STEM education and connected to degradation, dependency, and powerlessness in which STEM education policies and practices around access and agency are seldom determined *by* the students, but something given by others *to* students. Powell and colleagues (2022), for example, forward and double down on the so-called *truths* of mathematics education as determined by researchers and *valid* evidence while masking and invalidating the voices of multi-marginalized students. Within the realm of research, aside from classroom discourse, the perspective of disabled students is largely absent in research literature on equity in higher education STEM education. Furthermore, aside from publications from the chapter authors (e.g., Tan et al., 2022), there are no articles that use DisCrit or Disability Justice frameworks to analyze learning in the STEM classroom. Centering the perspectives of SDQTPOC can help to unmask and expose the normative processes that disable students in STEM classrooms, illuminate how they resist those normative processes, and contribute to reconceptualizations of equity by identifying factors that support access and retention in STEM pathways.

5.3.5 Intersections of Justice: Interdependence, Cross-Disability Solidarity, Collective Access, and Collective Liberation

Disability Justice offers an opportunity for a new frame to consider equity and social justice work in STEM education with ways in which we might pivot and adapt to center on collective care and intersections of justice. Oppression does not play out uniformly, but rather across the multitude of intersections of social identities and is often disguised through division and intentional pitting of one group against the other (e.g., using Asian Americans as a wedge against our siblings of color) (Museus & Iftikar, 2013). Many of us feel vulnerable ourselves while naming communities that require particular care and are more susceptible to COVID-19, including our elderly and people living with long-term, chronic illness. This is an opportunity to deepen our own understanding of vulnerability, disability, and chronic illness at both the individual and structural level. As authors, we think about accessibility and the goods, services, and rights we do and do not have access to today as womxn, People of Color, immigrants, and people who identify as LGBTQ2IA+. As Disability Justice activist Mia Mingus (2011) reminds us, "Accessibility is concrete resistance to the isolation of disabled people" (para. 3).

There is beauty and value in interdependence. However, a capitalist, ableist society focuses on individual achievement at the expense of what we can do as individuals. What would it mean to redefine success as a community endeavor rather than an individual one? In STEM education, success is measured through individual achievement and performance. What about relying on each other, being able to succeed together? As Synder and Mitchell (2001) suggest, disability studies show us how to problematize the very idea of ability as a singular achievement by an individual: to recognize this is not to deny the very real difference that actual physical disabilities make in the lives of individuals, nor to question the urgency of concrete political and ethical action on behalf of the disabled. It is rather to question the whole model of self-sufficiency, mastery, and perfect ability (not to mention perfectibility) that governs our image of the individual. Scholarship in STEM education has more recently begun to focus on interdependence and care (Padilla et al., in press; Yeh et al., 2021). Now more than ever, we must align ourselves in meaningful social and structural ways with people living with disabilities and deepen our work to ensure an understanding of ableism impacting all our movements for justice. Most prescient in this call for Disability Justice is advocating for our interdependency. Many of us are facing the realization that independence is truly a myth (e.g., Torres, 2020) that we cannot sustain ourselves and our communities alone. We cannot meet all our needs alone. We cannot stay safe alone.

In the sections that follow, we describe a case study involving the struggles of a disabled Student of Color in higher education mathematics courses. Utilizing DisCrit and Disability Justice frameworks, we then analyze this case study, which offers a poignant example of the intersections of ableism and racism in higher education.

5.4 Case Study: *Payan v Los Angeles Community College District*

In 2016, Roy Payan, who identifies as a blind Latino, was struggling with his required mathematics course. It was not because of his ability or motivation (he passed other courses with As and Bs; Vasquez, 2022) but because the math instructor, Mathematics Department, and administrators at Los Angeles Community College District (LACCD), the largest community college district in the state of California, made it virtually impossible for any blind student to successfully pass. Payan recalls,

> When I enrolled in one math class after another, they all used a textbook that was improperly digitized or not digitized at all and would have to be digitized one chapter at a time for me. These textbooks also came with an accompanying CD that had homework lessons, uncaptioned or fully articulated videos, student/instructor interactive chat rooms, and many other features that were completely inaccessible for persons with visual impairments or deafness. After informing the instructor and OSD [Office for Students with Disabilities] at the LACC campus, I was asked to talk with the Dean and Chair of the Math department. When I met with them, I was repeatedly told by both the Chair and Dean of the department, that they could not do anything about it. (Essig, 2022, para. 8)

Payan was not provided with accessible materials for his math class, despite his numerous requests and conversations with the powers-that-be as he went up the chain of command (Vasquez, 2022; R. Payan, personal communication, June 7, 2022).

In an attempt to navigate the situation, Payan transferred classes … multiple times. He came across ableist instructors at every turn. One instructor would not accept him in his math course because Payan was blind and would "slow the class down too much" (Vasquez, 2022). Another instructor refused to allow him to audio record lectures, while in another math class, Payan was given permission to record, but the instructor was not verbally explaining the formulas he was writing on the board during class. After demonstrating his experience as a blind person in this class to the instructor, Payan was told that his attendance was no longer welcome. There was one math instructor who went so far as to show her complete disdain for disabled students. Payan recollects,

> Another math professor stated that she could not understand why persons with disabilities insisted on attending college if we could not do the work. She went out of her way to humiliate me in front of the entire class by telling me that I was occupying a seat that a more worthy, not disabled student should have been allowed to occupy. This instructor went out of her way to make me feel unwelcome. (Essig, 2022, para. 13)

Ultimately, Payan was denied information in braille or audio and even tutors to read the material out loud. As this course was a gatekeeper for him to be able to transfer to a 4-year university, he had to come up with money for a tutor. Payan was not receiving financial aid and was living on his social security benefits; so, he was forced to hire a tutor on his already limited budget (R. Payan, personal communication, June 7, 2022).

Other blind students at LACCD, as Payan would find out, were not as fortunate. He talked to 15–20 disabled students about the inaccessibility of math textbooks, finding that each one of them had difficulties accessing the accommodations they had a right to by the American with Disabilities Act (ADA) and Section 504 of the Rehabilitation Act. Some of them spent years trying to obtain accessible materials, while others gave up and dropped out of school. Afterwards, Payan requested that the LACCD Math Department Chair contact the vendor of the book for an accessible copy of the MyMathLab and software because the current math materials were not compatible with screen readers. He was told that there were not enough of him (i.e., blind and Veteran students) to warrant that action (R. Payan, personal communication, June 7, 2022). In 2017, Payan and another plaintiff filed a lawsuit against the district.

During the litigation process, the LACCD claimed that any discrimination was unintentional. However, the plaintiffs were able to demonstrate multiple instances where the district failed to provide accommodations to its students. For example, LACCD was given $25 million dollars per year for almost 50 years (i.e., the passing of Section 504, Individuals with Disabilities Education Act (IDEA), and ADA) for compliance training and accommodations, and the district was unable to demonstrate how they allocated the funding (Essig, 2022, para. 11; R. Payan, personal communication, June 7, 2022). In addition, during the testimonies of the Adaptive Assistive Technologies (AAT) chair and the ADA compliance officer for LACCD, it was revealed that neither individual had credentials, certifications, classes, or other professional development on adaptive technologies or ADA (R. Payan, personal communication, June 7, 2022). In 2019, the plaintiffs were able to prove to the federal district court that "textbooks, handouts, websites, databases, and computer applications were not accessible to them because they did not function properly with screen-reading software" (DREDF, 2022, para. 3). Consequently, the judge ordered LACCD to address the barriers to access. At this point, Payan had already graduated from Los Angeles City College in 2018 and was an undergraduate public administration major at California State University, Los Angeles.

Rather than complying with the judge's orders, LACCD immediately appealed to the Ninth Circuit Court of Appeals, claiming that the plaintiffs should not be allowed to bring claims under the ADA or Section 504 using "disparate impact." According to the Disability Rights Education & Defense Fund (DREDF), "disparate impact" discrimination "means discrimination that is supposedly unintentional, such as when there is a neutral policy or practice that harms or excludes disabled people" (DREDF, 2022, para. 5). Payan and his litigation team won again, with the majority ruling that Section 504 and the ADA were created to address both intentional and unintentional discrimination "caused by 'thoughtless indifference' or 'benign neglect,' such as physical barriers to access public facilities" (DREDF, 2022, para. 6). There was one dissenting opinion by appellate judge, Judge Kenneth K. Lee, and LACCD announced in November 2021 and again in January 2022 that they intended to take their case to the Supreme Court to decide whether disabled plaintiffs could be allowed to use disparate impact under Section 504 and the ADA (DREDF, 2022). This attempt also failed, as the disability community rose up in arms in March 2022.

As of this writing, Roy Payan is still waiting for a settlement with the LACCD. In an interview with RespectAbility, he states,

> The only settlement I will accept is for them to agree to make all digitized systems completely accessible and to create a system to mandate that all professors undergo Accessible Document Conversion training such as Ally. The district must also be mandated to hire a credentialed Adaptive Technologies and Accessibility staff member. The staff member should test all software and digitized systems for compliance and JAWS compatibility. (Essig, 2022, para. 16)

Roy Payan is now finishing up his Masters' degree in public policy at the University of Southern California, and he intends to continue after graduation toward his doctorate in public policy to advocate for people with disabilities. Despite being able to move on with his own educational and professional aspirations, he is still in touch with six disabled friends who are still stuck at LACCD because they cannot fulfill math or science requirements due to inaccessibility (R. Payan, personal communication, June 7, 2022). He thinks about them and the thousands of disabled students in community colleges who are fighting against the odds to receive an education that they deserve.

5.4.1 Case Analysis

Using DisCrit and Disability Justice as intersectional lenses, we provide an analysis of Roy Payan's case to add to the understanding of oppression and liberation in STEM education. To begin, the case illustrates the importance of Payan's leadership as a disabled Latinx/Person of Color agitating for higher education transformation for him and his disabled and nondisabled peers. Disabled Students of Color are best positioned to lead changes as they offer unique knowledge and insights that educators simply do not possess (Annamma & Morrison, 2018). In Payan's case, he critically and quickly noted the inaccessible math curriculum at LACCD and took action. Being an older student, his life experience as a disabled Latinx certainly informed his activist stance, as he knew he needed to self-advocate as well as fight for the rights of his peers. Here, we are reminded of Piepzna-Samarasinha (2018) assertion that those most impacted by inaccessibility in our world are on the front-lines and must react as a matter of survival. In this way, Payan bore the brunt of initiating and sustaining countless communications with math instructors and college administrators as well as the pain of repeatedly encountering explicit and implicit ableist assaults throughout the process (DisCrit Tenet 4, 7, and Disability Justice: Sustainability and Change Through Leadership of Those Most Impacted).

Payan's case also reflects the compounding nature of injustices. As with many disabled Individuals of Color in the United States, Payan faced social and financial challenges as he encountered the material impact of being raced and disabled (DisCrit Tenet 3). For example, Payan's very limited income meant that hiring a math tutor, who was supposed to be provided by the LACCD, put him further on the cusp of vulnerability. Thus, in many instances, disabled Students of Color have no choice but to be pushed out of higher education due to these and many other disabling societal factors. The attitudes and actions of LACCD's educators and administrators ignore the compounding and systemic nature of injustices faced by multi-marginalized folx. Instead, their attitudes and actions are grounded in individualistic and deficit-laden approaches that blame disabled Students of Color for their own failures.

Similarly, the Payan case illustrates how racism and ableism are interdependent (DisCrit Tenet 1) and work to preserve Whiteness and ability as property (DisCrit Tenet 6 and Disability Justice: Intersectionality and Recognizing Wholeness). During his thorough investigation to provide evidence for his case, Payan himself noted that, although the LACCD has campuses throughout Los Angeles that include a few campuses in predominantly upper- and middle-class White communities of West Los Angeles and the San Fernando Valley, no White students with disabilities from those campuses filed complaints against the school district for discrimination. The discrimination complaints against LACCD were from disabled students primarily from Communities of Color who were marginalized by primarily White-identified American, Middle Eastern, or European instructors. As a vocal blind Latinx, Roy Payan's presence on a higher education campus, even a community college one, disrupts notions of Whiteness and normality through "smartness trespassing" (Carrillo & Rodriguez, 2016). He challenges what it means to be a degree-seeking student and a mathematician, as mathematics has been historically and currently viewed as a field for able-bodied, White men. By demanding accommodations that he deserves by law and being able to clearly articulate his access needs, he is a threat to established practices. Indeed Payan's mathematics professors, including the one who humiliated him by alluding to his unworthiness, act as police, judge, and jury who draw on the long history of racism and ableism in defining and protecting the so-called universal human—a universal which is more exclusive than inclusive of many minoritized and oppressed groups (Disability Justice: Intersections of Justice: Interdependence, Cross-Disability Solidarity, Collective Access, and Collective Liberation).

Instead, Payan "interrupt[s] a racialized gaze" that views Latinxs as "intellectually inferior" to Whites (Carrillo & Rodriguez, 2016, p. 241). In a recent email exchange, when asked if he felt that disability and race/ethnicity were intersectional factors in his case, he responded,

> During this lengthy legal battle, I have been and continue to be approached by hundreds of Students of Color with disabilities, yet only one white [student] amongst them. Do I believe that race has played a major role in the discrimination I faced as a person with a disability by LACCD, DEFINITELY!!! (R. Payan, personal communication, September 21, 2023, emphasis in original)

The continued attempts of the LACCD to ignore Payan's requests and to pass the "problem" up the chain of command demonstrate that academic ableism "renders disability as abject, invisible, disposable, less than human, while able-bodiedness is represented as at once ideal, normal, and the mean or default" (Dolmage, 2017, p. 7). Most disappointing are the able-bodied Administrators of Color on the LACCD Board of Directors who have bought into Whiteness and continue to perpetuate these oppressive, ableist practices. Now, the fact that the LACCD is determined to contest disparate impact to the detriment of the disability community rather than address their accessibility issues exhibits, how they are willing to use the legal system to deny the rights of disabled People of Color in their district (DisCrit Tenet 5).

5.5 Discussion

The case study analysis of *Payan v LACCD* using the frameworks of DisCrit and Disability Justice provides theoretical and methodological tools to not only identify structural issues related to racism and ableism but also suggest ways for higher education educators and researchers to move forward to address those issues. While DisCrit can help us to deconstruct systems of oppression connected to the intersection of race and disability, Disability Justice shifts power to legitimize the experiential knowledge of disabled students and community leaders and recognizes their ways of knowing and being as critical to understanding, analyzing, and contesting inequities. Together, these frameworks challenge our assumptions about what STEM education and educational research consider "normal" and what we say we value in STEM education, as they center the voices and multifaceted identities of SDQTPOC. By focusing on a case study of Roy Payan and his experience as a disabled Student of Color fighting for accommodations, we are able to bear witness to the challenge of fighting an ableist educational system, identify the permanence and intersectional nature of ableism with other forms of subordination, and are motivated to engage in the struggle. Yet, too few research studies that intersect STEM education and disability attend to the experiential knowledge and lived experiences of disabled students as pathways to finding solutions. Ultimately, Payan's story, as analyzed through a DisCrit and Disability Justice lens, inspires us, especially those of us who are racialized and disabled, to speak out because "the more of us who speak up, the more pressure there will be for higher education institutions to change" (Torres, 2021, p. 633).

More specifically, sites of transformations for STEM instructors can also be part of larger conversations in higher education and higher education research to prevent and reduce harms. For example, ongoing training as required by higher education institutions for all personnel are opportune sites of intersectional work. College instructors are often required upon hire and in regular intervals throughout their tenure to engage in trainings in areas, such as sexual harassment and student privacy. A discussion on the effectiveness of such training is beyond the scope of this chapter. What is clear is that higher education needs to prioritize these trainings, which signifies high importance with the purpose of learning, relearning, and perhaps unlearning. Thus, integrating intersectional approaches, such as DisCrit and Disability Justice, into the set of trainings that all higher education personnel are required to undergo can provide more robust learnings to move beyond disability-as-deficit. Intersectional approaches can also address the shortcomings of higher education's diversity, equity, and inclusion (DEI) efforts wherein faculty do not perceive disability as related to these efforts (Barnard et al., 2008). That is, Disability Justice must be approached as interconnected and bounded with the multitude of struggles toward liberation (Davis, 2016).

These trainings represent a step, but on its own they are inadequate. As part of the higher education's settler-colonial foundation and operation (Patel, 2021), STEM education must reckon with these while working to dismantle them. Inclusion of disabled Students of Color into what Martin Luther King, Jr. terms a *burning house* will inevitably be a losing proposition that Payan's case can attest to for the current state of STEM education. For example, Patel (2021) asserts that higher education structures of faculty tenure means that full professors who are mostly male, White, and able-bodied force out many multimarginalized junior faculty. Thus, if STEM education seeks better faculty representation, it must also dismantle promotion and retention structures that seek to maintain social hierarchies in higher education. That is, DEI efforts must not only aim to recruit and retain underrepresented Faculty of Color, but these efforts must also dismantle settler-colonial foundations and operations while building nurturing conditions for disabled students and Faculty of Color to thrive.

Decolonizing STEM education research must heed Yeh and colleagues' (2021) call for disobedience; for example, researchers and educators should resist traditional notions of what is mathematics and who can do mathematics. We point to higher education's refusal to provide even the most basic accommodations to folks like Roy Payan and so many other mathematics students as entrenched beliefs in whose bodies and minds are "valued" in mathematics. Research in higher education center and privilege male, White, and able-bodied ways of knowing and doing. Thus, resisting such prevailing practices in STEM education means centering those who have been most harmed and excluded in higher education (i.e., disabled Students of Color). This cannot mean, for example, sprinkling diverse topics and readings into existing STEM courses while still centering Whiteness (Patel, 2021). Rather, STEM courses must resist and disrupt settler colonialism's notions of domination, extraction, and commodification. Imagine if Roy Payan was enrolled in an institution of higher education that centered the Disability Justice principles of collective access and care, solidarity, and liberation. His and his peers' accommodations would be features of his mathematics course rather than perceived as an annoyance. Roy, his peers, and the mathematics course instructor would be working interdependently to rectify the settler colonialism's destruction of our environment rather than in competition to contribute to it.

References

Annamma, S. A., & Morrison, D. (2018). DisCrit classroom ecology: Using praxis to dismantle dysfunctional education ecologies. *Teaching and Teacher Education, 73*, 70–80. https://doi.org/10.1016/j.tate.2018.03.008

Annamma, S. A., Connor, D., & Ferri, B. (2013). Disability critical race studies (DisCrit): Theorizing at the intersections of race and disability. *Race Ethnicity and Education, 16*(1), 1–31.

Annamma, S. A., Ferri, B. A., & Connor, D. J. (2018). Disability critical race theory: Exploring the intersectional lineage, emergence, and potential futures of DisCrit in education. *Review of Research in Education, 42*, 46–71. https://doi.org/10.3102/0091732X18759041

Artiles, A. J. (2019). Fourteenth Annual Brown Lecture in Education Research: Reenvisioning equity research: Disability identification disparities as a case in point. *Educational Researcher, 48*(6), 325–335.

Barnard, L., Stevens, T., Siwatu, K. O., & Lan, W. (2008). Diversity beliefs as a mediator to faculty attitudes toward students with disabilities. *Journal of Diversity in Higher Education, 1*(3), 169–175. https://doi.org/10.1037/a0012707

Calabrese Barton, A., & Tan, E. (2020). Beyond equity as inclusion: A framework of "rightful presence" for guiding justice-oriented studies in teaching and learning. *Educational Researcher*, 0013189X2092736.

Carrillo, J. F., & Rodriguez, E. (2016). She doesn't even act Mexican: Smartness trespassing in the New South. *Race Ethnicity and Education, 19*(6), 1236–1246.

Chang, E. H., Milkman, K. L., Gromet, D. M., Rebele, R. W., Massey, C., Duckworth, A. L., & Grant, A. M. (2019). The mixed effects of online diversity training. *Proceedings of the National Academy of Sciences, 116*(16), 7778–7783. https://doi.org/10.1073/pnas.1816076116

Charlton, J. (2000). *Nothing about us without us: Disability oppression and empowerment*. University of California Press.

Cooper, K. M., & Brownell, S. E. (2016). Coming out in class: Challenges and benefits of active learning in a biology classroom for LGBTQIA students. *CBE—Life Sciences Education, 15*(3), ar37. https://doi.org/10.1187/cbe.16-01-0074

Copur-Gencturk, Y., Cimpian, J. R., Lubienski, S. T., & Thacker, I. (2019). Teachers' bias against the mathematical ability of female, Black, and Hispanic students. *Educational Researcher, 49*(1), 30–43. https://doi.org/10.3102/0013189X19890577

Crenshaw, K. W. (2017). *On intersectionality: Essential writings*. The New Press.

Davis, A. Y. (2016). *Freedom is a constant struggle: Ferguson, Palestine, and the foundations of a movement*. Haymarket Books.

Disability Rights Education & Defense Fund (DREDF). (2022, January). *Payan v LACCD explainer*. DREDF. https://dredf.org/2021/12/16/payan-v-laccd-explainer/

Dolmage, J. T. (2017). *Academic ableism: Disability and higher education*. University of Michigan Press.

Essig, M. (2022, February 15). *Fighting for students with disabilities: Spotlight on Roy Payan*. RespectAbility. https://www.respectability.org/2022/02/roy-payan-spotlight/

Few-Demo, A. L. (2014). Intersectionality as the "new" critical approach in feminist family studies: Evolving racial/ethnic feminisms and critical race theories. *Journal of Family Theory & Review, 6*(2), 169–183.

Hartman-Hall, H. M., & Haaga, D. A. F. (2002). College students' willingness to seek help for their learning disabilities. *Learning Disability Quarterly, 25*(4), 263–274. https://doi.org/10.2307/1511357

Joseph, N. M., Viesca, K. M., & Bianco, M. (2016). Black female adolescents and racism in schools: Experiences in a colorblind society. *The High School Journal, 100*(1), 4–25. https://doi.org/10.1353/hsj.2016.0018

Jensen, J. M., McCrary, N., Krampe, K., & Cooper, J. (2004). Trying to do the right thing: Faculty attitudes toward accommodating students with learning disabilities. *Journal of Postsecondary Education and Disability, 17*(2), 81–90.

Kim, W. H., & Lee, J. (2015). The effect of accommodation on academic performance of college students with disabilities. *Rehabilitation Counseling Bulletin, 60*(1), 40–50. https://doi.org/10.1177/0034355215605259

Mamiseishvili, K., & Koch, L. C. (2011). First-to-second-year persistence of students with disabilities in postsecondary institutions in the United States. *Rehabilitation Counseling Bulletin, 54*(2), 93–105. https://doi.org/10.1177/0034355210382580

Marshak, L., Wieren, T. V., Ferrell, D. R., Swiss, L., & Dugan, C. (2010). Exploring barriers to college student use of disability services and accommodations. *Journal of Postsecondary Education and Disability, 22*(3), 151–165.

Martin, D. B., Gholson, M. L., & Leonard, J. (2010). Mathematics as gatekeeper: Power and privilege in the production of knowledge. *Journal of Urban Mathematics Education, 3*(2), 12–24.

Mingus, M. (2011, February 12). Changing the framework: Disability justice—How our communities can move beyond access to wholeness. *Leaving Evidence*. https://leavingevidence.wordpress.com/2011/02/12/changing-the-framework-disability-justice/#:~:text=Accessibility%20is%20concrete%20resistance%20to,move%20beyond%20access%20by%20itself

Mirin, A. C., & Tan, P. (2022). Disability accommodations in college: Alarming discrimination in mathematics. In *The 24th Annual Conference on Research in Undergraduate Mathematics Education (RUME), Boston, MA*.

Morales, A. L. (2013). *Kindling: Writings on the body*. Palabrera Press.

Museus, S. D., & Iftikar, J. (2013). An Asian critical theory (AsianCrit) framework. *Asian American Students in Higher Education*, 18–29.

National Center for Education Statistics (NCES). (2018, May). Table 311.10. Number and percentage distribution of students enrolled in postsecondary institutions, by level, disability status, and selected student characteristics: 2015–16 [Data table]. In *Digest of Education Statistics*. U.S. Department of Education, Institute of Education Sciences. Retrieved January 25, 2022, from https://nces.ed.gov/programs/digest/d20/tables/dt20_311.10.asp

Newman, L., Wagner, M., Knokey, A. M., Marder, C., Nagle, K., Shaver, D., & Wei, X. (2011). *The post-high school outcomes of young adults with disabilities up to 8 years after high school: A report from the national longitudinal transition study-2* (NLTS2) (NCSER 2011–3005, National Center for Education Statistics, US Department of Education, Washington, DC).

Nieminen, J. H., & Pesonen, H. V. (2022). Anti-ableist pedagogies in higher education: A systems approach. *Journal of University Teaching & Learning Practice, 19*(4) https://ro.uow.edu.au/jutlp/vol19/iss4/08

Oliver, M. (2013). The social model of disability: Thirty years on. *Disability & Society, 28*(7), 1024–1026. https://doi.org/10.1080/09687599.2013.818773

Padilla, A., Tan, P., Bagger, A., & Nieminen, J.H. (in press). The politics of caring in mathematics education: Discursive and linguistic interrogations. *International Congress on Mathematic Education (ICME-15) Proceedings:* Sydney, Australia.

Patel, L. (2021). *No study without struggle: Confronting settler colonialism in higher education*. Beacon Press.

Powell, S. R., Benz, S. A., Mason, E. N., & Lembke, E. S. (2022). How to Structure and Intensify Mathematics Intervention. *Beyond Behavior, 31*(1), 5–15. https://doi.org/10.1177/10742956211072267

Piepzna-Samarasinha, L. L. (2018). *Care work: Dreaming disability justice*. Arsenal Pulp Press.

Prema, D., & Dhand, R. (2019). Inclusion and accessibility in STEM education: Navigating the duty to accommodate and disability rights. *Canadian Journal of Disability Studies, 8*(3), 121–142.

Rao, S., & Gartin, B. C. (2003). Attitudes of university faculty toward accommodations to students with disabilities. *Journal for Vocational Special Needs Education, 25*, 47–54.

Reinholz, D. L., Stone-Johnstone, A., White, I., Sianez, L. M., & Shah, N. (2020). A pandemic crash course: Learning to teach equitably in synchronous online classes. *CBE—Life Sciences Education, 19*(4), 1–13. https://doi.org/10.1187/cbe.20-06-0126

Reinholz, D. L., & Ridgway, S. W. (2021). Access needs: Centering students and disrupting ableist norms in STEM. *CBE-Life Sciences Education, 20*(3) https://www.lifescied.org/doi/10.1187/cbe.21-01-0017

Reinholz, D. L., & Wilhelm, A. G. (2022). Race-gender D/discourses in mathematics education: (Re)-producing inequitable participation patterns across a diverse, instructionally-advanced district. *Urban Education*. https://doi.org/10.1177/00420859221107614

Reinholz, D. L., Johnson, E., Andrews-Larson, C., Stone-Johnstone, A., Smith, J., Mullins, B., Fortune, N., Keene, K., & Shah, N. (2022). When active learning is inequitable: Women's participation predicts gender inequities in mathematical performance. *Journal for Research in Mathematics Education, 53*(3), 204–226. https://doi.org/10.5951/jresematheduc-2020-0143

Saia, T. (2019). *Disability as an identity: Disability cultural centers in higher education* [Doctoral dissertation, The University of Arizona]. ProQuest Dissertations and Theses Global.

Sins Invalid. (2019). *Skin, tooth, and bone: The basis of movement is our people* (2nd ed.). Sins Invalid.

Steele, C. M. (1997). A threat in the air: How stereotypes shape intellectual identity and performance. *American Psychologist, 52*(6), 613–629. https://doi.org/10.1037/0003-066X.52.6.613

Synder, S. L., & Mitchell, D. T. (2001). Re-engaging the body: Disability studies and the resistance of embodiment. *Public Culture, 13*(3), 367–389.

Tan, P., Padilla, A., & Lambert, R. (2022). A critical review of educator and disability research in mathematics education: A decade of dehumanizing waves and humanizing wakes. *Review of Educational Research*. https://doi.org/10.3102/00346543221081874

Tatum, B. D. (2000). The complexity of identity: "Who am I?". In M. Adams, W. J. Blumenfeld, H. W. Hackman, X. Zuniga, & M. L. Peters (Eds.), *Readings for diversity and social justice: An anthology on racism, sexism, anti-semitism, heterosexism, classism and ableism* (pp. 9–14). Routledge.

Taylor, S. R. (2021). *The body is not an apology: The power of radical self-love*. Berrett-Koehler Publishers.

Theobald, E. J., Hill, M. J., Tran, E., Agrawal, S., Arroyo, E. N., Behling, S., Chambwe, N., Cintrón, D. L., Cooper, J. D., Dunster, G., Grummer, J. A., Hennessey, K., Hsiao, J., Iranon, N., Jones, L., Jordt, H., Keller, M., Lacey, M. E., Littlefield, C. E., et al. (2020). Active learning narrows achievement gaps for underrepresented students in undergraduate science, technology, engineering, and math. *Proceedings of the National Academy of Sciences, 117*(12), 6476–6483. https://doi.org/10.1073/pnas.1916903117

Torres, L. E. (2020). Straddling death and (re)birth: A disabled Latina's meditation on collective care and mending in pandemic times. *Qualitative Inquiry, 27*(7), 895–904. https://doi.org/10.1177/1077800420960169

Torres, L. E. (2021). *Sobreviviendo sin sacrificando*—An intersectional DisCrit testimonio from a tired mother-scholar of color. *Race Ethnicity and Education, 24*(5), 623–636. https://doi.org/10.1080/13613324.2021.1918402

Treisman, U. (2013, April 19). *Iris M. Carl equity address: Keeping our eyes on the prize*. Presentation at the NCTM Annual Meeting, Denver, CO.

Vasquez, A. (Host). (2022, January 13). *Blind students fight for accessible college classes* [Audio podcast]. Education Beat by EdSource. https://edsource.org/podcast/blind-students-fight-for-accessible-college-classes

Vizenor, G. R. (2008). *Survivance: Narratives of Native presence*. University of Nebraska Press.

Yeh, C., Tan, P., & Reinholz, D. L. (2021). Rightful presence in times of crisis and uprisings: A call for disobedience. *Equity & Excellence in Education, 54*(2), 196–209.

Chapter 6
Disabled by Design: Dreaming the Future of DBER Through Intersectional Disobedience

Phillip A. Boda

Abstract Discipline-based education research (DBER) has come of age at the postsecondary level. A concerted prominence in this area of educational research for more just and equitable futures for all positional identities has, discouragingly, been often relegated to the margins. DBER scholars have traditionally focused on conceptual disciplinary learning as the penultimate benchmark to achieve equity; and yet, even among those measures, the notion of ableist structures, policy, and practices remains deeply entrenched in the ways such research is evaluated for its success, and how it can grow in transformative ways for all. Drawing on Intersectionality as critical social theory, Disability theorizing beyond individual accommodations, and epistemic disobedience, this chapter stakes the claim that we can Dream of futures yet untold in higher education for students pursuing STEM degrees and careers, if we are only courageous enough to relinquish our obsession with disciplinary commitments and work toward more relational praxis. The implications of this work extend beyond the postsecondary level and can be used to inform more just K-12 STEM futures for historically marginalized youth, while also contributing to critical design enquiries that seek to disrupt disciplinary hegemonic injustices.

Keywords Intersectionality · Disability · Exclusionary design · DBER · Epistemic disobedience

P. A. Boda (✉)
Department of Special Education, University of Illinois at Chicago, Chicago, IL, USA

G. L. Cochran, J. D. Adams (eds.), *Equity-Oriented, Discipline-Based STEM Education Research*, Critical Studies of Education 19,
https://doi.org/10.1007/978-981-96-9136-4_6

6.1 Dreaming Beyond Past and Present DBER: An Introduction

> [We must never refuse] to let go of the visionary—that which does not yet exist—and the radicalness of the imaginary as a space for what is yet unthinkable, at the edge of the possible. We say yes to Octavia Butler's brilliant speculativeness: we will dream our way out; we must imagine beyond the given [and our lifetimes]. (Davis et al., 2022, p. 29)

The abolition feminist authors and scholars of the recent work *Abolition. Feminism. Now* articulate above a need to go beyond "what is," and like my colleagues and I have challenged among racially just research methodologies in education (Boda et al., 2022), start Dreaming of "what if." This speculative approach—of engaging with the past and present, but not allowing them to define all possibilities for the future—is one that grounds a critical analysis of how we think about societal institutions, such as education, as well as their purpose in relation to past discrimination and present social positioning practices. We know that, in higher education in particular, the space to speak from an authentic lived reality in relation to how learning can be nurtured among historically marginalized students is often relegated to a secondary affective goal, and sometimes never considered valuable at all (Patton et al., 2022). This postsecondary witnessing of the purposes of higher education and its scholarship, then, beckons us to think about the aspirations of K-12 youth through paradigms that place our attention on notions of interdependence, self-determination, and "lifting as we rise" (Turner, 2022). Contextualizing this chapter in these theories of educational purpose and dismantling of cultural repertoires that value market value product over process-oriented relationality provide a critical approach to think with and through the perspectives of students in highly exclusive disciplines in higher education in ways that bring forth speculative possibilities to challenge the status quo (Warren, 2021).

In postsecondary education, the "hard sciences" are often taken for granted in terms of the professorial acumen to teach in favor of these "high-performing" academic scholars focusing on research more so than supporting the next generation of professionals in the field. Discipline-based education research (DBER), in response, originated as a push back to this persistent need to engage with training new professionals grounded in the principal improvement of academic achievement within Science, Technology, Engineering, and Math (STEM) disciplines at the postsecondary level, particularly among students who have been historically excluded from these "hard" disciplines (National Research Council, 2012). DBER emerged from this recognition that students and their professors in STEM degrees should develop more relational interactions beyond student–teacher or student–researcher to improve both the quality of teaching and, hopefully, begin to engage with how the broader culture of STEM education is designed to be exclusive to those with specific ways of knowing and being, which often neglects subjectivity and interpretative quality in these spaces because they stem from and become prevalent most when personal experience is considered pivotal for learning. The traditional DBER aim follows a market value product importance laden in higher education praxis more broadly to churn out as many graduates as possible in favor of garnering new research funds through this revolving door policy. This culture of STEM higher education has subsequently produced hegemonic student positions as valuable exemplars of who can do STEM and positioned others as non-STEM. In this context of exclusionary design that has been sustained through unquestioned historical precedence, the need to challenge what we think we know and how we know it is paramount.

Instead of taking an integrated STEM approach such as those used in K-12 schooling contexts (Roberts et al., 2022; Roehrig et al., 2021), DBER researchers have instead turned their focus on the improvement of traditional learning outcomes among these STEM disciplines primarily in isolation. This means that the lessons we are learning from K-12 spaces, while novel and seeping with possibility for Dreaming of futures untold, have not been well taken-up by those scholars working in postsecondary STEM education research in ways that could challenge the histories of exclusion we know to be ever-present and persistently impactful in these spaces. Recently, these have been nascent calls to develop more integrated STEM alliances among discipline-based education researchers and the type of cross-pollinating work that is beneficial across STEM disciplinary research at the higher education level (Henderson et al., 2017; Shipley et al., 2017). However, there remain hurdles to these efforts that often neglect systemic shifts in favor of individual change, which invariably upholds the cultural repertoires that have historically excluded subaltern positionalities by design (Reinholz et al., 2021), and that more "soft" interpersonal infrastructure building is needed (Fuqua et al., 2019). This reality speaks to STEM disciplinary cultures at the postsecondary level which valorize traditional content acquisition over more authentic STEM learning experiences (Santangelo et al., 2021), but also goes further to engage with how we can begin the process of "Dreaming our way out of it," to paraphrase Davis et al. (2022), and in doing so design STEM learning with our students such that we begin to disobey how the hegemony designed into these cultures is critically problematized.

Unfortunately, the reliance on market value products (i.e., STEM professionals that fit into historical precedent of student identities among these disciplines) is not all that surprising given the commitment to (post-)positivism at the higher education level in terms of the research deemed as valuable for DBER scholarship. While also focusing on isolated STEM disciplinary teaching and learning to improve postsecondary course design, instructor pedagogy, and facilitation of discussion to create "impact" in these courses as Diversity, Equity, and Inclusion efforts (cf. Davey et al., 2021; Freeman et al., 2014; Ruiz-Primo et al., 2011; Salehi et al., 2021). While these efforts are laudable broadly speaking among the capitalist rhetoric of universities as places of business rather than socially engaged learning for the public good, they neglect to critically make present the discursive and material realities where Whiteness and supremacist logics of exclusion are so deeply engrained in the structures, policy, and practices of higher education (Adejumo, 2021). This also denies a space for more speculative Dreams beyond the cultures that seek to sustain their power and challenge disciplinary hegemony (Boda et al., 2022). Additionally, it speaks to the silencing and smothering of dissent toward widely used theories of leadership and organizational structuring that have dominated how higher education perceives its role, as well as the way it sees its students. Ladkin and Patrick (2022) cohere with my points:

> Within transformational leadership theory, it is leaders who are full human beings with agency, a sense of purpose, high moral standards, and the ability to evoke positive regard from others. Followers on the other hand are passive, directionless and only capable of mediocrity without the leader's guidance. They are depicted as being less than fully rounded humans capable of exercising their will. (p. 220)

I note these efforts not in lament, but to bear witness to the ways in which traditional disciplinary cultures are sustained through a reliance on Western Modern epistemological approaches to systemic change. By focusing on content and teaching practices over the ways in which students from historically marginalized backgrounds could provide advantageous standpoints about how to radically change disciplinary cultures. Like Ladkin and Patrick (2022) above, as well as Adejumo (2021), disciplinary leadership in relation to STEM education research at the postsecondary level must reckon with these hegemonic cultures that exclude by design and characterize some positionalities being distinctly aligned with past and present structures, policies, and practices. This can start radically challenging the ways in which we research and inform how disciplines such as STEM in higher education require more than designing for greater content acquisition and conceptual learning. In sum, to afford place and space where historically marginalized students are centered rather than peripheral by design requires discipline-based education researchers working in STEM: define the center upon which all other positionalities are measured, assess the extent to which they themselves have the material histories to engage with this work in relational ways, and embark on a path toward Dreaming about changing these academic pathways to support students that are most vulnerable to these systemic policies and practices that embolden one center rather than many.

In this chapter, I Dream with and through a specific positionality that is invisible in the literature at both the postsecondary STEM and K-12 STEM levels: Students labeled with disabilities (cf. Boda, 2021; Prema & Dhand, 2019; Schneiderwind et al., 2021; Traxler & Blue, 2020). This population is of particular importance to STEM DBER scholarship given that Disability as an ideology can help unpack the designs among higher education spaces where students' needs are not being served well (cf. Boda, 2022; Hamraie, 2016; Reynolds, 2022). Based on the paucity of research in DBER and STEM education research more broadly in relation to this positionality (i.e., disabled identities being understood by the broader medical rhetoric of deficit and manifesting as socially positioned studenthoods of defect), this centering of the margins among inquiries that name themselves as diversity, equity, and inclusion (DEI) initiatives pushes forth a way to think not just about the challenges of the past, nor the present fiscal limitations from the market value commoditization of higher education. Instead, it asks that we Dream of future possibilities in DBER work where professors and students build new cultures of joy and feelings of relational belonging, subsequently nurturing the brilliance of students who have been historically marginalized because of their desire to work collaboratively, define their life trajectories on their own terms, and support those in their local community (Patton et al., 2022; Turner, 2022). This then also challenges epistemic obedience viewing STEM as "disciplinarily rigorous," which sustains ableist logics of disability/difference as deficit (Cech, 2021; Slaton, 2013) and rampant neglect of people's support needs in STEM (Reinholz & Ridgway, 2021). In this way, I argue that DBER scholars should work toward "what if" by leveraging intersectionality as critical social theory for design (Boda, 2022) and stake the claim that scholars can do more than change disciplinary teaching: They can help students Dream.

6.2 The (Lack of) Ideology Around Disability in Postsecondary STEM Education Research

Among a recent exploration to examine equity-oriented differences between student demographics in relation to the impact of active learning strategies in STEM fields, there has remained a commitment to enact change at the individual level for students with disabilities[1] solely as a function of accommodations (Nardo et al., 2022). This is not to say that students labeled with disabilities may not need these accommodations and support for their needs as learners. In response and articulating a new discourse around disability, I ask the questions: Who *doesn't* need support for their learning needs in higher education spaces? And what might this mean about how we view our scholarship working toward equity if we actively choose to neglect to theorize critically those identities from which universal design was borne? With such a lack of engagement of the ideology of disability among design scholarship (Hamraie, 2016)—and what Nusbaum and Steinborn coin as *ontological erasure* in education (2019)—what remains unexamined in thinking merely through a material realist paradigm[2] is that without an engagement with the ideologies that produce vulnerable subjects and subjectivities that more often experience inequity (Boda, 2022), little is done to engage with how those identities have been constructed as "in need" of those accommodations due to the ableist systemic designs in higher education. Indeed, broader articulations of institutional change in postsecondary STEM education research for students from historically marginalized demographics have noted the need for more critical theorizing as the primary focus to push systemic shifts and support the justification that individual learning can/has/will occur, rather than using theory to guide the epistemological approaches across research studies

[1] Identity-first language is purposefully chosen by Disabled and Disability Justice scholars, whereas person-first language is seen by many Disabled folks as a denial of individual voice and claimed identity among markers of marginalization used to self-determine. If I use students "labeled with disabilities," I am articulating how imposed positioning of students who have been subject to special education processes is enacted through a label.

[2] Material realism responds to specific sociology interactions and structuring of disability primarily focused on accommodations with little focus on the context of instruction (Allan, 2010).

(Reinholz et al., 2021). This work alludes to a greater need to examine DBER beyond content and practices to engage with epistemic cultures, but not just in ways that "tow the party line." This critical praxis invites disobedient approaches.

An important alignment with the work above is Lillywhite and Wolbring (2022) where they note how students labeled with disabilities are neglected in their presence among research opportunities, particularly in STEM disciplines, which articulates a new pathway to think about how to change the ways in which novel educational programs at the postsecondary level can impact the nature of who becomes STEM professionals and invariably disrupt this historical marginalization through broader systemic paradigm shifts. Through this example, we see these scholars not just theorizing who can/has/will become STEM professionals based on individual change and, alternatively, approached their enquiry to theorize systemic changes. In a similar vein, Pickard (2022) has to engage Disabled students as Whole,[3] knowledgeable participants in higher education to inform instructors' pedagogical approaches in ways that develop learning environments that are inclusive and humanizing. This invokes the Disability mantra of the Civil Rights movement era in the United States where Disabled people sought equity in society writ large (Charlton, 1998; Scotch, 2009). Drawing on these approaches, then, shifts the enquiry not on the individual Disabled student being "in need," but, pushes us to think about the proximity of our enquiry in relation to its impact on the ideology of disability that shapes and shades instructional designs that are inherently ableist. Dreaming with and through this scholarship, as well as my own work and my colleagues, I propose that we leverage the three dimensions laid out by Patricia Hill Collins around *Intersectionality as critical social theory* (2019) (intersectionality as *metaphor*; the use of intersectionality as a *heuristic*; and the *paradigm shift* in thinking that comes from taking up intersectionality as a form of reflexive, relational praxis) to illuminate/Dream how disobedient designs can inform equity in DBER.

6.3 Intersectionality as a Metaphor in DBER

When intersectionality as a term is used in (STEM) education research, it is most often (mis)used by leveraging it only as a *metaphor* to illustrate intersecting identities and describe the ways in which they are important in how we make sense of collective difference within and among positions of power. However, we must bear critical witness to the idea that this is not the primary articulation made by the originators of intersectionality as a theoretical perspective taken toward the world writ large in terms of understanding overlapping oppression. The position that intersectionality is not just about identity is something DBER and other social justice scholars working in STEM education must reckon with to fully engage intersectionality

[3] Berne et al. (2018) outline from Disabled peoples' viewpoints—their Lives-Hopes-Dreams—as being seen as Whole and valuable to inform the way we experience the world and its designs. This is Disability Justice as a Term.

as critical social theory. Indeed, Patricia Hill Collins in one of the first seminal texts to embody an intersectional sensibility but not explicitly named as such (*Black Feminist Thought*; 1991) argues we must first disrupt subordinate epistemic positioning: "The presence of subjugated knowledges means that groups are not equal in making their standpoints known to themselves or others. 'Decentering' the dominant group is essential and relinquishing privilege of this magnitude is unlikely to occur without struggle" (p. 236–237). She goes on in her recent book *Intersectionality as Critical Social Theory* to unpack how the premise of an intersectional sensibility in our work must resist how society has made rigid and flattened oppressed peoples' experiences as less than valuable sources of credible, reliable, and valid authors of their lived experiences, which is a testament to resistant knowledge projects that embody intersectionality (cf. Collins, 2017; May, 2015).

Kimberlé Crenshaw (2016) goes on to compliment these arguments about the theory of intersectionality as a critical social theorizing praxis when she points out that one very powerful component of intersectionality rests less on overlapping identities and more on how power is exerted among social and cultural locations. During her keynote at Women of the World 2016, she emphasized that intersectionality is most concerned with the "structures that make some identities the consequence of and vehicle for vulnerability." This articulation places importance on understanding lived realities as an integral praxis when conducting research to support people empowering themselves and being agents to self-determinate their life. In turn, this leveraging of intersectionality as metaphor, while the most frequently operationalized theorizing of the term, should be most engaged with as a way to broach reality as it exists right now in the present while taking into consideration historical precedence to understand it: that to disrupt discrimination and violence head-on, social justice scholarship must engage with the contexts that have led to "the exclusion of some people but not others," and that "if you want to know what discrimination is happening, you have to look at the context in which it is happening!" (Crenshaw, 2016). Here, we find the impetus of intersectionality as critical social theory: Context. It is by understanding the past and present of specific *contexts* in ways that articulate power and difference as being an interplay where we can begin to Dream with intersectionality through how it manifests in our approaches to ameliorate inequity among disciplinary epistemic commitments and their cultures.

For students labeled with disabilities in higher education, this idea of context comes forth when we begin theorizing difference as a metaphor for understanding the oppressive systems that exclude some but not others; that welcome some studenthoods but not others; and deny agency to those who are outside the historical purview of who can be named as a STEM professional.[4] Identifying what Collins (2019) calls "conceptual correspondences or relationships" (p. 30) by using an intersectional lens to name how markers of difference such as race and gender provide nuance to how we understand experiences in particular contexts is valuable.

[4] See Robertson et al. (2023) for an example of how race has been denied visibility by STEM professors.

But, for DBER, taking seriously the notion of intersectionality as a metaphor in the context of disabled students pursuing STEM degrees and careers being provided accommodations as the benchmark to define when "justice has been done" serves only to reify the "separate but equal" status emboldened through the label of disability seen elsewhere when describing colonial subjects as being raced, classed, and defined only by markers of difference (Boda, 2018, 2020, 2022; Cioè-Peña, 2021; Sins Invalid, 2019; Taylor, 2018; Vernon, 1999). Breaking through the metaphoric quality to push into "a place not simply to house experiences but also a way of working both politically and intellectually" (Collins, 2019, p. 33) requires that the legally mandated accommodations for students labeled with disabilities in higher education not happen just because a disabled student is enrolled in a STEM class, but goes further to think with and through intersectionality as critical social theory in ways to solve social problems in our fields. This further conceptualizing of intersectionality as a heuristic can then push DBER scholars beyond mere abstract metaphor to seek a systemic approach that cannot just identify intersectional oppression but Dreams of futures where the types of designs that have born and bred these architectures are dismantled.

6.4 Intersectionality as a Heuristic in DBER

If intersectionality as critical social theory is to be taken up by DBER scholars in ways that disrupt the ableist designs so deeply embedded in higher education, we must move beyond using intersectionality as a metaphor for describing overlapping oppression. In addition, we must engage intersectionality as a methodological and epistemological architecture that should be used to analyze "everyday experiences with an eye toward shaping action strategies in everyday life" and additionally "to solve social problems within an academic discipline" (Collins, 2019, p. 34). This *heuristic* approach to intersectionality helps rethink existing knowledge systems—past and present—to interrogate how problems were solved in the past by creating speculative questions of futures untold not based on the previously centered procedures and praxes. This move beyond metaphor to think heuristically also brings to bear previously unquestioned systems of power that define how identities are socially positioned as valuable/expendable in capitalist binary epistemic frames often used when attempting to unpack how oppression operates in a specific context. For students outside of the race/class/gender triad often used to understand intersectionality as a term in DBER and social justice scholarship more broadly, this heuristic implementation, echoing Crenshaw's seminal article title (1990), "maps the margins" among our disciplinary strongholds that have been used traditionally to engage with equity and ameliorate injustice. This *heuristic* take on intersectionality, then, is particularly valuable for disabled students given the active erasure of their humanity in society more broadly (Andrews et al., 2019; Reynolds, 2022), by education scholarship (Erevelles, 2018; Pugach et al., 2021), and in schools (Boda, 2021, 2022).

Leveraging intersectionality as a heuristic can thus evoke imaginative scholarship about the type of justice considered valuable "if research was primarily focused not on *serving* populations but participating in the radical possibility of helping *cultivate freedom* from the expectations of the majority" (Boda et al., 2022). I have argued in other work that the structural segregation between general and special education research delineates hierarchies of value—that disabled students are more medically (bodily) deficient than general education students by virtue of the label of disability not being welcomed at the social justice/multiculturalism table. This conceptual and disciplinary segregation (Boda, 2021) therefore serves only to sustain cultures of schooling that mirror the racist, classist, and hetero-sexist realities of educational debt accrued from historical-material precedents supported in America, as well as most other Global North and Westernized designs of who is "teachable," and who is "smart" (Leonardo & Broderick, 2011; Hatt, 2012). By witnessing the lack of engagement with the ideology disability within and among STEM education disciplinary research, even when thinking about social justice goals, we can view disability as having expansive power for transforming societal institutions such as schools. However, our approach, again, must not rest only on description of oppression, but demands that we move beyond the abstractive (and often redundant) metaphor that within markers of difference people experience oppression in ways that shed light on how to Dream beyond this reality. And when we seek to defy the rigid traditions that have built and sustained (post-)positivist obsessions within education research as sites of objectification and transaction rather than spaces of agency and liberation, such as those mono-categorical enquiries used in past DBER scholarship, we work toward cultivating *epistemic disobedience* (Mignolo, 2009).

6.5 Intersectional Disobedience and Hegemonic Epistemic Commitments

In 2009, Walter D. Mignolo put forth that academic knowledge-making commitments (*episteme*) should not be juxtaposed as opposite to personal meaning-making (*doxa*). Instead, he suggested an alternative to this binary by de-linking from the power vested in the colonial project because of the disciplinary traditions in philosophy, sociology, and epistemology that denied the lived experiences of colonized peoples around the world in ways that required more concerted attention if we are to pursue justice for all. In his work, by refusing binary categorizations of value that define who can be producers of knowledge, as well as whose social and cultural locations are there to be assembled by those at the center, Mignolo compliments arguments made by Disabled scholars and their kin. Indeed, Mignolo's articulation of *epistemic disobedience* also encourages a deconstruction of how to collect knowledge (methodology) and the treatment of personal interpretations of lived realities (epistemic commitments). When given power and authority to define truth about Self and Other, hierarchies of "other" pit mono-categorizations of difference (i.e., race or gender or disability, but never Both/And/All) against each other to fracture solidarity (Davis, 2016). Drawing from Leah Lakshmi Piepzna-Samarasinha (n.d.), a self-described "queer disabled femme writer, organizer, performance artist and educator of Burgher/Tamil Sri Lankan and Irish/Roma ascent" (para. 1), from her radical book *Care Work: Dreaming Disability Justice* (2018), we see a connection between Mingolo's epistemic aims and the nature of intersectionality as a heuristic used to change our collective sociopolitical worlds:

> [P]eople have a lot of different gifts to offer, and normal and ableds [*sic*] often assume not only that we [Sick and Disabled folks from multiply marginalized backgrounds such as being LGBTQ+] have nothing to offer, but that we can only be (patronizingly and abusively) cared for. But care doesn't have to be one way. It can become an ongoing responsive ecosystem, where what is grown responds to need. (p. 147)

Lakshmi Piepzna-Samarasinha above is embodying both an intersectional sensibility by understanding the metaphors of disability and difference as deficit laden among the rhetoric of our Western Modern world's obsession with capitalist profit, while also providing a lens through which these realities can be dismantled. Namely, Lakshmi Piepzna-Samarasinha (2018) employs intersectionality as a heuristic to think about how we can theorize deeply about *care networks, interdependence,* and *community*, and more specifically in ways that address the needs of people outside the purview of historical systemic hegemony. Hamraie and Fritsch (2019) in their *Crip Technoscience Manifesto* further this Dreaming with an intersectional heuristic and put forth that the "practices of critique, alteration, and reinvention of our material-discursive world" (p. 2) used by Disabled people constitute an embodiment of *crip technoscience*, encouraging us to think about these critical perspectives of ingenuity, survival, and imagination that lay bare the brilliance and, what many abled-bodied cannot Dream of, beauty that is present when claiming Disability as an identity. They argue that "crip technoscience claims disabled life as desirable life, as life worth living, and as a difference that matters … [which can] alter the very systems that produce disability or attempt to render us as broken" (2019, p. 20). Here, again, we find a connection between design of systems, the lived realities of those most relegated to the margins among enquiries pursuing equity, and pathways forward when thinking about how we construct knowledge about ourselves and with others. Disobeying the epistemic commitments that render disability in juxtaposition to ableist constructions of Self, thinking with Reynolds (2022), these scholars make present the necessity to move beyond "what is" and move toward "what if" (Boda et al., 2022) through an intersectional disobedience that values differential knowledge building.

To illustrate a pragmatics of this critical theorizing for DBER scholars wishing to unpack and dismantle the ableism through intersectional disobedience, Spektor and Fox (2020) provide a guided template to engage with Disabled people in ways that disrupt exclusionary designs by leveraging their lived experiences, centering their many Lives-Hopes-Dreams, and paying close attention to the ways in which speculative questioning plays an important role in the design process. In their work, they Dream about the roles and purposes of assistive technology *with* Disabled folks, and for DBER people this could plausibly transfer to the ways accommodations are provided using technology through a *crip technoscience* sensibility, as well as how the structural arrangements of classrooms, teaching practices, and assessment measurements are developed beyond the vacuous spaces of STEM disciplinary-focused content acquisition. They further conclude with a strong summary of why this work is important, and for whom, by using trans-disciplinary critical theorizing that incorporates multiple perspectives and goals, which, unfortunately, are not found in prominence among STEM DBER scholarship at the present time:

> Disabled bodies are commonly regarded as 'original cyborgs' in transhumanist futuring, outfitted with bionic limbs and intelligent canes in the hopes that they will be optimised—or, at the very least, normalised. Assistive technologies thus operate under a productivist ethos which conflates a working body with a valuable one. Far from being seamless, disabled individuals are forced to shoulder steep physical, social, and financial burdens to utilise assistive technologies. With and through their design concepts, our interviewees began to instead explore assistive devices that center the values of disability communities, rather than reinforce an optimisation rhetoric. (2020, p. 350)

Knowing that our futures are defined by what we do now, DBER scholars engaging in an equity ethos toward their work and the purpose of postsecondary education more broadly could learn much from inviting and engaging with their Disabled students, Disability communities, and their Disabled colleagues to support transformative changes to leadership, exclusionary structures, and oppressive practices that sustain disciplinary hegemony. But this takes a shift in their paradigms.

6.6 Intersectionality as a Paradigm Shift in DBER

Using Patricia Hill Collins (2019) definition of paradigmatic thinking (i.e., that which "involves having a model or provisional explanation in mind, a typical pattern of something, [and] a distinct set of concepts or thought patterns" (p. 41)), I argue that some of the tensions that may have emerged in the previous sections around theorizing beyond "what is"—beyond the paradigms currently employed by STEM disciplines and the DBER research professionals doing equity research—can be navigated by thinking through intersectionality to disrupt hegemony. Honoring Collins' words once more, her presentation of paradigm shifts and the value of intersectionality in helping toward these changes provides a bit more clarity for this approach:

> [Drawing on Kuhn's articulations, paradigm shifts] are significant because they describe what happens when traditional frameworks no longer sufficiently explain social realities and thus become ineffective. In this sense, the concept of a paradigm shift is especially important for intersectionality as a critical social theory in the making, because a paradigm shift identifies a significant turning point when established social theories lose their critical edge and when other social theories displace them. (2019, p. 43)

Lovingly, Collins' words on the importance of how intersectionality as a paradigm shifting theory urges scholars such as those working in DBER to think, as I presented above, where they have come to know how and why a particular discipline or field of study operates in the ways that are deemed "acceptable," and then determine to what extent researchers themselves become part and parcel of that hegemony, as well as if they seek intersectional disobedience to the epistemic commitments that have marginalized people in their fields. Using intersectionality as a metaphor and/or heuristic draws on and often seeks to push that "what is" in the favor of "what if" (Boda et al., 2022). Leveraging intersectionality as a paradigm shifting theory turns this critical gaze inward to the researchers, the theory itself, and adopts a reflexive stance toward the evolution of the term. Indeed, intersectionality used to critically theorize the world is an ever evolving and shifting practice which is defined, to reiterate Crenshaw's (2016) words above, by the context where discrimination is taking place. For Discipline-based Education Research, the context in which their inequity is born, bound, and bred emerges as a function of strong ableist and other oppressive commitments to the ways science, technology, engineering, and math have been traditionally subject by academic gatekeepers at the university and individual levels (Bayer Corporation, 2012; Rahming, 2022), primarily reserved for the "best and brightest" students perceived as such from broader societal expectations of "good" and "smart" identities (Hatt, 2012; Leonardo & Broderick, 2011), and frequently draw on the historical designs of White supremacy (Ladkin & Patrick, 2022), especially at elite universities (Jackson-Cole & Chadderton, 2021).

Dreaming beyond this reality of the present-day amalgam of exclusionary designs that go unquestioned because of the assumptions made about what constitutes STEM studenthood, using intersectionality as a paradigm-shifting process of critically theorizing these disciplinary histories and contexts brings to bare the importance of asking important questions brought forth at the beginning of this chapter but in different ways. Namely: Why do DBER research? And for whom is it designed to serve? I put forth another question, acting in mutiny to disciplinary theorizing that has plagued DBER work devoid of criticality and equity orientations, that can serve as a benchmark for engaging with intersectionality as a paradigm shift: Who decides when justice is served? Echoing my questions and perturbations to the hegemonic status quo of scholarly work in DBER spaces, Collins reminds us to keep a keen eye on how questions are often as important as the ways we think about finding answers: "When it comes to intersectionality as a critical social theory, asking the wrong questions is a far bigger problem than offering eloquent answers to what seem to be the right questions" (2019, p. 290). Here, I Dream through intersectionality to help DBER scholars think about the types of questions they are asking, and why *those questions* have been at the forefront of how we think about equity in the field. As stated before, Disabled positional identities are almost entirely erased from DBER scholarship aims and goals, and if they are at all they come in the form of comparative growth in relation to the *ableist center of higher education*. Indeed, I am nuancing Leonardo and Broderick's term the *normative center of schooling* (2011) to re-center this term in STEM higher education spaces where we know that students labeled with disabilities as seen as "being served" merely by obtaining their legally mandated accommodations (Nardo et al., 2022), which is crucial when witnessing Disabled peoples' lack of support needs being noted at all in STEM education policy (Gordián-Vélez, 2022). However, Friedensen et al. (2021) note that the support needs for students labeled with disabilities in STEM pathways at the postsecondary level do not start, nor do they end, with accessibility to accommodations; and, rather, that able-bodied Dreams are not all that dissimilar.

In their work, Friedensen et al. (2021) point to an ecological shifting that is needed to fully support students labeled with high incidence disabilities in STEM pathways. For clarity, high incidence disabilities are those that account for about 80% of all disability labels, and include autism, intellectual, emotional, and physical disabilities, which are further categorized by the extent of support need such that these disabilities would be most frequently found in general education contexts. Therefore, when applying selection criteria in their study, Friedensen et al. (2021) have ostensibly excluded low incidence disabilities that often come with multiple, overlapping, and complex supports that many scholars believe constitute a need to segregate outside of these general education contexts. Putting to the side these limitations, DBER scholars can learn much in terms of how to shift their paradigms about what successful STEM learning looks and feels like at the postsecondary level by considering how these students with high incidence disability labels articulate their needs for science, technology, engineering, and mathematics in higher education. While Cech (2021) showcases the exclusionary designs sustained within engineering postsecondary pathways, and Lillywhite and Wolbring (2022) describe how students with disabilities are rarely given the opportunity to participate in research across disciplines, including STEM, Friedensen et al. (2021) provide more nuance and a critical take on how to think about these overlapping identities in relation to non-Disabled social justice scholarship (i.e., intersectionality as a metaphor), as well as how the current structures, policies, and practices are not designed to meet these students' support needs (i.e., intersectionality as a heuristic). Leveraging this text in the context of pursuing equity in DBER, then, can help us think how to ask new questions and develop new modes of enquiry beyond content learning.

Friedensen et al. (2021) first note that these exclusionary designs start with the transition to college itself, which for any university student new to a campus is a struggle and a process of acclimation. Here, I am illuminating the nature of this articulated support need in a comparative way to illustrate that the needs of those students labeled with disabilities in STEM pathways do not emerge solely from a site that is defined by the disability itself. This thereby further critically theorizes and, subsequently, problematizes the ideology that disability rests primarily in the individual and less in the constructed spaces they frequent, occupy, and survive. Friedensen et al. (2021) then showcase how peer and professor behavior and attitudes toward teaching greatly impact the ways Disabled students view their acceptance and belonging in STEM spaces. This second point, again, illustrates a paradigm shift of how we think about equity in DBER for STEM disciplines given that these perspectives by Disabled students in higher education have been consistently found in the literature voiced from their non-Disabled peers (cf. Dancy et al., 2016; Hughes & Kothari, 2021; Reyes, 2011; Suran, 2021; Trenshaw, 2022). Finally,

Friedensen et al. (2021) eloquently expose the cracks in STEM learning environment design at the postsecondary level which they report Disabled students don't feel like they "fit into" many times in these spaces, and that the organizational structures—often held over without question from historical precedence—severely affected how they could Live, Learn, and develop an Identity in STEM. They do note that some of these organizational issues come from accommodations not being serviced well, but by and large the critiques provided by these Disabled voices on the ground articulate a space for radically changing the system and its supportive procedures entirely. And when we know that poor students labeled with disabilities in K-12 spaces receive less STEM courses than their able-bodied counterparts (Peterson, 2021), the nature of an equity-based DBER intervention cannot effectively be situated solely in a classroom space. These are systemic cultures that have sustained exclusionary designs over decades in higher education. To disrupt these mechanisms of oppression, we must think beyond the paradigms we have used.

This brings us full circle to think about using intersectionality as a paradigm-shifting way to critically theorize the world, while also being cognizant and careful about how using the term is concomitant with understanding how power and difference manifest in context, even if the term is not used (Cho et al., 2013). Therefore, for DBER scholars pursuing equity-oriented research in STEM pathways at the postsecondary level, this means that the questions once asked about how to support and sustain students, and particularly Disabled students in these contexts, require re-framing, re-articulation, and/or re-creation entirely. In the pragmatic sense, this can take form by asking some of the following questions that are not currently well taken-up in the literature within the field: To what end do DBER approaches to STEM success support Disabled students Becoming STEM professionals?; What mechanisms have been more productive and/or destructive at affording Disabled students a sense of Belonging in STEM pathways?; and How can we foster learning environments that sustain cultures in STEM learning where Disabled students feel Beloved among STEM peers, professors, and researchers? These questions are fundamentally different from asking content acquisition enquiries around whether a student has "learned" a specific concept, to what extent did they develop argument sophistication, and in what ways we could support retention in STEM pathways (devoid of critical and society impact overlaps that compound the exclusive social positioning). There are not currently answers to the questions I posed above, and little outside of quantitative analyses for the latter questions I gave. I urge DBER scholars to engage with the ideology of Disability that they currently hold, seek out ways to understand how that ideology was constructed based on their personal experiences and exposure to broader theory, and then truly perceive that Disabled students, along with other historically marginalized positionalities in STEM, have something to contribute to their growth. Without such a radical paradigm shift that actively disobeys the epistemic commitments of (post-) positivistic scholarship in DBER among STEM pathways, there is no shift in the ways that systems and their corresponding policies have engrained hegemony that translates into practices.

6.7 Conclusive Dreaming and Speculative Innovation

In this chapter, I have unpacked the origins and goals of Discipline-based Education Research (DBER) in Science, Technology, Engineering, and Mathematics (STEM) pathways at the postsecondary level, as well as provided nuanced perspectives toward higher education leadership in relation to how reform has traditionally not been designed to meet the criticality of equity-oriented scholarship more broadly. I then showcased the active erasure of disability among these endeavors in both postsecondary and K-12 contexts, providing an articulation (Clarke, 2015) whereby to embody an equitable praxis in these contexts requires listening and critically theorizing the ways subaltern positional identities are marginalized by design, thereby seeking to draw on intersectionality as critical social theory to disrupt the lack of engagement around power, agency, and self-determination by thinking with Disabled scholars and Disabled activism. Through the cognitive architecture set down by Collins (2019) around intersectionality as a knowledge building project that demands trans-disciplinary theorizing and diverse material responses, I further expounded on how intersectionality as metaphor is often the most (mis)used typology of the term, as well as how using this metaphoric abstraction of intersectional identity comes to fruition namely by drawing on intersectional sensibilities being directly connected to using the term as a heuristic to challenge previous Discourses that dominate fields of study, and can invariably stagnate equity-oriented research. I finished by diving into intersectionality as a paradigm-shifting form of critical theorizing social injustice for Disabled students in DBER scholarship such that we begin to engage in the needed changes across systems and minds.

These arguments are not taken lightly in the face of research that shows how Disabled students have differential access to STEM disciplines at the K-12 level (Peterson, 2021), and drawing connections between other markers of difference such as race when thinking about how to Dream beyond "what is" important for equity work in both DBER and K-12 education research from past precedence and current theorizing of disability (cf. Boda, 2022; Reinholz & Ridgway, 2021; Reinholz et al., 2021; Suran, 2021; Turner, 2022). This shifting of paradigms also takes time; but we must also ask *whose time* is being subject to deferral of equity, and *who* is mostly served by the activities we engage with as equity-oriented scholars in DBER. The value and promise of leveraging intersectionality as critical social theory in higher education contexts is limited only by our resistance to Dream beyond what has been deemed "acceptable" research aims, methods, and analyses. This chapter serves as a guide forward to Dream of futures untold.

References

Adejumo, V. (2021). Beyond diversity, inclusion, and belonging. *Leadership, 17*(1), 62–73.

Allan, J. (2010). The sociology of disability and the struggle for inclusive education. *British Journal of Sociology of Education, 31*, 603–619.

Andrews, E. E., Forber-Pratt, A. J., Mona, L. R., Lund, E. M., Pilarski, C. R., & Balter, R. (2019). # SaytheWord: A disability culture commentary on the erasure of "disability". *Rehabilitation Psychology, 64*(2), 111.

Bayer Corporation. (2012). Bayer facts of science education XV: A view from the gatekeepers—STEM department chairs at America's top 200 research universities on female and underrepresented minority undergraduate STEM students. *Journal of Science Education and Technology, 21*(3), 317–324.

Berne, P., Morales, A. L., Langstaff, D., & Invalid, S. (2018). Ten principles of disability justice. *WSQ: Women's Studies Quarterly, 46*(1), 227–230.

Boda, P. A. (2018). Exclusion from participation in science: Confessions from an ally on the other side of the fence. In M. Koomen, S. Kahn, C. L. Atchison, & T. A. Wild (Eds.), *Towards inclusion of all learners through science teacher education* (pp. 301–311). Brill Sense.

Boda, P. A. (2020). 'More than an educator but a political figure': Disability and critical pedagogy in teacher education. In S. R. Steinberg & B. Down (Eds.), *The SAGE handbook of critical pedagogies* (pp. 869–883). Sage Publishers Ltd.. https://doi.org/10.4135/9781526486455.n81

Boda, P. A. (2021). The conceptual and disciplinary segregation of disability: A phenomenography of science education graduate student learning. *Research in Science Education, 51*(6), 1725–1758.

Boda, P. A. (2022). Identity making as a colonization process, and the power of disability justice to cultivate intersectional disobedience. *Education Sciences, 12*(7), 462.

Boda, P. A., Nusbaum, E. A., & Kulkarni, S. S. (2022). From 'what is' toward 'what if' through intersectionality: Problematizing ableist erasures and coloniality in racially just research. *International Journal of Research & Method in Education*, 1–14.

Cech, E. A. (2021, July). Engineering's systemic marginalization and devaluation of students and professionals with disabilities. In *2021 ASEE Virtual Annual Conference Content Access*.

Charlton, J. I. (1998). Nothing about us without us. In *Nothing about us without us*. University of California Press.

Cho, S., Crenshaw, K. W., & McCall, L. (2013). Toward a field of intersectionality studies: Theory, applications, and praxis. *Signs: Journal of Women in Culture and Society, 38*(4), 785–810.

Cioè-Peña, M. (2021). Raciolinguistics and the education of emergent bilinguals labeled as disabled. *The Urban Review, 53*(3), 443–469. https://doi.org/10.1007/s11256-020-00581-z

Clarke, J. (2015). Stuart Hall and the theory and practice of articulation. *Discourse: Studies in the Cultural Politics of Education, 36*(2), 275–286.

Collins, P. H. (1991). *Black feminist thought: Knowledge, consciousness, and the politics of empowerment*. Routledge.

Collins, P. H. (2017). Intersectionality and epistemic injustice. In *The Routledge handbook of epistemic injustice* (pp. 115–124). Routledge.

Collins, P. H. (2019). *Intersectionality as critical social theory*. Duke University Press.

Crenshaw, K. (1990). Mapping the margins: Intersectionality, identity politics, and violence against women of color. *Stanford Law Review, 43*, 1241.

Crenshaw, K. (2016, March 15). *On intersectionality* [Video]. YouTube. https://www.youtube.com/watch?v=-DW4HLgYPlA&t=190s

Dancy, M. H., Rainey, K., Mickelson, R., Stearns, E., & Moller, S. (2016). Influences of teaching style and perceived care of instructor on retention of underrepresented groups in STEM. *PERC Proc*.

Davey, T., Salazar Luces, J. V., & Davenport, R. (2021). Individual-centred approaches to accessibility in STEM education. *Education Sciences, 11*(10), 652.

Davis, A. Y. (2016). *Freedom is a constant struggle: Ferguson, Palestine, and the foundations of a movement*. Haymarket Books.

Davis, A. Y., Dent, G., Meiners, E. R., & Richie, B. E. (2022). *Abolition. Feminism. Now* (Vol. 2). Haymarket Books.

Erevelles, N. (2018). Toward justice as ontology: Disability and the question of (in) difference. In *Toward what justice?* (pp. 67–83). Routledge.

Freeman, S., Eddy, S. L., McDonough, M., Smith, M. K., Okoroafor, N., Jordt, H., & Wenderoth, M. P. (2014). Active learning increases student performance in science, engineering, and mathematics. *Proceedings of the National Academy of Sciences, 111*(23), 8410–8415.
Friedensen, R., Lauterbach, A., Kimball, E., & Mwangi, C. G. (2021). Students with high-incidence disabilities in STEM: Barriers encountered in postsecondary learning environments. *Journal of Postsecondary Education and Disability, 34*(1), 77–90.
Fuqua, J. L., Phillips, J. A., Bargagliotti, A., & Herreiner, D. (2019, January). Facilitators and outcomes of STEM-education groups working toward disciplinary integration. In *Proceedings of the Physics Education Research Conference (PERC)* (pp. 131–134).
Gordián-Vélez, W. J. (2022). Policy position paper: Ensuring the inclusion of people with disabilities in STEM education and careers. *Journal of Science Policy & Governance, 20*(02).
Hamraie, A. (2016). Universal design and the problem of "post-disability" ideology. *Design and Culture, 8*(3), 285–309.
Hamraie, A., & Fritsch, K. (2019). Crip technoscience manifesto. *Catalyst: Feminism, Theory, Technoscience, 5*(1), 1–33.
Hatt, B. (2012). Smartness as a cultural practice in schools. *American Educational Research Journal, 49*(3), 438–460. https://doi.org/10.3102/0002831211415661
Henderson, C., Connolly, M., Dolan, E. L., Finkelstein, N., Franklin, S., Malcom, S., et al. (2017). Towards the STEM DBER alliance: Why we need a discipline-based, STEM-education research community. *Journal of Geoscience Education, 65*(3), 215–218.
Hughes, B. E., & Kothari, S. (2021). Don't be too political: Depoliticization, sexual orientation, and undergraduate stem major persistence. *Journal of Homosexuality*, 1–28.
Jackson-Cole, D., & Chadderton, C. (2021). White supremacy in postgraduate education at elite universities in England: The role of the gatekeepers. *Whiteness and Education, 8*, 1–19.
Ladkin, D., & Patrick, C. B. (2022). Whiteness in leadership theorizing: A critical analysis of race in Bass' transformational leadership theory. *Leadership, 18*(2), 205–223.
Leonardo, Z., & Broderick, A. A. (2011). Smartness as property: A critical exploration of intersections between whiteness and disability studies. *Teachers College Record, 113*(10), 2206–2232.
Lillywhite, A., & Wolbring, G. (2022). Undergraduate disabled students as knowledge producers including researchers: Perspectives of disabled students. *Education Sciences, 12*(2), 77.
May, V. M. (2015). *Pursuing intersectionality, unsettling dominant imaginaries*. Routledge.
Mignolo, W. D. (2009). Epistemic disobedience, independent thought and decolonial freedom. *Theory, Culture & Society, 26*(7/8), 159–181. https://doi.org/10.1177/0263276409349275
Nardo, J. E., Chapman, N. C., Shi, E. Y., Wieman, C., & Salehi, S. (2022). Perspectives on active learning: Challenges for equitable active learning implementation. *Journal of Chemical Education*.
National Research Council. (2012). *Discipline-based education research: Understanding and improving learning in undergraduate science and engineering*. S. R. Singer, N. R. Nielsen, & H. A. Schweingruber (Eds.). The National Academies Press. http://www.nap.edu/catalog.php?record_id=13362
Nusbaum, E. A., & Steinborn, M. L. (2019). A "visibilizing" project: "Seeing" the ontological erasure of disability in teacher education and social studies curricula. *Journal of Curriculum Theorizing, 34*(1).
Patton, L. D., Jenkins, T. S., Howell, G. L., & Keith, A. R., Jr. (2022). Unapologetically Black creative educational experiences in higher education: A critical review. *Review of Research in Education, 46*(1), 64–104.
Peterson, S. (2021). The interaction of disability status and family income level as a predictor of STEM achievement for youth in US high schools. *Rehabilitation Professional, 29*(2).
Pickard, B. (2022). The importance of collaboration: Valuing the expertise of disabled people through social confluence. In *Developing expertise for teaching in higher education* (pp. 69–84). Routledge.
Piepzna-Samarasinha, L. L. (2018). *Care work: Dreaming disability justice*. Arsenal Pulp Press.

Piepzna-Samarasinha, L. L. (n.d.). *Leah Lakshmi Piepzna-Samarasinha: Writer, disability/transformative justice worker*. https://brownstargirl.org/

Prema, D., & Dhand, R. (2019). Inclusion and accessibility in STEM education: Navigating the duty to accommodate and disability rights. *Canadian Journal of Disability Studies, 8*(3), 121–141.

Pugach, M. C., Matewos, A. M., & Gomez-Najarro, J. (2021). Disability and the meaning of social justice in teacher education research: A precarious guest at the table? *Journal of Teacher Education, 72*(2), 237–250.

Rahming, S. G. (2022). The STEM glass ceiling: The influence of immigration status on STEM trajectories of Afro-Caribbean women (a narrative approach). *Journal of International Students, 12*(1), 156–174.

Reinholz, D. L., & Ridgway, S. W. (2021). Access needs: Centering students and disrupting ableist norms in STEM. *CBE—Life Sciences Education, 20*(3), es8.

Reinholz, D. L., White, I., & Andrews, T. (2021). Change theory in STEM higher education: A systematic review. *International Journal of STEM Education, 8*(1), 1–22.

Reyes, M. E. (2011). Unique challenges for women of color in STEM transferring from community colleges to universities. *Harvard Educational Review, 81*(2), 241–263.

Reynolds, J. M. (2022). Theories of disability. In *The disability bioethics reader* (pp. 30–38). Routledge.

Roberts, T., Maiorca, C., Jackson, C., & Mohr-Schroeder, M. (2022). Integrated STEM as problem-solving practices. *Investigations in Mathematics Learning, 14*, 1–13.

Robertson, A. D., Vélez, V., Hairston, W. T., & Bonilla-Silva, E. (2023). Race-evasive frames in physics and physics education: Results from an interview study. *Physical Review Physics Education Research, 19*(1), 010115.

Roehrig, G. H., Dare, E. A., Ring-Whalen, E., & Wieselmann, J. R. (2021). Understanding coherence and integration in integrated STEM curriculum. *International Journal of STEM Education, 8*(1), 1–21.

Ruiz-Primo, M. A., Briggs, D., Iverson, H., Talbot, R., & Shepard, L. A. (2011). Impact of undergraduate science course innovations on learning. *Science, 331*(6022), 1269–1270.

Salehi, S., Ballen, C. J., Trujillo, G., & Wieman, C. (2021). Inclusive instructional practices: Course design, implementation, and discourse. In *Frontiers in education* (p. 395).

Santangelo, J., Hobbie, L., Lee, J., Pullin, M., Villa-Cuesta, E., & Hyslop, A. (2021). The (STEM) 2 network: A multi-institution, multidisciplinary approach to transforming undergraduate STEM education. *International Journal of STEM Education, 8*(1), 1–15.

Schneiderwind, J., Johnson, J. M., Clifton, A., & Bourelle, K. (2021, January). Hidden in plain sight: Students with disabilities in STEM education. In *New Perspectives in Science Education 2021 Conference Proceedings*.

Scotch, R. K. (2009). "Nothing about us without us": Disability rights in America. *OAH Magazine of History, 23*(3), 17–22.

Shipley, T. F., McConnell, D., McNeal, K. S., Petcovic, H. L., & John, K. E. S. (2017). Transdisciplinary science education research and practice: Opportunities for GER in a developing STEM discipline-based education research alliance (DBER-A). *Journal of Geoscience Education, 65*(4), 354–362.

Sins Invalid. (2019). *Skin, tooth, and bone: The basis of movement is our people* (2nd ed.). Primedia eLaunch. https://www.sinsinvalid.org/disability-justice-primer

Slaton, A. E. (2013, June). Body? What body? Considering ability and disability in STEM disciplines. In *2013 ASEE Annual Conference & Exposition* (pp. 23–247).

Spektor, F., & Fox, S. (2020). The 'working body': Interrogating and reimagining the productivist impulses of transhumanism through crip-centered speculative design. *Somatechnics, 10*(3), 327–354.

Suran, M. (2021). Keeping Black students in STEM. *Proceedings of the National Academy of Sciences, 118*(23), e2108401118.

Taylor, A. (2018). Knowledge citizens? Intellectual disability and the production of social meanings within educational research. *Harvard Educational Review, 88*(1), 1–25. https://doi.org/10.17763/1943-5045-88.1.1

Traxler, A., & Blue, J. (2020). Disability in physics: Learning from binary mistakes. In *Physics education and gender* (pp. 129–152). Springer.

Trenshaw, K. (2022). Local minima and maxima in trans-STEM affirmations. In *Queering STEM culture in US higher education* (pp. 94–110). Routledge.

Turner, J. D. (2022). Freedom to aspire: Black children's career dreams, perceived aspirational supports, and Africentric values. *Race Ethnicity and Education, 25*(1), 128–153.

Vernon, A. (1999). The dialectics of multiple identities and the disabled people's movement. *Disability & Society, 14*(3), 385–398. https://doi.org/10.1080/09687599926217

Warren, C. A. (2021). *About centering possibility in Black education*. Teachers College Press.

Chapter 7
Political Disability Identity and DBER Disability Politics

Liam McDermott

Abstract When discussing disability, it is critical that researchers and educators acknowledge the inherent politicality of the subject. Disability is political. Education is political. Our bodies, our minds, and how they are policed and how they are liberated within the classroom, among other spaces, are political. In this chapter, I present Political Disability Identity (PDI), a useful tool and framework for understanding disability in a political space. Using PDI aids us in understanding the power dynamics within the classroom and provides us a strengths-based understanding of disability and identity through the ways which individuals and communities develop and wield political agency. By drawing on real-world examples and insights from the literature, this chapter illuminates the role of disability politics in shaping educational research agendas and in fostering environments that prioritize diversity, equity, inclusion, and justice. I discuss PDI as a framework for understanding disability in DBER spaces, and provide recommendations for conducting education research in a strengths-based, political disability way.

Keywords Disability · Diversity, equity, and inclusion · Disability justice · Disability education · Disability politics · Accessibility

To invigorate and hold space for Eduardo Bonilla-Silva's words, I take up Disability Justice[1] (DJ) to infuse: "When whiteness [ableism] becomes normative, it works like God, in mysterious ways" (Zuberi & Bonilla-Silva, 2008). When we confront

[1] In conversation with the words and works of Sami Schalk (2022), Judy Heumann and Joiner (2021), Alice Wong (2018) Bailey and Mobley (2018), Audre Lorde (2017), among others.

L. McDermott (✉)
University of Connecticut—Avery Point, Groton, CT, USA
e-mail: vtb24001@uconn.edu

G. L. Cochran, J. D. Adams (eds.), *Equity-Oriented, Discipline-Based STEM Education Research*, Critical Studies of Education 19,
https://doi.org/10.1007/978-981-96-9136-4_7

that ableism is especially political; that it is so incredibly entrenched in our laws and social norms; that it is readily accepted by society and incorporated into other forms of oppression such as racism and hetero-sexism, we are met with a nuanced, oftentimes invisible challenge (Wolbring, 2008; McRuer, 2006, 2018; Goodley, 2014; Leigh & Brown, 2020; Bell, 2012), a challenge which must be overcome if we are to make academia a welcoming, productive space for everyone.

So ubiquitous is ableism, especially in academic circles, that it is often excused as pragmatism (Kimball et al., 2016). Intellectual and disciplinary citizenship is repeatedly denied to disabled students and professionals under the guise of academic rigor or productivity (Kimball et al., 2016; D'Amico, 2016). In fighting against or within an ableist system for their own right to exist, disabled individuals are both internally and externally silenced (Lourens, 2018) or are removed entirely from their field (Olsen et al., 2020). It is therefore critical that we, disabled academics, and our allies continuously cultivate a space in academia which empowers us to make good change, good trouble, and good work. That we understand where useful political agency and power is developed and nurtured, and effectively wield this agency and power to challenge and dismantle oppressive academic ableist systems. Regarding this, Political Disability Identity (PDI) has recently begun being used in educational contexts (Smith & Mueller, 2022; McDermott et al., 2024). Developed by Putnam (2005), it can be understood twofold. First, PDI is disability identity formation in a political space. Second, PDI is disability identity formation through the development of political agency. This dual nature of PDI makes it an incredibly interesting framework for study, a very powerful tool for understanding how social change can be achieved for disabled students in higher education. 19.4% of undergraduates are disabled (NCES), and more and more disabled students enter college each year. It is therefore imperative that we, as researchers, center disabled students in our work and make their voices heard, as those with privilege cannot always fully understand the experiences of marginalized groups (Hartsock, 2003). It is also critical that we consistently aim our research at enabling and creating positive social change (Aguirre et al., 2017). Furthermore, it is important that we encourage positive disability identity development in students (Jain, 2019). PDI as a framework for research is one way to do all three. In order to use PDI in research, however, it is important we fully understand PDI and its philosophical background, as well as how to effectively wield it as a tool. Keeping this in mind, in the subsequent handbook chapter, I will do the following: establish political disability in the classroom, establish why a PDI framework is effective for disability studies in Discipline-Based Education Research (DBER), and discuss how to use PDI's guiding themes in research.

7.1 Positionality

Before diving head-on into PDI and its relevance to DBER, it is important that I position myself and the identities which make up me, the author of this chapter. I am a white, queer, disabled physics education researcher. I am both neurodivergent and

I am hard-of-hearing, wearing assistive devices (hearing aids). In my studies, both primary, secondary, and higher, I have experienced many forms of institutional and direct ableism and homophobia. It is this repetitive and tiring experience of academic violence which steered me in the direction of DBER and Physics Education Research, and what drives me to share and discuss a PDI framework for use in DBER.

7.2 Political Disability in the Classroom

Political, the act of doing politics, is a negotiation of power and resources among individuals and groups (Aguirre et al., 2017). This way of understanding politicalness allows us to define certain aspects of ourselves and our environment as political and gives us space to play in and examine these aspects under this lens. In this definition, disability is political (Jarman et al., 2002; McRuer, 2006; Goodley, 2014). Ableism is so incredibly baked-into how society, both in and out of how academia operates; the language used when talking about disability, the way we interact with others and the environment, and the way we interact epistemologically with our work all are constructed with a particular "correct" able-bodied viewpoint in mind (Brilmyer, 2018). A disabled person must negotiate resources (in the form of accessibility and accommodations) and power, or lack thereof, given to them without their input. To understand this, consider the following scenario:

> A student identifying with attention deficit hyperactivity disorder (ADHD) begins taking an introductory biology course. The lectures for this course are scheduled for 1 h 30 min. The professor has designed the course such that the lectures take up all 1 h 30 min, with some time at the end for questions. This course structure is incompatible with this student's learning, as since they have ADHD, they perform deeper analysis of individual topics covered in class, making them lose focus if the lecture advances too quickly past certain topics. Because of this, they go talk to the professor to ask them to slow down. The professor suggests the student should simply "try to focus better, as the topics are critical to the class."

In this scenario, the student is subject to political choices made for them without their input by a person wielding more power than them. Specifically, the professor wields this power of a professor-student dynamic and refuses to negotiate their power difference with the student. They do this by dismissing the student's concerns about accessibility. This is despite the student attempting to gain power by asking the professor directly to alter their teaching practices. Now, the student must negotiate resources with the professor, either through gaining more power through employing the Office of Disability Services, or through enacting coping methods in class. Further, many disadvantageous political choices were enacted toward the student before they even stepped into class. For instance, the professor initially designed the course to take the full hour and 30 min. Whether consciously or unconsciously, the professor made this choice to exclude the student instead of considering that a lecture could be more accessible with intermittent extended breaks for questions and catching up on material covered (Rich, 2009). Even higher up, the act of setting an 1 h 30 min time limit on classes by administrators is another choice deliberately ignoring this student's needs for success.

In the spirit of the scenario, the classroom is one of the first places we encounter where we have to negotiate power structures and relations in order to gain resources and status (Candela, 1998; Oral, 2013; Le et al., 2016). The classroom itself becomes a political sphere by which power and resources are negotiated. The choices educators make for students, deliberate or not, are intrinsically political and have real and lasting ramifications on their lives in and out of the classroom (Lavy & Sand, 2015). This goes double for students from marginalized backgrounds, like disabled students. It is thus important that educators are cognizant of the politicality of their decisions and deliberately make choices which encourage disabled students to occupy for themselves positions of equal power to their non-disabled peers and teachers; welcoming us to the negotiation table and including us in the discourse (McCoy et al., 2017). The choices of educators—who to include in the canon of our subjects, whose messages to frame as "valid," what to allow and not allow in class, what we encourage and what we discourage in class—are all political choices. Choices that affect power structures and resource allocation in the classroom will affect class culture. Educators have a unique opportunity to acknowledge the political power we have within their political spheres, shaping and cultivating the space in which disabled students learn and develop their sense of self, professionally and otherwise.

The question arises: How can educators center their students and encourage their agency in and out of the classroom? They can do this by encouraging and developing a PDI in their students. We know that by establishing a PDI, students are collectively uplifted and begin to thrive in their field (Jain, 2019). A good place to start this establishment as an educator is through understanding how learners develop a PDI and how the choices educators make affect this development. We have already achieved the first step to this, establishing the classroom as a political space which political actors (educators and learners) navigate. The next step is to delve deeper into PDI and understand what exactly it says about disability and research.

7.3 From a Political Classroom to a Political Disability Identity

As a social identity, it is impossible to disentangle disability rights and educational institutions (Titchkosky, 2014; Scotch, 1989; Pelka, 2012). Universities and other educational institutions have often been the breeding ground of political movements regarding disability in the United States (Altbach & Cohen, 1990). This is incredibly salient when considering the mass student-led protests in the 1960s and 1970s which culminated in legislation promoting accessibility in schools and other public institutions, as well as wider support in the self-determinism of disabled people regardless of how far they deviate from the "norm" (Putnam, 2001). This publicization of the politics of disability led to further academic discussion of how we conceptualize disability (Dolmage, 2017). Specifically, in 1993, Harlan Hahn developed a minority model of disability that conceptualized disability not as a fault of the individual, but as a construct of society. Previously, disability was often seen in American discourse as something which an individual had to overcome, due to some perceived moral failing. Disability was a dishonorable, sinful thing, something other and, if not deserving of eradication, deserving of being hidden and segregated (Andrews, 2019). Later, these moral failings, and how we viewed them, shifted to an external reference frame. This "new" idea about disability indicated that disability was not due to personal failings but due to failings of disease, external evils on an otherwise pure human body (Kattari et al., 2017). Disability came from impairment, the medical model of disability.

It should be noted that the moral model and the medical model existed simultaneously; one did not replace the other. They also are not things of the past. The moral, medical, minority, social, and diversity models, among others, all still exist. One need only be disabled and go to the doctor or listen to political discourse on addiction to see that the moral and medical models are still alive and well (Retief & Letšosa, 2018).

Hahn's work sparked a lineage of disability studies research through his conception of the minority model (Hahn, 1993). Specifically, his work was expanded upon by Michelle Putnam (2005) and combined with other disability studies work (Gill, 1997; Schur, 1998; Watson, 2002; Gilson et al., 1997) to develop her PDI framework. This framework establishes six domains of PDI development, three external and three internal. The internal domains are: *self-worth*, *discrimination,* and *pride*. These domains are internal thoughts and beliefs about disability and a person's relationship with disability and are caused by external sociopolitical forces. The external domains are: *common cause*, *policy alternatives*, and *engagement in political action.* These domains are externalizable and beliefs and actions regarding disability and society, and are caused by internal beliefs about disability. Figure 7.1 serves to define these individual domains. Developing these domains aids us in "positive sense of identity, membership in disability community, and connection to self-advocacy" (Smith & Mueller, 2022). In order to encourage development in these domains, we have to know what they look like in students.

These internal and external domains reinforce each other. Internalizing beliefs about ableism and disability promotes external domains such as taking political action or collectivizing. External performance such as communicating and organizing with other disabled people and non-disabled allies offers us new perspectives on disability as an identity and the experiences of others, strengthening internal domains. In communicating and organizing, we bring others who have a weaker sense of PDI, or those without a PDI entirely, into our political space, introducing to them new thoughts about disability and how ableism affects themselves and those they care about, starting them on the path to reinforcing their own PDI. In fact, I posit that non-disabled allies, in gaining and strengthening their own PDI, may come to positively question their own relationship with ability and ableism. Just as learning about gender and sexuality, developing identity from an allyship standpoint

Domain	How it could be expressed
Self-Worth	A person's understanding that being disabled does not make them lesser or deficient. In fact, they hold equal value in society to everyone else.
Discrimination	A person's understanding that they experience discrimination, oppression, or violence because of how society is structured and how people view disability.
Pride	A person's sense of belonging to a larger disabled community. A person's belief that belonging to this community is a good thing.
Common Cause	A person's understanding that other disabled people share similar experiences to theirs, and due to these similarities, they have similar political interests, forming a political body.
Policy Alternatives	A person's belief that the negative experiences disabled people share could and should be change through changing policy.
Political Engagement	A person's individual or group engagement in politics to enact policy alternatives. This can be through self-advocacy, protest, holding public office, etc.

Fig. 7.1 PDI domains and their definitions

promotes questioning of cis-heterosexual normativity and can cause educators and learners to challenge the performative aspects of their own identity (Pennell, 2020); learning about disability, developing a PDI through allyship, may help non-disabled educators and learners challenge the performative aspects of their ability. By reinforcing our own PDI through our performance, we open the door for people outside the community to see themselves in the "other," as it were, developing their own internal PDI domains, and strengthening our collective political disability movement (Habermas & Cronin, 2005) (Fig. 7.2).

7.4 Doing Political Disability Research

PDI, and research which is philosophically guided by its ideals, has four key themes (McDermott et al., 2024). They are, *The Social Model, Multidimensionality, Variance in Outcomes,* and *Fluidity/Inevitability*. Its themes are central to PDI (McDermott et al., 2024). These themes represent the fundamental philosophy from which PDI research is built. When using PDI in research, I suggest using all the themes when designing methodology and analyzing data. How do we do that? I, below, outline each theme, defining them and examining each's methodological and epistemological strengths to using each in research.

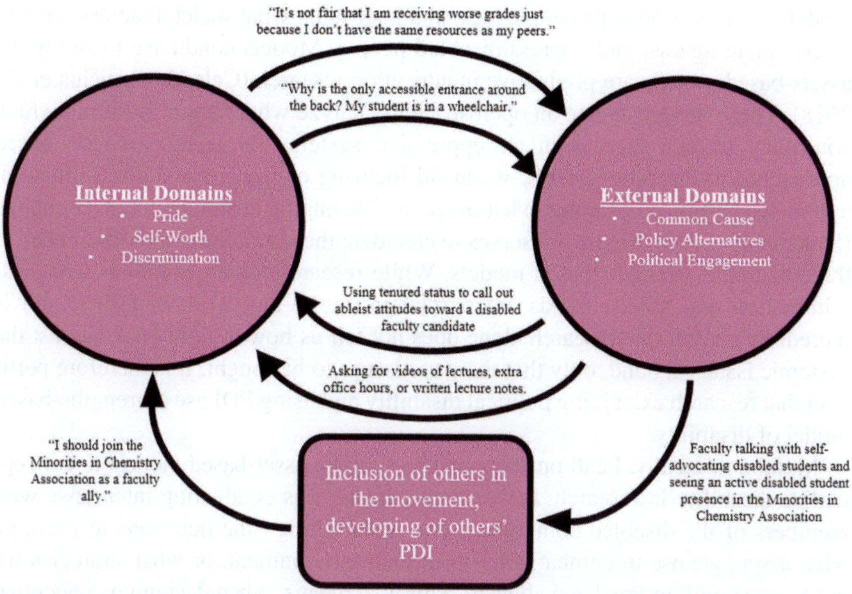

Fig. 7.2 An illustration of how a person's experiences with PDI internal domains influence their performing external domains, reinforcing internal domains, and how seeing performance of external domains can bring others into this cycle of strengthening PDI

7.5 Social Model

The social model, also called the minority model, forms the philosophical foundation of PDI. PDI is first and foremost grounded in Hahn's (1993) work, enmeshing disability-focused scholarship with his minority model to form the six domains of disability identity formation. Standing in direct opposition to the medical model, which defines disability as a negative result of impairment, this model situates disability as a result of sociopolitical forces. Otherwise saying, disability is a social construct which oppresses people with sensory, cognitive, or physical differences which society deems as "deficient." This is particularly useful when it comes to discussing recommendations for creating an accessible classroom. Instead of conceptualizing issues disabled students have in the classroom as the result of personal deficit which requires students to be "fixed" or to work harder, conceptualizing these issues as the result of incompatibility of classroom strategies with students' needs empowers educators and their students to enact structural change (Barnes, 2019).

Upon further thought however, I believe the social model of disability is not the end-all-be-all model for PDI. Instead, I believe *any* model which encourages strengths-based or assets-based (non-deficit) approaches will marry well into research using PDI. This includes the social model and expands to include the diversity model, socio-ecological model, and many other models (Dwyer, 2022). These models focus less on what people lack, and instead on what societal factors serve to discriminate against and oppress disabled people. Models conducive to strengths/assets-based models are productive in education research (Celedón-Pattichis et al., 2018). These models create an open space to analyze what factors we bring which contribute to our success in an oppressive system. By using strengths-based approaches to disability DBER, we avoid focusing on deficits and discrimination, and instead begin to examine what assets and strengths enable us to enact change (Sauer et al., 2020). Figure 7.3 serves to elucidate the similarities and differences of three example strengths-based models. While research which examines what discrimination and ableist trends are normal in day-to-day work in DBER and is incredibly useful, that research alone does not tell us how to fight back against the systemic issues at hand, only that there are issues to be fought. It is therefore pertinent that research examining political disability and using PDI use a strengths-based model of disability.

Methodologically, I call on researchers to utilize asset-based models to conceptualize disability in research. For example, if one was conducting interviews with members of the disabled community, one would design the questions to examine what assets we use to combat ableism in their environment, or what strategies we use to cope with internalized ableism. This also means, when designing conceptual frameworks for research, one should use frameworks which pull from asset- or strength-based philosophies. Examples of these frameworks are: Community Cultural Wealth (Yosso, 2005), PDI (Putnam, 2005), International Family Strengths Model (Defrain & Asay, 2007), and many others. This utilization of strengths-based models extends to formulation of recommendations from research. Recommendations should come from the viewpoint of things to increase agency and focus on building our strengths/assets to achieve change.

It is also important, when designing a study, to ask what will happen once research is complete. How will the findings and recommendations gleaned from research affect the populations involved? To understand this, it is critical to include members of the disabled community in the research design and in the dissemination process (Aguirre et al., 2017).

Additionally, each of these strengths-based models contains the important facet that the world becomes accessible when all barriers are broken. Regardless of how we conceptualize disability, the goal of research using these models should be to

Social Model	Diversity Model	Socio-Ecological Model
• Disability stems from socio-political forces against people with natural variations deemed "bad". • Disability is a social construct. • Disability is not intrinsic to the person. • The world becomes accessible when all barriers are broken.	• Violence against disabled people stems from socio-political forces against people with natural variations deemed "bad". • Disability is a social construct. • Disability is an inherent quality of the person. • The world becomes accessible when all barriers are broken.	• Disability comes from interactions between the person and the environment. • Disability is a unique experience. • Disability is not intrinsic to the person. • The world becomes accessible when all barriers are broken.

Fig. 7.3 Descriptions of three strengths/assets-based models of disability

break down barriers. We need to be conscious of our political choices at all stages of research. It is not enough to simply point out and identify what barriers exist. We as researchers have an obligation to make good trouble with our work.

7.6 Multidimensionality

Disability, as an identity, is intrinsically and extrinsically multidimensional. Intrinsically, one can be disabled in multiple ways (multiply disabled). For instance, one can be hard-of-hearing and have chronic pain (like the author). This is epistemologically important; disability identities intersect with other disability identities. Thus, one disabled person's experience will likely be different, potentially in drastic ways, from another disabled person's experience, even if they share an axis of disability. To use myself as an example again, my experiences having ADHD are very different from another person's experiences with having ADHD because my chronic pain means I have a much lower focus threshold than a person without chronic pain.

Extrinsically, disability is simply one axis of oppression a person can experience (Gillborn, 2015). Disability intersects with race, gender, sexuality, class, etc. As an example, a Black disabled person experiences disability very differently from a White disabled person. This can have very important implications for identity development. In her thesis examining disability identity development of an achondroplasic woman in the 1930s, Saskatchewan, Bray (2010) provide an instance of this using PDI to conclude that due to the intersection of class and disability, her ancestor did not develop a sense of being marginalized due to disability, nor did she see herself as disabled.

Methodologically, it is my recommendation that political disability research and research using PDI use frameworks which allow for an intersectional approach to studying oppression. It is also critical to make sure that when designing research, we take care to be open to different standpoints, using critical methodologies and being open to qualitative methods to ascertain the full scope of disabled people's experiences.

7.7 Variance in Outcomes

A corollary of the multidimensional nature of disability is that change will not be one-size-fits-all. As the adage goes, if you've met one disabled person, you've met one disabled person. The marginalization one disabled person experiences is going to be different than the marginalization another disabled person experiences due to differences in how they interact with their environment, and how society interacts with their intersecting identities. For instance, I require different accessibility requirements than other people with ADHD due to my experience with rejection sensitive dysphoria (psychosomatic pain and paralysis caused by the fear of failure) and my non-neurotypical automaticity (the ability to learn and absorb information without focusing on learning). Extended time on exams is not a useful accommodation for me, instead flexible due dates on homework assignments is a far more useful accommodation. Comparatively, many students with ADHD require extended time on exams, while flexible due dates would be detrimental to their educational success. Similarly, the additional resources necessary for success in academia will vary between disabled people. A first-generation White cis-male autistic student will navigate academia differently than a Latinx cis-female autistic student whose parents graduated college. Both students may need additional resources provided to them, however if both students receive *exactly* the same resources say, extra time on exams and participation in affinity groups, they may have drastically different outcomes. The first-generation student may be quickly lost to the mire of navigating college without additional academic counseling, whereas the other student may flourish given a space among peers to voice concerns about their college experience.

This is not to say that large overarching change and broad accommodation is not needed or is not helpful. Instead, suggestions for broad changes should be made with individual variance in outcomes in mind. While enacting flexible due dates for assignments will help many disabled and non-disabled students alike, it will not help everyone, and potentially hinder certain students' success.

Furthermore, researchers who study disability so often neglect, ignore, or reject the interplay that racism, classism, and hetero-sexism have with ableism (Bailey & Mobley, 2018) and thus the variance in outcomes which comes from these intersections. This outright harmful academic tradition consequently ignores vast swaths of our community. Keeping our work in conversation with Crenshaw (2017) and Hill Collins (2022), among others, is critical to establishing a better, more inclusive political disability research. We don't have to stray far from DBER to find the tools necessary for this.[2] We need not reinvent this intersectional wheel to affirm our support for everyone in our incredibly diverse community through our research.

Methodologically, research should be designed conscious of individual needs. Introducing a wide variety of accessibility options is a very good start to this. This can be in the form of offering an audio version of written materials, breaking down questions into smaller chunks, holding virtual interviews, etc. Perhaps the most important addition to research design when dealing with human subjects is to add "if there is anything we/I can do to make this study more accessible to you, or accommodate you in your participation in this study, please let us/me know" in recruitment materials. This allows us to accommodate on the individual level and make us aware of any political choices we make unconsciously regarding disability. Additionally, recommendations drawn from research should be made conscientious of this variance. Echoing PDI's *common cause* domain, accommodation the individual uplifts us all. In recommendations, it is important that we stress that one-size-fits-all is not a productive strategy for accessibility. That, while certain accommodations may be beneficial to the majority of students, they will likely be detrimental to others. It is critical that we are conscious of the individual and promote change informed through individualized radical kindness.

[2] See Cochran et al. (2020).

7.8 Fluidity/Inevitability

Disability is temporally, spatially, and situationally contextual. Due to how society is presently constructed, throughout a single day we may encounter numerous situations in which we are disabled, even without claiming a disability label. For example, a blind person traversing the streets of town may be disabled by a lack of clear and accessible signals in the form of truncated domes and detectible warning pavers. Further, a lack of curb cuts and braille signage on businesses may prove to be disabling. However, when at home, where this hypothetical blind person has a clear and curated mental map of their surroundings, they are no longer disabled. By crossing a physical threshold, this person is non-disabled.

One logical continuation of this idea that disability is contextual is that there must exist some people who are never disabled by society. If disability is temporally, spatially, and situationally contextual, there simply must be a way of living life, whether through deliberate action, pure luck, or some combination of the two, where one *never* enters this context, never crosses the threshold. However, this is not the case. Being able-bodied is a farce, as able-bodiedness is unattainable (McRuer, 2006). Disability is inevitable; everyone will experience or has experienced being disabled at some point in their life. It reasons then, that if everyone will be disabled at some point in their life, disability justice affects everyone. Political disability DBER is research which has ramifications for every single person in the field.

It is therefore critical that we are crafting our theses with this in mind. Methodologically, this looks like framing disability justice as nonbinary and wide reaching. Shifting our framing from disabled and non-disabled people (a binary) to disabled people, people who are not currently disabled, and the oppressive ableism looming over everyone, will bring more people into disability justice movements and create a more inclusive space for those seriously affected by ableism but do not yet politically conceive of themselves as disabled. This also means broadening the scope of research to include less "traditional" disabled people in our DBER. It is worth examining the effect ableism has on, for instance, a pregnant scientist navigating their field, or students with OCD, or to be perhaps a tad more radical, able-bodied people in STEM. These people experience ableism daily, and there are countless new questions to be asked about their experiences and the effect a heavily-baked-in-ableism has on how they navigate their fields, which would otherwise remain hidden if they were not included in the fluid and inevitable nature of disability.

7.9 A Brief Case for Developing Non-disabled PDIs

Ableism affects us all. While I speak from a standpoint of having experienced routine, normalized academic and "regular" ableism in my day-to-day life, having it shape my worldview, my relationships, my research, and my teaching, I also speak from a standpoint where I may uniquely examine how the same ableism I experience affects and shapes the individuals, environments, and communities around me.

Not only will everyone experience disability, and thus experience ableism, at some point in their life, non-disabled people experience ableism every day. This ableism is a detriment to their lives too, and similarly affects their relationships with themselves, their loved ones, their communities, and their environments.

How often do folks from the United States avoid seeking better employment, or are kept from becoming family caretakers or starting their own businesses due to fear of losing private health insurance, the fear of becoming disabled? It turns out, often (Chute & Wunnava, 2018). In a university context, students, due to systemic and structural expectations to be the ideal abled and productive student and their lack of social support, are increasingly experiencing mental health complications (Wörfel et al., 2016). Graduate students, too, are experiencing increasing burnout and mental illness from the same ableist sources, something which is mitigated by satisfaction with advisor choice (Allen et al., 2022), something which (I speculate) may be due to advisor demands of students and expectations vis-à-vis breaks and productivity.

While I have not addressed this in this chapter, I certainly imagine that a PDI can be developed in non-disabled students and faculty. In fact, a strong PDI development may be incredibly beneficial to non-disabled people, helping them avoid burnout, and contextualizing their working habits through an understanding of compulsory-ablebodiedness.[3] Moreover, unpacking compulsory-ablebodiedness and the normative nature of ableism through a strong PDI in non-disabled people creates stronger allies, something of which we always are in dire need. On this front, I call for more research into how non-disabled people (allies and otherwise) develop or strengthen their PDI. Further, I call for more research on how disabled/non-disabled coalitions, through their PDI and their wielding of political agency can better dismantle ableist systems in academia.

7.10 Questions to Ask When Doing Political Disability Research

Now that we have a better understanding of the methodological and epistemological implications of using these PDI themes to guide our research, I feel it would be most useful to boil them down into seven questions. Some of these are similar to Aguirre et al.'s work (2020) on doing political mathematics research, however here I expand on them in relation to a broader DBER and more specific political disability. These questions should be asked at all stages of research. Even when our research is but a twinkle in our eye and we are simply formulating a research question, we should still be asking ourselves these questions. These are questions designed to facilitate political acts of disability research. Without answering them, we become vulnerable to unconscious political acts which could have far-reaching ramifications on our research and our community. I also provide sub-questions and commentary to better understand these questions and how to formulate answers to them.

[3] For more reading on compulsory ablebodiedness, see McRuer's *Crip theory cultural signs of queerness and disability* (2006).

1. How am I including disabled people in my research?
    (a) If it's only as subjects, how am I certain the issues which matter to the disabled community at large are being discussed?
    Ways to remedy this could be hiring community members as consultants or reaching out and including community members as principal investigators/authors.
2. How am I centering disabled people in my research?
    (a) Am I considering my own positionality in my interpretation of data?
    (b) As I make recommendations, how am I ensuring these recommendations uplift our community?
    (c) Am I cognizant of what assumptions I make about "proper" communication looks and sounds like? (Kerschbaum & Price, 2017).
3. What strengths-based methods and conceptual-frameworks am I using?
    (a) Would a layperson reading the findings of this research assume discrimination or political violence to be the sole experiences of the study population?
4. How will the findings be used in effecting the lives of disabled people?
    (a) In the spirit of W.E.B Du Bois and Zuberi and Bonilla-Silva, the same statistics can be used just as well to uphold the status-quo. Am I being neutral, or even liberal with my methods? Or am I using my research to challenge the status-quo and break barriers (Du Bois, 1899; Zuberi & Bonilla-Silva, 2008).
5. How am I engaging members of the disabled community after this research is completed?
    (a) Will this research be used by community members to uplift others?
6. Am I reading outside my field?
    (a) Am I including the work of disabled scholars and the work of scholars who use their work to talk back to ableist structures?
7. How am I engaging my colleagues with this work?
    (a) Am I challenging their practices and encouraging their political agency to disrupt ableist systems?

References

Aguirre, J., Herbel-Eisenmann, B., Celedón-Pattichis, S., Civil, M., Wilkerson, T., Stephan, M., Pape, S., & Clements, D. H. (2017). Equity within mathematics education research as a political act: Moving from choice to intentional collective professional responsibility. *Journal for Research in Mathematics Education, 48*(2), 124–147. https://doi.org/10.5951/jresematheduc.48.2.0124

Allen, H. K., Lilly, F., Green, K. M., Zanjani, F., Vincent, K. B., & Arria, A. M. (2022). Graduate student burnout: Substance use, mental health, and the moderating role of advisor satisfaction. *International Journal of Mental Health and Addiction, 20*(2), 1130–1146.

Altbach, P. G., & Cohen, R. (1990). American student activism. *The Journal of Higher Education, 61*(1), 32–49. https://doi.org/10.1080/00221546.1990.11775090

Andrews, E. E. (2019). *Disability as diversity: Developing cultural competence*. Oxford University Press.

Bailey, M., & Mobley, I. A. (2018). Work in the intersections: A black feminist disability framework. *Gender & Society, 33*(1), 19–40. https://doi.org/10.1177/0891243218801523

Barnes, C. (2019). Understanding the social model of disability: Past, present and future. In *Routledge handbook of disability studies* (pp. 14–31). Routledge.

Bray, C. (2010). Domestic citizenship and Disability in Saskatchewan in the 1930's.

Bell, C. M. (2012). *Blackness and disability: Critical examinations and cultural interventions*. Michigan State University Press.

Brilmyer, G. (2018). Archival assemblages: Applying disability studies' political/relational model to archival description. *Archival Science, 18*, 95–118. https://doi.org/10.1007/s10502-018-9287-6

Candela, A. (1998). Students' power in classroom discourse. *Linguistics and Education, 10*(2), 139–163. https://doi.org/10.1016/s0898-5898(99)80107-7

Celedón-Pattichis, S., Borden, L. L., Pape, S. J., Clements, D. H., Peters, S. A., Males, J. R., Chapman, O., & Leonard, J. (2018). Asset-based approaches to equitable mathematics education research and practice. *Journal for Research in Mathematics Education, 49*(4), 373–389. https://doi.org/10.5951/jresematheduc.49.4.0373

Chute, B. W., & Wunnava, P. V. (2018). Is there a link between employer-provided health insurance and job-mobility? Evidence from NLSY79. *Open Journal of Human Resource Management, 1*(1), 38–52.

Cochran, G. L., Boveda, M., & Prescod-Weinstein, C. (2020). Intersectionality in STEM education research. In *Handbook of research on STEM education* (pp. 257–266). Routledge.

Collins, P. H. (2022). *Black feminist thought: Knowledge, consciousness, and the politics of empowerment*. Routledge.

Crenshaw, K. W. (2017). *On intersectionality: Essential writings*. The New Press.

D'Amico, M. M. (2016). Community college workforce development in the student success era. In *Higher education: Handbook of theory and research* (pp. 217–273).

DeFrain, J., & Asay, S. M. (2007). Epilogue: A strengths-based conceptual framework for understanding families world-wide. *Marriage & Family Review, 41*(3–4), 447–466.

Dolmage, J. (2017). *Academic ableism: Disability and higher education*. University of Michigan Press.

Du Bois, W. E. B. (1899) [1996]. *The Philadelphia Negro: A social study*. University of Pennsylvania Press.

Dwyer, P. (2022). The neurodiversity approach(es): What are they and what do they mean for researchers? *Human Development, 66*(2), 73–92. https://doi.org/10.1159/000523723

Gill, C. (1997). Four types of integration in disability identity development. *Journal of Vocational Rehabilitation, 9*, 39–46.

Gillborn, D. (2015). Intersectionality, critical race theory, and the primacy of racism: Race, class, gender, and disability in education. *Qualitative Inquiry, 21*(3), 277–287.

Gilson, F., Tulser, A., & Gill, C. (1997). Ethnographic research in disability identity: Self-determination and community. *Journal of Vocational Rehabilitation, 9*, 7–17.

Goodley, D. (2014). *Dis/ability studies: Theorising disablism and ableism*. Routledge.

Habermas, J., & Cronin, C. (2005). *The inclusion of the other: Studies in political theory*. Polity Press.

Hahn, H. (1993). The political implications of disability definitions and data. *Journal of Disability Policy Studies, 4*(2), 41–52. https://doi.org/10.1177/104420739300400203

Hartsock, N. C. (2003). In S. Harding & M. B. Hintikka (Eds.), *The feminist standpoint: Developing the ground for a specifically feminist historical materialism*. Kluwer Academic Publishers.

Heumann, J., & Joiner, K. (2021). *Being Heumann: An unrepentant memoir of a disability rights activist*. Random House.

Jain, N. R. (2019). Political disclosure: Resisting ableism in medical education. *Disability & Society, 35*(3), 389–412. https://doi.org/10.1080/09687599.2019.1647149

Jarman, M., Lamp, S., Mitchell, D., Nepveux, D., Nowell, N., & Snyder, S. (2002). Theorising disability as political subjectivity: Work by the UIC disability collective on political subjectivities. *Disability & Society, 17*(5), 555–569.

Kattari, S. K., Lavery, A., & Hasche, L. (2017). Applying a social model of disability across the life span. *Journal of Human Behavior in the Social Environment, 27*(8), 865–880.

Kerschbaum, S. L., & Price, M. (2017). Centering disability in qualitative interviewing. *Research in the Teaching of English, 52*(1), 98–107.

Kimball, E. W., Wells, R. S., Ostiguy, B. J., Manly, C. A., & Lauterbach, A. A. (2016). Students with disabilities in higher education: A review of the literature and an agenda for future research. In *Higher education: Handbook of theory and research* (pp. 91–156).

Lavy, V., & Sand, E. (2015). *On the origins of gender human capital gaps: Short and long term consequences of teachers' stereotypical biases* (No. w20909). National Bureau of Economic Research.

Le, A. T., LaCost, B. Y., & Wismer, M. (2016). International female graduate students' experience at a Midwestern University: Sense of belonging and identity development. *Journal of International Students, 6*(1), 128–152.

Leigh, J. S., & Brown, N. (2020). Internalised ableism: Of the political and the personal. In *Ableism in academia: Theorising experiences of disabilities and chronic illnesses in higher education*. UCL Press. E-ISBN 978-1-78735-497-5.

Lorde, A. (2017). *A burst of light and other essays*. Ixia Press.

Lourens, H. (2018). Driving in unheard silence: Disability and the politics of shutting up. *Journal of Health Psychology, 23*(4), 567–576.

McCoy, D. L., Luedke, C. L., & Winkle-Wagner, R. (2017). Encouraged or weeded out: Perspectives of students of color in the STEM disciplines on faculty interactions. *Journal of College Student Development, 58*(5), 657–673.

McDermott, L. G., Mosley, N. A., & Cochran, G. L. (2024). Diverging nonlocal fields: Operationalizing critical disability physics identity with neurodivergent physicists outside academia. *Physical Review Physics Education Research, 20*(1), 010111.

McRuer, R. (2006). *Crip theory cultural signs of queerness and disability*. New York University Press.

McRuer, R. (2018). *Crip times: Disability, globalization, and resistance*. New York University Press.

Olsen, J., Griffiths, M., Soorenian, A., & Porter, R. (2020). Reporting from the margins: Disabled academics reflections on higher education. *Scandinavian Journal of Disability Research, 22*(1), 265–274.

Oral, Y. (2013). "The right things are what I expect them to do": Negotiation of Power Relations in an English classroom. *Journal of Language, Identity & Education, 12*(2), 96–115. https://doi.org/10.1080/15348458.2013.775877

Pelka, F. (2012). *What we have done; An oral history of the Disability Rights Movement*. University of Massachusetts Press.

Pennell, S. M. (2020). Queer theory/pedagogy and social justice education. In *Handbook on promoting social justice in education* (pp. 2291–2311).

Putnam, M. M. (2001). Individual disability identity: A beginning exploration into its nature, structure, and relevance for political coalition building. University of California, Los Angeles.

Putnam, M. (2005). Conceptualizing disability: Developing a framework for political disability identity. *Journal of Disability Policy Studies, 16*(3), 188–198.

Retief, M., & Letšosa, R. (2018). Models of disability: A brief overview. *HTS Teologiese Studies/Theological Studies, 74*(1).

Rich, L. P. (2009). *Prompting self-monitoring with assistive technology to increase academic engagement in students with attention—Deficit/Hyperactivity disorder symptoms* (Order No. 3358336). Available from ProQuest Dissertations & Theses Global. (304893494). https://login.proxy.libraries.rutgers.edu/login?qurl=https%3A%2F%2Fwww.proquest.

com%2Fdissertations-theses%2Fprompting-self-monitoring-with-assistive%2Fdocview%2F304893494%2Fse-2%3Faccountid%3D13626

Sauer, J. S., Rossetti, Z., de Lourdes, B., & Serpa, M. (2020). Forging connections in family engagement. In *Affirming disability: Strengths-based portraits of culturally diverse families. Essay*. Teachers College Press.

Schalk, S. (2022). *Black disability politics*. Duke University Press.

Schur, L. (1998). Disability and the psychology of political participation. *Journal of Disability Policy Studies, 9*, 4–31.

Scotch, R. K. (1989). Politics and policy in the history of the Disability Rights Movement. *The Milbank Quarterly, 67*, 380. https://doi.org/10.2307/3350150

Smith, I., & Mueller, C. O. (2022). The importance of disability identity, self-advocacy, and disability activism. *Inclusive Practices, 1*(2), 47–54.

Titchkosky, T. (2014). *Disability, self, and Society*. University of Toronto Press.

Watson, N. (2002). Well, I know this is going to sound very strange to you, but I don't see myself as a disabled person: Identity and disability. *Disability & Society, 17*, 509–527.

Wolbring, G. (2008). The politics of ableism. *Development, 51*(2), 252–258. https://doi.org/10.1057/dev.2008.17

Wong, A. (2018). *Resistance and hope*. Alice Wong.

Wörfel, F., Gusy, B., Lohmann, K., Töpritz, K., & Kleiber, D. (2016). Mental health problems among university students and the impact of structural conditions. *Journal of Public Health, 24*, 125–133.

Yosso, T. J. (2005). Whose culture has capital? A critical race theory discussion of community cultural wealth. *Race ethnicity and education, 8*(1), 69–91.

Zuberi, T., & Bonilla-Silva, E. (2008). *White logic, White Methods: Racism and methodology*. Rowman & Littlefield.

Chapter 8
Addressing Biases Through Decoloniality and Belonging in STEM: A Cross-Disciplinary Dialogue

Jamilah Dei-Sharpe, Kirsten Crandall, Jessica Ford, Nathalie Jreidini, Ezgi Ozyonum, and Hadjara Sidibé

Abstract This chapter illustrates the application of a decoloniality, equity, diversity, inclusion and belonging (DEDIB) framework to assess and dismantle the oppressive systems impacting marginalized groups in science, technology, engineering, and medicine (STEM) research and education. Insights from a syllabus deconstruction workshop co-hosted by the Decolonial Perspectives and Practices Hub and STEMM Diversity @ McGill determined actionable items that instructors can use to unpack the historical and theoretical context of the STEM fields in their course content and design. First, we illustrate the historical and contemporary biases

Jamilah Dei-Sharpe, Kirsten Crandall, Jessica Ford, Nathalie Jreidini, Ezgi Ozyonum and Hadjara Sidibé contributed equally with all other contributors.

J. Dei-Sharpe (✉)
Department of Sociology and Anthropology, Concordia University, Montreal, QC, Canada
e-mail: info@jamilahds.com

K. Crandall
Department of Biology, University of Ottawa, Ottawa, ON, Canada

Department of Biology, McGill University, Montreal, QC, Canada

Redpath Museum, McGill University, Montreal, QC, Canada

J. Ford · N. Jreidini
Department of Biology, McGill University, Montreal, QC, Canada

Redpath Museum, McGill University, Montreal, QC, Canada

E. Ozyonum
Department of Education, Concordia University, Montreal, QC, Canada

H. Sidibé
Département de neurosciences, Université de Montréal, Montreal, QC, Canada

Centre de recherche du Centre hospitalier de l'Université de Montréal (CRCHUM), Montreal, QC, Canada

G. L. Cochran, J. D. Adams (eds.), *Equity-Oriented, Discipline-Based STEM Education Research*, Critical Studies of Education 19,
https://doi.org/10.1007/978-981-96-9136-4_8

in the STEM fields and how a DEDIB framework can be used as a solution to overcome these prejudices. Second, we present three actionable DEDIB approaches that can be adopted in STEM syllabi and classrooms: (1) pluralizing forms of knowledge, (2) combating systemic racism, and (3) increasing inclusivity and accessibility in learning. We conclude with final reflections and actionable commitments for STEM instructors. By exposing educators and students to a DEDIB framework, we aim to enrich STEM pedagogy to advance ethical and sustainable STEM education and research by diversifying learning, knowing, and practicing in the field.

Keywords Decoloniality · Equity · Diversity · Inclusion · Belonging · STEM · Discipline-Based Education Research

A growing consensus has emerged that systemic racism and colonialism are the roots of inequality within higher education (Itchuaqiyaq & Matheson, 2021; Pandey, 2020; Perrin-Stowe et al., 2023; Skopec et al., 2021; Talanquer, 2014; Winberg et al., 2019). More specifically, **STEM** disciplines remain embedded in a complicated past, where historical **Westernized** and **Anglocentric** views continue to predominate modern science, a bias that is underacknowledged in the current curriculum (Quijano & Ennis, 2000). All fields, especially those in STEM, are being called upon to address their role in reproducing hierarchical knowledge that shape education and research in universities and beyond (Pandey, 2020). As a response to these social justice calls, institutions are focusing on educational **Equity**, **Diversity**, and **Inclusion (EDI)** efforts, with policies and task forces addressing the barriers and isolation experienced by marginalized students and faculty (Frances et al., 2017).

In higher education, EDI initiatives seek to foster "inclusive learning and working environments" that are representative, fair, and respectful (Henderson et al., 2017). These spaces strive to accept differences and offer opportunities for diverse people to excel (Roberts, 2020; Wilson-Kennedy et al., 2020). This work focuses on using accessible frameworks that can be adopted by STEM educators and researchers to identify and address **systemic biases**. **Syllabus deconstruction** workshops explore the possibilities of restructuring the syllabus with an anti-colonial approach in the course objectives, readings, assessments, and student leadership. Through a STEM-based syllabus deconstruction co-hosted by the **Decolonial Perspectives and Practices Hub** and **STEMM Diversity @ McGill**, we uncovered that STEM educators and researchers can be redirected toward systematic and ideological change when EDI is accompanied by emphasizing "**decolonizing**" and "**belonging**" principles, or a **Decolonizing, Equity, Diversity, Inclusion, and Belonging (DEDIB) framework** (Fig. 8.1; Roberts, 2020).

In this chapter, early-career scientists from the Arts and Sciences provide a cross-disciplinary perspective on the importance of a DEDIB framework and how the framework can be applied. We first unpack the historical context and biases of the STEM fields and how the DEDIB framework is positioned to combat these biases. Next, we present how three DEDIB approaches can be adopted into the STEM fields and syllabi: (1) pluralizing forms of knowledge, (2) combating systemic

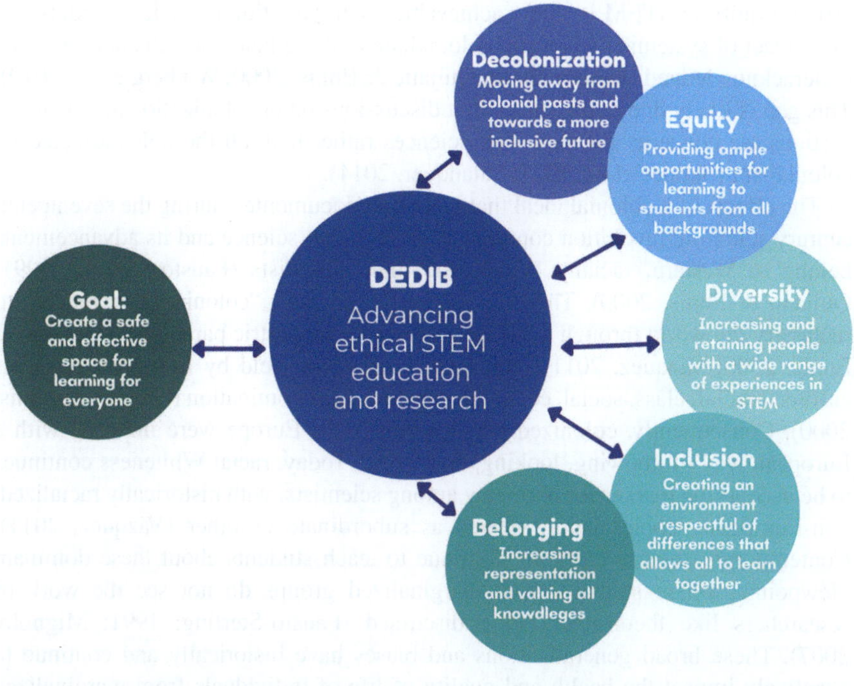

Fig. 8.1 Conceptual diagram of the Decolonization, Equity, Diversity, Inclusion & Belonging (DEDIB) framework to be used in postsecondary education in the STEM fields

racism, and (3) increasing inclusivity and accessibility in learning. We aim to enrich STEM pedagogy by exposing educators and students to how a DEDIB framework can advance ethical STEM education and research by diversifying learning through appropriate practices.

8.1 Part 1: Discipline-Based Education Research, Decolonization, and DEDIB Principles

8.1.1 Discipline-Based Education Research and Decolonization

For decades, **Discipline-Based Education Research** has dedicated itself to improving teaching and learning in the STEM fields through developments in knowledge and practices (Talanquer, 2014). This research seeks to address teaching and learning challenges by focusing on the content, context, and cultures that practitioners are experiencing (Henderson et al., 2017), with goals centered on making STEM more accessible and interactive to diverse students (Talanquer, 2014). However, a general improvement across STEM disciplines in educational research requires a greater emphasis on creating equitable educational environments (Henderson et al.,

2017). Equity in STEM is solely achievable when attention is made to understand the impact of systemic racism and colonialism in these fields, which continue to be underacknowledged (Pandey, 2020; Quijano & Ennis, 2000; Winberg et al., 2019). This gap is likely due to the stance that discussions on decolonization and its intersections are concerns for the social sciences rather than all the fields affected by colonialism (Skopec et al., 2021; Talanquer, 2014).

The effects of a colonial ideal that were first documented during the seventeenth century scientific revolution continue to enforce that science and its advancements belong to Western, racially White, and male scientists (Fausto-Sterling, 1991; Quijano & Ennis, 2000). These researchers used their "coloniality of power" to reclassify the world through a "white gaze" or "Eurocentric paradigm" (Quijano & Ennis, 2000; Vázquez, 2011). This matrix of power held by European colonists enforced racial, class, social, economic, and political domination (Quijano & Ennis, 2000). Consequently, colonized peoples outside of Europe were imposed with a European way of knowing, looking, and living. Today, racial Whiteness continues to be used as the marker for normalcy among scientists, with historically racialized, non-European peoples being viewed as subordinate or other (Vázquez, 2011). Contemporary science curricula continue to teach students about these dominant viewpoints, where students from marginalized groups do not see the work of researchers like themselves being discussed (Fausto-Sterling, 1991; Mignolo, 2007). These broad generalizations and biases have historically and continue to negatively impact the health and quality of life of individuals from marginalized groups. Welcoming a decolonial lens to learning environments can provoke and enliven difficult, but essential dialogues surrounding our biased patterns that impede interconnectedness, equity, inclusion, and diversity within the STEM fields.

8.1.2 The Politics of Decolonization

Decolonization is a complex concept, political process, and movement that does not have one definition (Itchuaqiyaq & Matheson, 2021; Sium et al., 2012). Decolonial efforts act to "replace and reshape institutional, political, and social structures that impede in the livelihood, engagement, opportunities, and survival of colonized peoples, lands, identities and existences" (Itchuaqiyaq & Matheson, 2021), where Indigenized and racialized communities work to combat different forms of settler and **epistemic colonialism** (Dei, 2017; Sium et al., 2012).

The exploitation, appropriation, seizure, and control of Indigenous lands and communities continue to occur through the complex system of **settler colonialism** (Tuck & Yang, 2012). European settlement on Turtle Island continues to displace and subjugate Indigenous communities through systemic acts of assimilation, erasure, control, and policing (Simpson, 2014; Tuck & Yang, 2012). Indigenous peoples and supporting communities must advance Indigenous sovereignty by exposing complicit non-Indigenous settlers who maintain and benefit from these systems. By centering Indigenous stories, histories, and practices in policy and education, decolonial efforts can bring awareness to mobilize systemic change (Itchuaqiyaq & Matheson, 2021; Simpson, 2014). Decolonization can be conceived as a global project led by diverse Indigenous peoples to reclaim their lands and revive their cultures (Dei, 2017; Sium et al., 2012).

Increased attention has been placed on decolonizing efforts for the Black/African peoples who were colonized, racialized, enslaved, and displaced on the Indigenous lands of Turtle Island through the transatlantic slave trade (Dei, 2017; hooks, 1992). Enslaved peoples were sold as capital and forced to be laborer's building "civilization" on stolen Indigenous lands (Dei, 2017; hooks, 1992). Today, Black peoples continue to be subjugated by systemic acts of control, policing, and other human rights violations (Dei, 2017; Maynard, 2017). Presenting Blackness with an Afro-Indigenous perspective may help guide deeper understandings beyond slavery and displacement toward a heritage of rich communal, spiritual, and land-based education, cross-Indigenous coalition-building and educational research (Dei, 2017).

During the twentieth century, Hispanic, Indigenous, Black/African, and Global South scholars began assessing the colonial patterns and power relations within the academy, and calling this phenomenon also known as "epistemic colonialism"[1] (Freire, 2000; Grosfoguel, 2013; hooks, 1992; Lugones, 2010; Maldonado-Torres, 2007; Mignolo, 2007; Quijano & Ennis, 2000; Spivak, 1988). Western-centric knowledge and student-teacher power dynamics are framed as colonial tools to assimilate, erase, and subordinate the contributions of Indigenous and racialized peoples in educational settings[2] (Freire, 2000; Mignolo, 2007). However, the return of decolonial politics within educational institutions is critiqued for reducing decolonization to a "metaphor," where generalized experiences of marginalized groups lead to changes in thoughts and ideas rather than supporting efforts to return Indigenous land through rematriation (Tuck & Yang, 2012).

Our work adopts the both/and stance to decolonization as formulated by Sium et al. (2012), where structural and epistemological objectives should be pursued concurrently. This coexistence emphasizes systems change to support Indigenous sovereignty, honor the rights and freedoms of marginalized peoples, and reimagine knowledge production. Therefore, coalition building between academia and activism is necessary for the larger cause (Sium et al., 2012) for anti-colonial education and as a tool to propel decolonizing futures. We value educational institutions as a site to prime the next generation to further grassroots decolonization projects. Here, we take steps toward a non-metaphorical adoption of decoloniality within STEM education by seeking to transform knowledge to incite social and material change for Indigenous and other colonized racial communities.

[1] Decolonial and post-colonial scholars invested in epistemic decolonization are connected to a complex literary tradition of "subverting the gaze" and empowering the marginalized class as taken up by Du Bois (1898), Fanon (1952), Mahatma Gandhi, and other visionaries championed for founding critical and emancipatory theory (Ashar, 2015).

[2] Key theories and concepts surrounding epistemic decolonization include "banking education" (Freire, 2000), the "subaltern," as well as the coloniality of power, being, or gender (Lugones, 2010; Maldonado-Torres, 2007; Quijano & Ennis, 2000).

8.1.3 DEDIB Principles: Unpacking EDI and Belonging in STEM

The movements for EDI and belonging are alternative efforts that can complement and push forward decolonizing objectives from different perspectives, resulting in a unique environment for decoloniality to grow. We have developed a heuristic drawing from our educational and teaching experiences on how DEDIB principles can work independently and together (Fig. 8.1). This model provides pedagogical tools that improve STEM education and create a sense of belonging in the classroom. Decolonizing tools target institutional structural change, while EDI tools target emotions (belonging) and institutional cultural change. Educators should engage with the DEDIB principles as distinct and irreducible components, which are directed toward a shared goal to disrupt, reimagine, and restructure the current ways of teaching, learning, and being in the classroom. Decolonizing education is dedicated to reconstructing the current system as it exists (Sium et al., 2012), while EDI is recognized as a commitment to creating equitable, inclusive, and diverse learning environments (Eaton, 2022). Educational spaces should provide equal access to opportunities and resources for marginalized and non-marginalized individuals, where differences are respected and represent the **intersectional** backgrounds of its members (Eaton, 2022; Wilson-Kennedy et al., 2020). Efforts to address belonging are dedicated to understanding and creating solutions for the emotions and feelings of individuals experiencing biases in STEM (Goodenow & Grady, 1993; Roberts, 2020; Ruttenberg-Rozen et al., 2021).

STEM educators and researchers engage with EDI principles to address the "biases, microaggressions and marginalization" impacting student outcomes (Wilson-Kennedy et al., 2020). Since 2012, Discipline-Based Education Research has represented "inclusive science" as a necessary and central goal for advancements in STEM. Engagement with EDI continues to deepen and target social justice in response to the tragic events of 2020 instigating an outcry to redress "legacies of systemic racism" (Wilson-Kennedy et al., 2020). In response, scholars and practitioners have worked to diversify literature and histories of **underrepresented groups**. The inequalities faced by women in STEM are currently the largest body of literature, while studies on Peoples of Color in STEM continue to be an underacknowledged gap in the field (Wilson-Kennedy et al., 2020). In Chemistry, some professionals have embraced EDI as a tool to generate more "opportunities [for] women and minorities." However, Black Americans were consistently shown to receive 1–2% of doctoral degrees and 6–8% of bachelor's degrees in Chemistry between 1997 and 2018 (Wilson-Kennedy et al., 2020). An EDI approach emphasizes the acquired skills of diverse students and how to disrupt marginalization they may experience within and outside the STEM classroom. Therefore, more nuanced understandings rather than a lack of interest or cognitive deficiencies are required to explain the low participation rate of racialized students in STEM (Wilson-Kennedy et al., 2020).

Belonging is often associated with a feeling of "interconnectedness with others," being "accepted," and "valued" (Goodenow & Grady, 1993; Roberts, 2020). A sense of belonging is an underrepresented, yet crucial factor in the interpersonal relationships and psychological well-being of students in STEM (Williams et al., 2021). A greater sense of belonging significantly improves the student's perception of their own capabilities by fostering a growth mindset and impacts how they are perceived by their instructors and peers (Williams et al., 2021). A growth mindset, which relates to both personal and intellectual growth, could potentially buffer environmental stress due to stereotypes and discrimination (Williams et al., 2021). Few studies investigate undergraduate student belonging and well-being in comparison to studies on recruitment and retention in postsecondary STEM programs (Williams et al., 2021). There has also been growing concern on the well-being of Women of Color in STEM, whose treatment has been characterized as "inhospitable" (Ruttenberg-Rozen et al., 2021). Here, DEDIB principles may address the lack of equity, diversity, inclusion, and belonging in classrooms, as well as contributing persistent systemic issues.

8.1.4 Applying DEDIB to the Course Syllabus

The incorporation of decolonizing principles and belonging with EDI efforts in postsecondary institutions is intended to ensure that educators and researchers are cultivating educational environments that address the roots of historical and systemic problems. This call to action emphasizes the use of the course syllabus as a tangible tool to apply the DEDIB framework in the STEM fields in an accessible and sustainable manner. The syllabus significantly determines the boundaries and parameters of the curriculum through learning objectives and anticipated outcomes. The flexibility of the syllabus may allow modifications, such as the removal and replacement of teaching and learning approaches that perpetuate biases.

We deepened our understanding of addressing biases and incorporating DEDIB principles into STEM syllabi and classrooms through a co-facilitated syllabus deconstruction workshop with the Decolonial Perspectives and Practices Hub and STEMM Diversity @ McGill in April 2022. The syllabus deconstruction workshop[3] was formulated into two sections using cross-disciplinary orientations: (1) identifying colonial patterns in STEM and sharing uncomfortable experiences in the classroom and (2) deconstructing a mock syllabus in real time including the course description and objectives, grading/assessments, and reading list. Participants brainstorm how to re-formulate the syllabus to include greater student–faculty collaboration, marginalized knowledges, land acknowledgements, lived experiences, and diverse assessments. Scientists unpacked biases in STEM through experience

[3] Syllabus deconstruction workshops are popular sessions at the DPPH for exploring the subversive possibilities of the course syllabus to disrupt the exclusionary curriculum in the Arts and Sciences.

sharing during a moderated panel. Participants were then given a STEM syllabus and asked to integrate DEDIB principles using critical reflection questions.

The reflection questions provoked ways that syllabi can maintain exclusionary scientific curricula, which may be combatted when embracing DEDIB principles. We learned that syllabi are an imperfect tool with epistemological and pedagogical authority in the classroom, which can normalize Western-centric standards and colonial ideology in STEM. However, syllabi may also act as a portal of growth and alternative possibilities for decolonization. A DEDIB-centered STEM syllabus incorporates plural knowledges, land acknowledgements, racially sensitive content, accessibility needs, linguistic diversity, and promotes interconnectedness. Syllabi should be used as a *compass* that can train future generations of STEM researchers to be compassionate, empathetic, and community informed.

8.2 Part 2: DEDIB Approaches Applicable to STEM Syllabi and Classrooms

8.2.1 Pluralizing Knowledge in STEM Discussions

Postsecondary institutions have been major players in perpetuating coloniality across systems of knowledge (Alvares & Faruqui, 2011; Andreotti et al., 2015; Miller, 2007). Although many disciplines including STEM have started contemplating decoloniality (e.g., Alvares & Faruqui, 2011; Battiste, 2013; Bhambra et al., 2018), these discussions have largely focused on research practices and theories, with few interrogating curriculum and pedagogy (Sanchez, 2018). Many racialized students and instructors feel outcast from their field, as they recognize that current curricula focus predominantly on White, anglocentric researchers (Fig. 8.2a; Chemers et al., 2011; Peters, 2015; Wong et al., 2021).

DEDIB pedagogical practices across disciplines can be actualized in four main ways: (1) examining and critiquing knowledge positionality in educational spaces, (2) constructing an inclusive curriculum moving away from dominant knowledge systems, (3) adopting relational teaching and learning practices through collaborations with external communities, and (4) connecting institutions with sociopolitical movements (Shahjahan et al., 2022). Scientific concepts taught in higher education are often considered factual, but students should be encouraged to be more open-minded by challenging and improving these concepts. Consequently, the teacher becomes a facilitator of activities, where students take responsibility for their learning (Gomez & Lee, 2015; Koehler et al., 2015). Diversifying STEM education encourages the pluralization of knowledge production outside the Western-centric viewpoint by including Afro-diasporic, Eastern, Indigenous, cross-disciplinary, anecdotal, and oral knowledges (Fig. 8.2c; Ford et al., 2021; Mignolo, 2007; Silko, 2012).

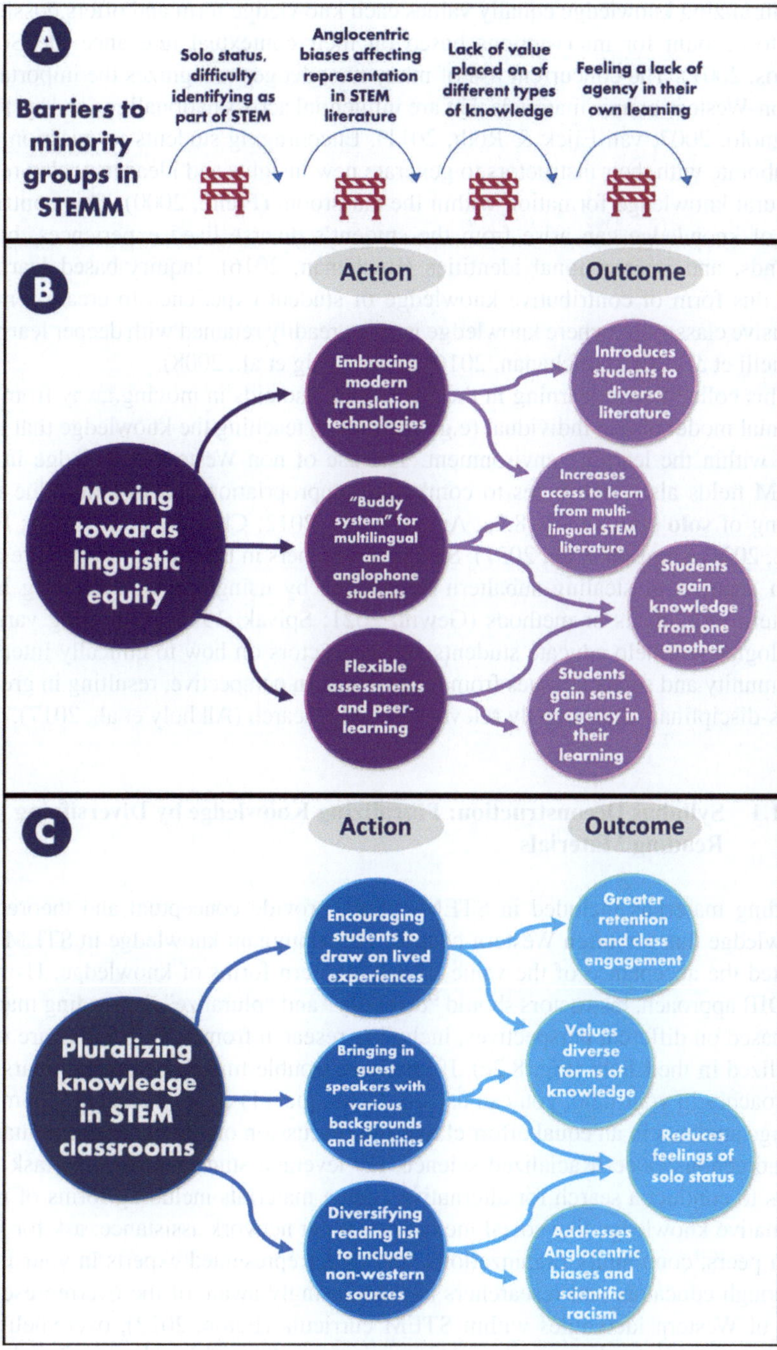

Fig. 8.2 Barriers faced by marginalized STEM students, and what STEM educators can do to address them. (**a**) Common barriers to learning experienced by marginalized STEM students. (**b**) Tangible actions and outcomes resulting from increasing linguistic equity in STEM classrooms. (**c**) Tangible actions and outcomes resulting from pluralizing knowledge in STEM classrooms

Pluralizing knowledge equally values each knowledge form and offers possibilities to account for interventions based on their contextual relevance (de Sousa Santos, 2007). The concurrent use of many knowledges recognizes the importance of non-Western course materials that are influential across nationalities and systems (Mignolo, 2007; van Eijck & Roth, 2011). Encouraging students to question and collaborate with their instructors to generate new insights and ideas may also result in plural knowledge formation within the classroom (Freire, 2000). This contribution of knowledge can arise from the student's diverse lived experiences, backgrounds, and intersectional identities (Buchanan, 2016). Inquiry-based learning uses this form of contributive knowledge of student experience to create a more inclusive classroom, where knowledge is more readily retained with deeper learning (Bonelli et al., 2016; Buchanan, 2016; Rosenzweig et al., 2008).

This collaborative learning in the classroom also aids in moving away from the colonial model of one individual (e.g., instructor) teaching the knowledge that they hold within the learning environment. The use of non-Western knowledge in the STEM fields also contributes to combatting appropriation and isolation due to a feeling of **solo status** (Fig. 8.2a; Archer et al., 2012; Chemers et al., 2011; Ford et al., 2021; Simpson et al., 2021). STEM researchers in the Western world are also often accused of stealing subaltern knowledge by using and not crediting non-Western individuals or methods (Gewin, 2021; Spivak, 1999). Exploring various ideologies will help educate students and instructors on how to ethically interpret community and societal issues from a non-Western perspective, resulting in greater cross-disciplinary and socially relevant STEM research (Alkholy et al., 2017).

8.2.1.1 Syllabus Deconstruction: Pluralizing Knowledge by Diversifying Reading Materials

Reading materials included in STEM syllabi provide conceptual and theoretical knowledge that are often Western-centric. This dominant knowledge in STEM has limited the acceptance of the value of non-Western forms of knowledge. Using a DEDIB approach, instructors should "diversify" and "pluralize" the reading materials based on different perspectives, including research from scientists that are marginalized in their field (Fig. 8.2c). If you have trouble finding diverse scholarship/approaches in your field, you can use our technique: (1) create a teaching moment: engage students in an equal effort classroom discussion on the underrepresentation of Indigenous/Queer/Racialized science, (2) leverage student resources: task students to conduct a search for alternative course materials including forms of non-normative knowledge like social media, or (3) get network assistance: ask for help from peers, community organizations, and underrepresented experts in your field. Although educators and researchers are increasingly aware of the overrepresentation of Western ideologies within STEM curricula (Eaton, 2022), overwhelming feelings of labor burden may deter instructors from diversifying their reading list to create a new framework within their classrooms (Perrin-Stowe et al., 2023).

At the Decolonial Perspectives and Practices Hub, two approaches may be used to diversify materials: mixed method and radical method. The mixed method focuses on representing materials of diverse forms of knowledge concurrently with Western-centric knowledge, which demonstrates the equal importance of non-Western discoveries and innovations. The radical method involves engaging with marginalized forms of knowledge and scientists in STEM, while subverting the "traditional" Western perspective. We acknowledge the fatigue and apprehension experienced by instructors with additional labor burdens or discomfort. In addition, a concern of "reverse elitism" may occur, where individuals may be hesitant to integrate marginalized knowledge due to a resulting stigmatization from Western viewpoints, or foundational field beliefs. Yet, incorporating marginalized knowledge in classrooms enriches knowledge transfer and possibilities for STEM instructors and students.

This change in curricula may be especially difficult and labor intensive in a course that the instructor teaches routinely with a pre-designed structure. Instructors may also have not received formal training to effectively implement this change, as it demands a concerted effort and a substantial shift in established practices. Therefore, small steps should be taken for instructors that are reluctant to incorporate diverse reading materials in their classroom. The first step can include researching your field related to anti-coloniality, anti-racism, and decolonization. Through time, the work of underrepresented communities or individuals may be slowly integrated to pluralize knowledge. To further incorporate collaboration, an activity may be assigned to students to research and share marginalized scholars that align with the course themes, further welcoming knowledge transference and collaboration within STEM classrooms.

8.2.2 Acknowledging and Discussing Biases in STEM Research and Course Content

Marginalized groups are not only disproportionately excluded from conducting research, but they have also been harmed by intentional and unintentional biases in STEM research (Webb et al., 2022). Women, LGBTQ2SIA+, Black, Brown, Indigenous, and People of Color continue to be omitted from studies in STEM (Cascio et al., 2021; García et al., 2017; Gould & Ashton-Smith, 2011; Quinn & Madhoo, 2014; Seydel, 2021; Spates, 2012; Thomas, 2004; Wendler et al., 2006). Research unintentionally perpetuates the norm of a world surrounding the White, cisgender heterosexual male further oppressing these marginalized groups. For example, women and People of Color continue to be underdiagnosed with autism spectrum disorders, and sexual orientation may in fact affect several physiological responses including stress responses (Gould & Ashton-Smith, 2011; Juster et al., 2015; Quinn & Madhoo, 2014; Spates, 2012; Thomas, 2004; Webb et al., 2022). In addition, Women of Color are greatly affected by these biases and further marginalized due to their intersection of gender and race (Spates, 2012; Thomas, 2004).

Therefore, scientists should become more inclusive in their research practices, especially when collaborating with diverse marginalized individuals, groups, or model systems outside their local community. Research can only be beneficial if these impacted communities are valued and given space to collaborate as knowledge holders and sharers (Silko, 2012). Using various forms of knowledge (e.g., Thompson et al., 2019), hiring linguistic and non-linguistic interpreters to correctly understand community members or beings, and open mindedness are some steps toward making science research more inclusive. These solutions represent steps for inclusive research practices that can continue to be used in classrooms and syllabi, where students can acknowledge and discuss biases present in STEM research.

8.2.2.1 Racial Biases Impacting STEM Research

Race is a sociological construct and not a biological reality (Smedley & Smedley, 2005; Webb et al., 2022; Yudell et al., 2016). The phenotypic or externally observable differences that we consider as indicators of race include skin color, hair texture, and body composition, which appear on a continuum for humans that simply covaries with our ancestral latitude (Webb et al., 2022). However, the collection of phenotypes and cultural indicators that we associate with one's "race" still has power and deeply affects the lived experiences of individuals within the society.

Assumptions of inherent differences between racial categories have long fueled STEM research (Webb et al., 2022). **Social Darwinism** and **scientific racism** have fueled discriminatory scientific questioning and hierarchization, leading to the oppression of marginalized groups through exploitation, segregation, and exclusion from power and cultural imperialism (Brandt, 1978; Dennis, 1995; Pandey, 2020; Saini, 2019; Webb et al., 2022). Many innovations in the STEM fields have advanced at the expense of these marginalized groups, including the beginnings of gynecology or the research conducted on enslaved peoples and inhumanely imprisoned people in concentration camps (Berger & Sarnyai, 2015; Freire, 2000; Sartin, 2004).

As a result, medical mistrust has been sown in several communities, which has yet to be acknowledged and addressed in healthcare and STEM research. Many of these affected communities continue to avoid medications or medical research due to mistrust, resulting in a vast avoidable health disparity (Freimuth et al., 2001). This exclusion of racial minorities in STEM research and, more specifically, medicine directly harms individuals and potentially encourages the development of biased medical technologies (Lester et al., 2020; Louie & Wilkes, 2018; Webb et al., 2022).

These persistent exclusionary biases in research may dissuade STEM researchers from using inclusive practices by adopting a "colorblind" approach (Juster et al., 2015), which can prevent the discovery of biological mechanisms and the optimization of technology in marginalized groups. For example, several physiological devices were not initially designed to handle phenotype variability, leading to a "phenotypic bias" and systematic data erasure from people with darker skin (Juster et al., 2015). Biases in biomedical optics have been identified with the failure to account for skin tone variability during diagnostics and the overgeneralized assumption of successful interpretation for all skin tones (Phan et al., 2021; Sardar et al., 2001). As a result, devices measuring oxygen saturation in the blood are systematically higher in individuals with darker skin due to the greater light absorption, potentially resulting in misdiagnosis and worse health outcomes (Fawzy et al., 2022).

STEM researchers are slowly beginning to recognize the need to create more inclusive innovations and technologies. They have the responsibility to design their research in an inclusive manner to ensure that the evaluated effect is true and not the result of confounding variables such as phenotypic bias or lived experiences (Berger & Sarnyai, 2015; Bird et al., 2021; Carter et al., 2017; Harnett et al., 2019; Kadambi, 2021; Webb et al., 2022). Instructors must acknowledge and assess the impact of biases present in historical and current STEM research in their curriculum to decrease the isolation faced by marginalized individuals in the classroom (Leibnitz et al., 2022). Decolonizing educational content would enable equity-oriented viewpoints to be voiced, leading to greater ethical and community-based research that challenges racial stigmas (Fiscella & Williams, 2004).

8.2.2.2 Interpersonal Relationships to Limit Biases

Using a DEDIB approach, STEM researchers may discover new concepts or innovations by trying to empathize with the experiences, emotions, and mental states of affected communities. Research shows that individuals generally show more empathy for people from the same ethnic group (Hein et al., 2010; Xu et al., 2009; Zuo & Han, 2013). In contrast, limited empathy occurs when an acquaintance on the same team is from a different ethnic group (Cao et al., 2015). Strong interpersonal relationships with individuals of different ethnic groups, such as neighbors, colleagues, and friends, are the only way to increase empathy and eliminate racial biases (Cao et al., 2015). Therefore, a more diverse team of individuals is not enough to combat our biases; strong interpersonal connections are required. Spaces including learning environments need to be optimized for individuals to feel safe and authentic, while allowing deeper interpersonal connections. This sense of belonging can also encourage students to enter and remain within STEM (Rainey et al., 2018).

A sense of belonging among STEM students is correlated with the number of individuals "like them" who also are in their major (Rainey et al., 2018). Having the presence of like students in a STEM learning environment as normative rather than an aberration conveys to students that they belong in the space (Rainey et al., 2018). Students leaving their STEM program report having a low sense of belonging, which has a greater impact among individuals from underrepresented groups (Rainey et al., 2018). Even when students sensed they belonged in STEM, they sometimes expressed feelings that others did not feel they belonged in the learning environment due to not fitting stereotypes (Rainey et al., 2018). A better understanding is required to determine the ways in which the educational and cultural climates prevent marginalized groups from feeling a sense of belonging and persisting in the STEM fields (Rainey et al., 2018).

STEM researchers from underrepresented groups continue to be disproportionately pushed out of their fields due to greater barriers such as funding access. For example, Black STEM researchers receive less funding than their White counterparts, with the rate of National Institutes of Health funding for Black researchers at approximately 55% of those of White researchers of similar academic standing (Erosheva et al., 2020; Ginther et al., 2011; Stevens et al., 2021). Underrepresented researchers may also have less access to positions in higher education, limiting their representation in the classroom and impacting the sense of belonging amongst STEM students. Therefore, marginalized STEM researchers should be fairly and proportionately included due to their impact and innovation in the learning environment, inclusive technologies, and methods.

8.2.2.3 Syllabus Deconstruction: Attending to Biases in the Course Description

The course description of STEM syllabi sets the stage for peer-to-peer rapport and relationality between students and the instructor. The course description traditionally shares key concepts, historical context, and learning objectives. Using a DEDIB approach, STEM educators should consider the biases present in the course description and themes related to marginalized groups and subsequently devise collaborative, reflexive learning objectives. The process of fostering collaboration touches on the DEDIB principles of anti-colonial power relations, acceptance, trust, and equitable co-operative engagements.

Alternative seating arrangements in the classroom should be used to increase collaboration and participation. Circular seating arrangements were the most common recommendation to instill a sense of belonging drawing from the African, Asian, and Indigenous practices of "healing circles." This circularity allows knowledge transference and interpersonal relationality, which dissuades power relations through equal standing amongst members. This pattern increases critical student engagement by engaging with peers for teaching and guidance, where instructors are no longer the sole knowledge holder. Although there is a trend favoring the incorporation of movable chairs and tables within educational environments, it remains indisputable that conventional classroom setups endure. Typically, securely affixed furniture remains the norm, thereby constricting the space arrangement. Consequently, instructors can focus on designing class activities to increase student-to-student interaction rather than physical spatial configurations.

Land acknowledgments are a decolonial practice that should be incorporated in STEM classrooms in an intentional and meaningful manner. Beyond a standardized statement in syllabi, land acknowledgements should be interconnected with course materials throughout the semester. Indigenous land ethics and practices can be valued for providing an alternative lens on the organisms and biological environment of inquiry. This connection can be further explored through readings by Indigenous scholars and considered at a deeper level in exercises or course assignments. Course descriptions and objectives are also a space to present and define the limits, biases, and knowledge gaps within the course content and in STEM. Instructors should avoid framing the course material as having an unfalsifiable and omnipotent account of the course topic and field.

Instructors should be transparent regarding the primary use of Western-centric knowledge in the course, where alternative forms of knowledge may be limited yet are present beyond the course. This transparency creates a space for students to reflect on other approaches and to explore potential biases in STEM. The course description should include a list of terminology that students are invited to question, expand, or critique, thereby welcoming critical thinking in the STEM classroom. This first step can include deconstructing foundational terminology in STEM like "science," "research," and "objectivity." Consequently, the STEM classroom can further evolve and be enriched by student knowledge. With more diverse individuals in classrooms, instructors can leverage prior lived experiences and knowledge to incorporate alternative forms of knowledge (Ozyonum, 2023).

8.2.3 A Decolonizing Approach to Inclusivity and Accessibility in STEM

STEM education tends to be uniformed and standardized, leaving little opportunity for students who do not fit into this normative mold to shine (Gao et al., 2020; Xie et al., 2015). Accessibility and inclusivity in STEM education can involve removing barriers associated with teaching and learning styles, language, and course content. Using a DEDIB approach, accessibility and inclusivity in STEM classrooms emphasizes unlearning colonial patterns that impede the contributions and interpersonal relationships of diverse faculty and students (Johnson, 2011; Roberts, 2020; Weissmann et al., 2019).

8.2.3.1 Promoting Adaptive Curriculum

Teaching materials and their implementation in STEM classrooms should be designed in response to the diverse needs and backgrounds of the instructors and students (Debarger et al., 2013). A DEDIB approach suggests employing a **Universal Design for Learning** framework in STEM classrooms, which emphasizes multiple ways of presenting curricula to engage and support different types of learners (Basham & Marino, 2013). This framework challenges the idea that "one size fits all" promoting diverse teaching techniques of curricula through multiple means of (1) representation, (2) engagement, and (3) expression to increase student accessibility (Izzo & Bauer, 2015; Schreffler et al., 2019).

Multimodal teaching uses multiple means of representation for course content (e.g., class notes, articles, podcasts, videos, music, etc.) to increase accessibility to learning materials (Izzo & Bauer, 2015). However, certain forms of media, such as journals and news articles, may be inaccessible behind paywalls. Therefore, a mixture of accessible media may cater to a student's needs by explaining course concepts in ways that are most beneficial to their learning. Greater classroom engagement provides multiple opportunities for students to interact with the course content (Izzo & Bauer, 2015). Instructors are encouraged to actively engage students through inquiry-based discussions rather than relying on lectures (Rosenzweig et al., 2008). This active learning benefits students with difficulty focusing on content during lectures or that fall behind while taking notes. For example, students may actively learn mathematics concepts through independent critical thinking followed by group discussions, resulting in better content retention and higher achievement in the course (Murray & Brookover, 2012). Group discussions may also lead to the knowledge transference because of diverse cognitive styles and increase interpersonal relationships (Mitchell et al., 2017).

Finally, students can express what they have learned using different assessment tools. Multiple choice questions are the most used assessment method in higher education (DiBattista & Kurzawa, 2011). Yet, standardized testing has been shown to be an inefficient means of quantifying student learning. This form of testing does not accurately showcase a student's achievements, as it provides unreliable markers of their level of knowledge (Bhattacharyya et al., 2013). Alternatively, offering various kinds of assessments in course curricula (e.g., multimedia projects or written essays) will allow students with different learning styles to demonstrate their comprehension. Instructors are then better able to determine whether students have grasped the essential course contents (Izzo & Bauer, 2015). The Universal Design for Learning framework challenges instructors to rethink and restructure adaptive curricula, allowing greater flexibility to serve a diverse learning population (Rose et al., 2002).

From a DEDIB perspective, the use of this framework does include a few limitations. This framework focuses on the standards and norms typically found in Western-centric classrooms. A greater global generalization of DEDIB principles is needed to go beyond tokenistic inclusion in higher education instead pursuing learning innovations, which include incorporating the geographical context of learning environments, power relations, post-colonial dialogues, and decolonial curricula (Fovet, 2020). Transforming this framework would help move away from standardized learning practices rooted in colonial ideologies, allowing more STEM students to excel in higher education.

8.2.3.2 Considering Learning Barriers

Anglocentrism is a predominant issue in the accessibility of the academic and educational systems in postsecondary institutions (Fig. 8.2a). Researchers from non-Western regions as well as those who speak English as a second language or additional language must overcome significant obstacles when learning or collaborating (Curry & Lillis, 2017). Structural biases in the knowledge production system add a barrier for non-native English speakers when participating in academic activities (Minakova & Canagarajah, 2020), especially for undergraduate and graduate students in STEM (Huang, 2010).

Many academic journal articles that are presented in STEM classrooms use English as the predominant language (Curry & Lillis, 2017). Non-English journal articles are rarely cited in English publications and are typically not presented as part of the reading materials in STEM programs (Davis & Malfitano, 2017). Linguistic privilege most prominently manifests through publication bias, with the promotion and dissemination of monolingual scientific research (Hyland, 2016). With the increasing need for graduate students to publish early in their STEM careers, it is important to instill pedagogical training for instructors and supervisors to support their students (Lee & Aitchison, 2009).

This linguistic bias can also impact the retention of course content and understanding course assessments of non-native English students in English STEM classrooms. Therefore, linguistic equity can be achieved by embracing modern translation technologies to provide polylingual institutional websites and by providing technologies with correction software when assessing non-native English students (Taylor, 2021). Having flexible assessments, such as peer-learning instead of oral presentations, also further encourages students to create a sense of agency where they are responsible for their own learning, while limiting linguistic obstacles (Fig. 8.2a; Gomez & Lee, 2015; Rissanen, 2014). This sense of agency, learning, and engagement in students via course content and assessments is largely controlled by emotion and motivation in the classroom. Motivation not only relates to an individual's values and goals (the "why"), but also relates to the individual's competence (the "can I") (Linnenbrink-Garcia et al., 2016).

As motivation depends on educational context, increasing the retention of underrepresented individuals in higher education could be achieved by applying the **self-determination theory**, a well-established concept in education psychology. Self-determination theory states that three psychological needs dictate a student's internal motivation: autonomy, competency, and relatedness (Linnenbrink-Garcia et al., 2016; Ryan & Deci, 2000). For example, research in Biology education has shown that self-determination theory plays a role in the positive impact of team-based learning (Jeno et al., 2018), as well as retention in STEM programs in higher education (Findley-Van Nostrand & Pollenz, 2017; Kuchynka et al., 2019). Therefore, STEM instructors should strive to limit common barriers for accessible learning in addition to helping students create a sense of agency and competence to foster their belonging in the educational space.

8.2.3.3 Syllabus Deconstruction: Accessible and Inclusive Assessments

The grading scheme measures student success and participation through assessments such as written assignments, quizzes, presentations, laboratory exams, and written final exams. The grading scheme also grounds the hierarchical relationship between the students and instructors. A DEDIB approach to the grading scheme calls instructors to develop flexible and meaningful measurements that attend to different learning styles and abilities beyond memorization (Fig. 8.2b; Freire, 2000). Reconsidering the grading scheme of a course is challenging and often feels impossible because of the institutional mandates and standardized grading structure.

Instructors can break the status quo of assessments with creativity and identify what can be modified. For example, instructors should consider avoiding heavily weighted final tests, such as the common 50–60% final exam in STEM courses. This greater weight on final exams sets the precedent that individual exercises that require repetition and memorization are more valued in the STEM fields. Alternatively, several different activities with similar weights should be employed as assessments, which are collaborative, creative, and equitably value different learning styles of students. Assessments could include group or paired student projects and the acceptance of diverse media assignments or projects such as oral presentations, podcasts, or videos. These kinds of assessments help translate the scientific knowledge of students in the classroom in accessible and interactive formats. Instructors can also incorporate course members in the assessment process by assigning students to peer-review each other's work as a class exercise. These assessments can then be factored into the student's final grade. The inclusion of student assessors ensures that the instructor's perspective is cross-referenced with other perspectives in the learning environment, resulting in a nuanced and more equitable evaluation focusing on comprehension rather than performance.

To combat linguistic biases in the classroom, instructors can integrate a lesson demonstrating the prevalence of Anglocentrism in STEM, where a language interpreter can be hired as a special guest to unpack a relevant non-English text. If there are budget limitations, instructors can share the history of multilingualism within the field, ask multilingual student volunteers to unpack relevant texts, or utilize the multilingual subtitle function in virtual classrooms for students to receive the lecture in their native language. Alternatively, instructors can create a student-buddy system, where anglophone students are paired with bilingual or multilingual "buddy" students who can help translate a text in the non-English language. Finally, a more interactive and alternative solution would be to have instructors invite Indigenous and racialized STEM researchers as guest speakers to their classroom to provide an intersectional perspective through the dissemination of oral histories and knowledge.

8.3 Further Reflections and Commitments

This chapter demonstrates the cross-disciplinary potential of decolonizing, equity, diversity, inclusion, and belonging principles as an antidote for exclusionary education. The STEM fields are not exempt from the oppressive historical and contemporary impacts of colonialism. Exposing students to diverse thoughts and practices will encourage them to unlearn systemic inequalities and prime them to become ethical, sustainable, inventive scientists and practitioners, which will inform their knowledge beyond the classroom. We call on STEM educators and researchers to begin identifying and addressing biases by examining the syllabus as a concrete step toward DEDIB in STEM.

The syllabus should be reformulated as a navigating tool for sharing knowledges while maintaining course outcomes. Educators can engage with DEDIB principles to reform their syllabus to incite a collaborative classroom with ethical engagement, collective rules, student empowerment, balanced student-instructor power dynamics, and acknowledgments between the class and course topic with the land. The syllabus is an unfinished product that promotes uncomfortable, but necessary dialogue on field-specific biases between students and instructors through a collective agreement.

Our cross-disciplinary dialogue and role as early-career educators and scientists fills the gap within Discipline-Based Education Research for knowledge exchange across expertise and disciplines in the Arts and Sciences. We provided application-based evidence that DEDIB principles promote critical awareness of biases and reformative strategies of action that warrant greater inquiry in the STEM fields. DEDIB principles can work to ethically advance STEM education by dethroning the idealized Western-centric scientist, further reconfiguring the standards of who can contribute to science.

Incorporating decolonization and belonging to the resurgent call for EDI in postsecondary institutions orients efforts toward absolving the root of the problem and creating more sustainable solutions. We hope to inspire further inquiry into solution-based practices related to decolonization, equity, diversity, inclusion, and belonging in STEM that can improve Indigenous and marginalized lives. We ask that you consider the potential for DEDIB to inspire a new generation of empathetic and multivocal scientists that can move us closer to STEM fields where we all belong.

Glossary[4]

Anglocentrism Considering the English language or English-speaking culture to be more important or better, or of expecting everything to be in English.

Belonging A sense or determination that one has an affinity to a place or situation. For example, feeling that one has agency and acceptance in an education or research space.

Decolonial Perspectives and Practices Hub (DPPH) A non-profit organization that began as a student initiative at Concordia University, dedicated to co-creating decolonial teaching and learning strategies, experiential learning programs, and community collaborations that challenge systemic inequities in academia.

Decolonization Decolonizing higher education is dedicated to reimagining and reconstructing the current system as it exists today, moving away from colonial

[4] These terms are sociological concepts that can be defined in a variety of ways, with meanings that can change over time. The definitions provided here have been written by the co-authors and are relevant to this chapter.

patterns of power and ways of viewing the world, and toward a more plural and inclusive future.

DEDIB framework Decolonizing, Equity, Diversity, Inclusion, and Belonging—A framework in which EDI is accompanied by "decolonizing" and "belonging" principles and can be used to assess and dismantle oppressive systems impacting marginalized groups.

Discipline-Based Education Research (DBER) A research field with the goal of making STEM education more accessible to diverse students through evidence-based knowledge, practices, and assessments to improve teaching and learning.

Diversity Having or comprising people of different identities in a group in an education setting or otherwise (diverse).

EDI Equity, Diversity, and Inclusion

Epistemic colonialism This concept criticizes the perceived universality of Western knowledge and Western ways of knowing. Also known as decolonization of knowledge.

Equity The identification and elimination of barriers to equal and just treatment, access, and opportunity that prevent the full participation of some individuals or groups (equitable).

Inclusion The act or practice of involving or integrating people who historically have been and/or continue to be excluded because of their identity (inclusive).

Intersectional Identifying with or belonging to more than one (usually) underrepresented group. For example, a person who is both Black and a woman may declare to have an intersectional identity.

Scientific racism A pseudoscientific belief that humans can be divided into biological groups called "races," where one race is superior or inferior to another, and that false empirical evidence exists to support or justify racism or racial inferiority/superiority. Scientific racism can also include the purposeful exclusion of groups in scientific or medical studies.

Self-determination theory A theory of human motivation and personality based on human basic needs, usually including competence (ability to be effective at a task or in an environment), autonomy (ability to have a sense of control over one's life), and relatedness (ability to have close relationships with others).

Settler colonialism A structure that perpetuates the displacement and/or removal of Indigenous people or cultures to replace them with a society built by the settlers.

Social Darwinism A highly criticized group of theories that apply biological concepts such as natural selection and "survival of the fittest" to claim that certain human populations, societies, and politics are innately better.

Solo status A feeling or observation that one is the singular member of a particular identity in a group setting or space.

STEM Science, Technology, Engineering, and Math

STEMM Diversity @ McGill A student-run initiative housed at the Redpath Museum of McGill University promoting equity, diversity, inclusion, and representation in science, technology, engineering, math, and medicine (STEMM).

Systemic bias A bias is a preconceived idea of a person, place, thing, or circumstance. It can be conscious or unconscious and is often influenced by society and considered to be unfair. When human systems, such as institutions, are biased toward a particular outcome, it is referred to as a systemic bias.

Underrepresented groups A subset of the population that includes a smaller percentage of individuals than that of the general population, leading to an insufficient and/or disproportionately low representation of that group in that population. Underrepresented groups can vary across different areas and can include, but are not limited to, race, gender, sexual orientation, and physical or mental disabilities, and can include visible and invisible differences.

Universal Design for Learning An educational design or practice that encourages and guides the formation of learning spaces that are inclusive of individual learning differences.

Westernized To adopt or be influenced by the cultural, economic, or political systems of Europe and North America, or to give them preference.

References

Alkholy, S. O., Gendron, F., McKenna, B., Dahms, T., & Ferreira, M. P. (2017). Convergence of indigenous science and Western science impacts students' interest in STEM and identity as a scientist. *Ubiquitous Learning: An International Journal, 10*(1), 1–13. https://doi.org/10.18848/1835-9795/CGP/v10i01/1-13

Alvares, C. A., & Faruqui, S. S. (2011). *Decolonising the university: The emerging quest for non-Eurocentric paradigms*. Penerbit Universiti Sains Malaysia.

Andreotti, V. D. O., Stein, S., Ahenakew, C., & Hunt, D. (2015). Mapping interpretations of decolonization in the context of higher education. *Decolonization: Indigeneity, Education and Society, 4*(1), Article 1.

Archer, L., DeWitt, J., Osborne, J., Dillon, J., Willis, B., & Wong, B. (2012). Science aspirations, capital, and family habitus: How families shape children's engagement and identification with science. *American Educational Research Journal, 49*(5), 881–908. https://doi.org/10.3102/0002831211433290

Ashar, M. (2015). Decolonizing what? Categories, concepts and the enduring 'not yet.'. *Cultural Dynamics, 27*(2), 253–265. https://doi.org/10.1177/0921374015585231

Basham, J. D., & Marino, M. T. (2013). Understanding STEM education and supporting students through Universal Design for Learning. *TEACHING Exceptional Children, 45*(4), 8–15. https://doi.org/10.1177/004005991304500401

Battiste, M. (2013). *Decolonizing education: Nourishing the learning spirit*. Purich Publishing.

Berger, M., & Sarnyai, Z. (2015). "More than skin deep": Stress neurobiology and mental health consequences of racial discrimination. *Stress, 18*(1), 1–10. https://doi.org/10.3109/10253890.2014.989204

Bhambra, G. K., Gebrial, D., & Nişancıoğlu, K. (2018). *Decolonising the university*. Pluto Press.

Bhattacharyya, S., Junot, M., & Clark, H. (2013). Can you hear us? Voices raised against standardized testing by novice teachers. *Creative Education, 4*(10), 633–639. https://doi.org/10.4236/ce.2013.410091

Bird, C. M., Webb, E. K., Schramm, A. T., Torres, L., Larson, C., & de Roon-Cassini, T. A. (2021). Racial discrimination is associated with acute posttraumatic stress symptoms and predicts future posttraumatic stress disorder symptom severity in trauma-exposed Black adults in the United States. *Journal of Traumatic Stress, 34*(5), 995–1004. https://doi.org/10.1002/jts.22670

Bonelli, L., Coons, L., & Holt, P. M. (2016). *The impact of the EDI framework and inquiry-based learning strategies on student achievement in middle school [Lipscomb University]*. ProQuest Dissertations and Theses (1850989146). ProQuest Dissertations & Theses Global.

Brandt, A. M. (1978). Racism and research: The case of the Tuskegee syphilis study. *The Hastings Center Report, 8*(6), 21. https://doi.org/10.2307/3561468

Buchanan, S. (2016). Exploring the lived experience of middle school students engaged in inquiry based learning. In S. Kurbanoğlu, J. Boustany, S. Špiranec, E. Grassian, D. Mizrachi, L. Roy, & T. Çakmak (Eds.), *Information literacy: Key to an inclusive society* (pp. 490–498). Springer International Publishing.

Cao, Y., Contreras-Huerta, L. S., McFadyen, J., & Cunnington, R. (2015). Racial bias in neural response to others' pain is reduced with other-race contact. *Cortex, 70*, 68–78. https://doi.org/10.1016/j.cortex.2015.02.010

Carter, R. T., Johnson, V. E., Roberson, K., Mazzula, S. L., Kirkinis, K., & Sant-Barket, S. (2017). Race-based traumatic stress, racial identity statuses, and psychological functioning: An exploratory investigation. *Professional Psychology: Research and Practice, 48*(1), 30–37. https://doi.org/10.1037/pro0000116

Cascio, M. A., Weiss, J. A., & Racine, E. (2021). Making autism research inclusive by attending to intersectionality: A review of the research ethics literature. *Review Journal of Autism and Developmental Disorders, 8*(1), 22–36. https://doi.org/10.1007/s40489-020-00204-z

Chemers, M. M., Zurbriggen, E. L., Syed, M., Goza, B. K., & Bearman, S. (2011). The role of efficacy and identity in science career commitment among underrepresented minority students. *Journal of Social Issues, 67*(3), 469–491. https://doi.org/10.1111/j.1540-4560.2011.01710.x

Curry, M. J., & Lillis, T. (2017). Problematizing English as the privileged language of global academic publishing. In *Global academic publishing: Policies, perspectives and pedagogies* (pp. 1–20). Multilingual Matters.

Davis, J. A., & Malfitano, A. P. S. (2017). Supporting research dissemination across languages and cultures: Working beyond Western and anglocentric lines. *Canadian Journal of Occupational Therapy, 84*(4–5), 204–206. https://doi.org/10.1177/0008417417745843

de Sousa Santos, B. (2007). Beyond abyssal thinking: From global lines to ecologies of knowledge. *Review (Fernand Braudel Center), 30*(1), 45–89.

Debarger, A. H., Choppin, J., Beauvineau, Y., & Moorthy, S. (2013). Designing for productive adaptations of curriculum interventions. *Teachers College Record, 115*(14), 298–319. https://doi.org/10.1177/016146811311501407

Dei, G. J. S. (2017). *Reframing Blackness and Black solidarities through anti-colonial and decolonial prisms*. Springer International Publishing. https://doi.org/10.1007/978-3-319-53079-6

Dennis, R. M. (1995). Social darwinism, scientific racism, and the metaphysics of race. *The Journal of Negro Education, 64*(3), 243–252. https://doi.org/10.2307/2967206

DiBattista, D., & Kurzawa, L. (2011). Examination of the quality of multiple-choice items on classroom tests. *The Canadian Journal for the Scholarship of Teaching and Learning, 2*(2). https://doi.org/10.5206/cjsotl-rcacea.2011.2.4

Du Bois, W. E. B. (1898). The Study of the N* Problems. *The Annals of the American Academy of Political and Social Science, 11*, 1–23. JSTOR.

Eaton, S. E. (2022). New priorities for academic integrity: Equity, diversity, inclusion, decolonization and Indigenization. *International Journal for Educational Integrity, 18*(1), 10. https://doi.org/10.1007/s40979-022-00105-0

Erosheva, E. A., Grant, S., Chen, M.-C., Lindner, M. D., Nakamura, R. K., & Lee, C. J. (2020). NIH peer review: Criterion scores completely account for racial disparities in overall impact scores. *Science Advances, 6*(23), eaaz4868. https://doi.org/10.1126/sciadv.aaz4868

Fanon, F. (1952). *Black skin, white masks*. Grove Press.

Fausto-Sterling, A. (1991). Race, gender and science. *Transformations: The Journal of Inclusive Scholarship and Pedagogy*, 2(2), 4–12.

Fawzy, A., Wu, T. D., Wang, K., Robinson, M. L., Farha, J., Bradke, A., Golden, S. H., Xu, Y., & Garibaldi, B. T. (2022). Racial and ethnic discrepancy in pulse oximetry and delayed identification of treatment eligibility among patients with COVID-19. *JAMA Internal Medicine, 182*(7), 730. https://doi.org/10.1001/jamainternmed.2022.1906

Findley-Van Nostrand, D., & Pollenz, R. S. (2017). Evaluating psychosocial mechanisms underlying STEM persistence in undergraduates: Evidence of impact from a six-day pre-college engagement STEM academy program. *CBE—Life Sciences Education, 16*(2), 1–15. https://doi.org/10.1187/cbe.16-10-0294

Fiscella, K., & Williams, D. R. (2004). Health disparities based on socioeconomic inequities: Implications for urban health care. *Academic Medicine, 79*(12), 1139–1147. https://doi.org/10.1097/00001888-200412000-00004

Ford, J., Jreidini, N., Crandall, K. E., Sanderson, S., & Xu, C. C. Y. (2021). Promoting equity and inclusion with student-driven initiatives. *Trends in Ecology and Evolution, 36*(12), 1063–1066. https://doi.org/10.1016/j.tree.2021.08.013

Fovet, F. (2020). Universal Design for Learning as a tool for inclusion in the higher education classroom: Tips for the next decade of implementation. *Education Journal, 9*(6), Article 6. https://doi.org/10.11648/j.edu.20200906.13

Frances, H., Dua, E., James, C. E., Kobayashi, A., Li, P., Ramos, H., & Smith, M. S. (2017). *The equity myth: Racialization and Indigeneity at Canadian Universities*. UBC Press.

Freimuth, V. S., Quinn, S. C., Thomas, S. B., Cole, G., Zook, E., & Duncan, T. (2001). African Americans' views on research and the Tuskegee Syphilis study. *Social Science and Medicine, 52*(5), 797–808. https://doi.org/10.1016/S0277-9536(00)00178-7

Freire, P. (2000). *Pedagogy of the oppressed* (30th anniversary ed.). Continuum.

Gao, X., Li, P., Shen, J., & Sun, H. (2020). Reviewing assessment of student learning in interdisciplinary STEM education. *International Journal of STEM Education, 7*(1), 24. https://doi.org/10.1186/s40594-020-00225-4

García, A. A., Zuñiga, J. A., & Lagon, C. (2017). A personal touch: The most important strategy for recruiting Latino research participants. *Journal of Transcultural Nursing: Official Journal of the Transcultural Nursing Society, 28*(4), 342–347. https://doi.org/10.1177/1043659616644958

Gewin, V. (2021). How to include Indigenous researchers and their knowledge. *Nature, 589*, 315–317. https://doi.org/10.1038/d41586-021-00022-1

Ginther, D. K., Schaffer, W. T., Schnell, J., Masimore, B., Liu, F., Haak, L. L., & Kington, R. (2011). Race, ethnicity, and NIH research awards. *Science (New York, N.Y.), 333*(6045), 1015–1019. https://doi.org/10.1126/science.1196783

Gomez, K., & Lee, U.-S. (2015). Situated cognition and learning environments: Implications for teachers on- and offline in the new digital media age. *Interactive Learning Environments, 23*(5), 634–652. https://doi.org/10.1080/10494820.2015.1064447

Goodenow, C., & Grady, K. E. (1993). The relationship of school belonging and friends' values to academic motivation among urban adolescent students. *The Journal of Experimental Education, 62*(1), 60–71. https://doi.org/10.1080/00220973.1993.9943831

Gould, J., & Ashton-Smith, J. (2011). Missed diagnosis or misdiagnosis? Girls and women on the autism spectrum. *Good Autism Practice, 12*(1), 34–41.

Grosfoguel, R. (2013). The structure of knowledge in westernized universities: Epistemic racism/sexism and the four genocides/epistemicide of the long 16th century. *Human Architecture: Journal of the Sociology of Self-Knowledge, 11*(1), 73–90.

Harnett, N. G., Wheelock, M. D., Wood, K. H., Goodman, A. M., Mrug, S., Elliott, M., Schuster, M., Tortolero, S., & Knight, D. C. (2019). Negative life experiences contribute to racial differences in the neural response to threat. *NeuroImage, 202*, 116086. https://doi.org/10.1016/j.neuroimage.2019.116086

Hein, G., Silani, G., Preuschoff, K., Batson, C. D., & Singer, T. (2010). Neural responses to ingroup and outgroup members' suffering predict individual differences in costly helping. *Neuron, 68*(1), 149–160. https://doi.org/10.1016/j.neuron.2010.09.003

Henderson, C., Connolly, M., Dolan, E. L., Finkelstein, N., Franklin, S., Malcom, S., Rasmussen, C., Redd, K., & St. John, K. (2017). Towards the STEM DBER alliance: Why we need a discipline-based STEM education research community. *International Journal of STEM Education, 4*(1), 14. https://doi.org/10.1186/s40594-017-0076-1

hooks, b. (1992). *Black looks: Race and representation*. South End Press.

Huang, J. C. (2010). Publishing and learning writing for publication in English: Perspectives of NNES PhD students in science. *Journal of English for Academic Purposes, 9*(1), 33–44. https://doi.org/10.1016/j.jeap.2009.10.001

Hyland, K. (2016). Academic publishing and the myth of linguistic injustice. *Journal of Second Language Writing, 31*, 58–69. https://doi.org/10.1016/j.jslw.2016.01.005

Itchuaqiyaq, C. U., & Matheson, B. (2021). Decolonizing decoloniality: Considering the (mis)use of decolonial frameworks in TPC scholarship. *Communication Design Quarterly, 9*(1), 20–31. https://doi.org/10.1145/3437000.3437002

Izzo, M. V., & Bauer, W. M. (2015). Universal Design for Learning: Enhancing achievement and employment of STEM students with disabilities. *Universal Access in the Information Society, 14*(1), 17–27. https://doi.org/10.1007/s10209-013-0332-1

Jeno, L. M., Danielsen, A. G., & Raaheim, A. (2018). A prospective investigation of students' academic achievement and dropout in higher education: A self-determination theory approach. *Educational Psychology, 38*(9), 1163–1184. https://doi.org/10.1080/01443410.2018.1502412

Johnson, D. R. (2011). Examining sense of belonging and campus racial diversity experiences among women of color in STEM living-learning programs. *Journal of Women and Minorities in Science and Engineering, 17*(3), 209–223. https://doi.org/10.1615/JWomenMinorScienEng.2011002843

Juster, R.-P., Hatzenbuehler, M. L., Mendrek, A., Pfaus, J. G., Smith, N. G., Johnson, P. J., Lefebvre-Louis, J.-P., Raymond, C., Marin, M.-F., Sindi, S., Lupien, S. J., & Pruessner, J. C. (2015). Sexual orientation modulates endocrine stress reactivity. *Biological Psychiatry, 77*(7), 668–676. https://doi.org/10.1016/j.biopsych.2014.08.013

Kadambi, A. (2021). Achieving fairness in medical devices. *Science, 372*(6537), 30–31. https://doi.org/10.1126/science.abe9195

Koehler, C., Binns, I. C., & Bloom, M. A. (2015). The emergence of STEM. In *STEM road map* (1st ed., pp. 13–22). Routledge.

Kuchynka, S., Findley-Van Nostrand, D., & Pollenz, R. S. (2019). Evaluating psychosocial mechanisms underlying STEM persistence in undergraduates: Scalability and longitudinal analysis of three cohorts from a six-day pre-college engagement STEM academy program. *CBE—Life Sciences Education, 18*(3), ar41. https://doi.org/10.1187/cbe.19-01-0028

Lee, A., & Aitchison, C. (2009). Writing for the doctorate and beyond. In *Changing practices of doctoral education* (1st ed., pp. 99–111). Routledge.

Leibnitz, G. M., Gillian-Daniel, D. L., Greenler, R. M. C. C., Campbell-Montalvo, R., Metcalf, H., Segarra, V. A., Peters, J. W., Patton, S., Lucy-Putwen, A., & Sims, E. L. (2022). The inclusive professional framework for societies: Changing mental models to promote diverse, equitable, and inclusive STEM systems change. *Frontiers in Sociology, 6*, 784399. https://doi.org/10.3389/fsoc.2021.784399

Lester, J. C., Jia, J. L., Zhang, L., Okoye, G. A., & Linos, E. (2020). Absence of images of skin of colour in publications of COVID-19 skin manifestations. *The British Journal of Dermatology, 183*(3), 593–595. https://doi.org/10.1111/bjd.19258

Linnenbrink-Garcia, L., Patall, E. A., & Pekrun, R. (2016). Adaptive motivation and emotion in education: Research and principles for instructional design. *Policy Insights from the Behavioral and Brain Sciences, 3*(2), 228–236. https://doi.org/10.1177/2372732216644450

Louie, P., & Wilkes, R. (2018). Representations of race and skin tone in medical textbook imagery. *Social Science and Medicine, 202*, 38–42. https://doi.org/10.1016/j.socscimed.2018.02.023

Lugones, M. (2010). Toward a decolonial feminism. *Hypatia, 25*(4), 742–759.

Maldonado-Torres, N. (2007). On the coloniality of being. *Cultural Studies, 21*(2–3), 240–270. https://doi.org/10.1080/09502380601162548

Maynard, R. (2017). *Policing Black lives: State violence in Canada from slavery to the present*. Fernwood Publishing.

Mignolo, W. D. (2007). Delinking: The rhetoric of modernity, the logic of coloniality and the grammar of de-coloniality. *Cultural Studies, 21*(2–3), 449–514. https://doi.org/10.1080/09502380601162647

Miller, D. (2007). The objectives of science. *Philosophia Scientae, 11*(1), 21–43. https://doi.org/10.4000/philosophiascientiae.314

Minakova, V., & Canagarajah, S. (2020). Monolingual ideologies versus spatial repertoires: Language beliefs and writing practices of an international STEM scholar. *International Journal of Bilingual Education and Bilingualism, 26*, 1–14. https://doi.org/10.1080/13670050.2020.1768210

Mitchell, R., Boyle, B., O'Brien, R., Malik, A., Tian, K., Parker, V., Giles, M., Joyce, P., & Chiang, V. (2017). Balancing cognitive diversity and mutual understanding in multidisciplinary teams. *Health Care Management Review, 42*(1), 42–52. https://doi.org/10.1097/HMR.0000000000000088

Murray, E., & Brookover, J. (2012). Universal Design for Learning in the mathematics classroom. In *Universal Design for Learning in the classroom: Practical applications* (pp. 71–89). Guilford Press.

Ozyonum, E. (2023). Let's Talk About Decolonial Internationalization. *YU-WRITE: Journal of Graduate Student Research in Education, 2*(1), Article 1. https://doi.org/10.25071/28169344.17

Pandey, S. (2020). Ready, primed, go: Ending the racism pandemic in science. *The Plant Cell, 32*(12), 3660–3661. https://doi.org/10.1105/tpc.20.00783

Perrin-Stowe, T. I. N., Horner, M., Coon, J. J., Lynch, L. R., de Flamingh, A., Alexander, N. B., Golebie, E., Swartz, T. M., Bader, A. C., & Halsey, S. J. (2023). "Where do I even start?" Recommendations for faculty diversifying syllabi in ecology, evolution, and the life sciences. *Ecology and Evolution, 13*(1), e9719. https://doi.org/10.1002/ece3.9719

Peters, M. A. (2015). Why is my curriculum white? *Educational Philosophy and Theory, 47*(7), 641–646. https://doi.org/10.1080/00131857.2015.1037227

Phan, T., Rowland, R., Ponticorvo, A., Le, B. C., Wilson, R. H., Sharif, S. A., Kennedy, G. T., Bernal, N., & Durkin, A. J. (2021). Characterizing reduced scattering coefficient of normal human skin across different anatomic locations and Fitzpatrick skin types using spatial frequency domain imaging. *Journal of Biomedical Optics, 26*(2), 026001. https://doi.org/10.1117/1.JBO.26.2.026001

Quijano, A., & Ennis, M. (2000). Coloniality of power, eurocentrism, and Latin America. *Nepantla: Views from South, 1*(3), 533–580.

Quinn, P. O., & Madhoo, M. (2014). A review of attention-deficit/hyperactivity disorder in women and girls: Uncovering the hidden diagnosis. *The Primary Care Companion for CNS Disorders, 16*(3), PCC.13r01596. https://doi.org/10.4088/PCC.13r01596

Rainey, K., Dancy, M., Mickelson, R., Stearns, E., & Moller, S. (2018). Race and gender differences in how sense of belonging influences decisions to major in STEM. *International Journal of STEM Education, 5*(1), 10. https://doi.org/10.1186/s40594-018-0115-6

Rissanen, A. (2014). Active and peer learning in STEM education strategy. *Science Education International, 25*(1), 1–7.

Roberts, L. W. (2020). Belonging, respectful inclusion, and diversity in medical education. *Academic Medicine, 95*(5), 661–664. https://doi.org/10.1097/ACM.0000000000003215

Rose, D. H., Meyer, A., Strangman, N., & Rappolt, G. (2002). *Teaching every student in the digital age: Universal Design for Learning*. Association for Supervision and Curriculum Development; WorldCat.org.

Rosenzweig, G. M., Carrodegaus, M., & Lucky, L. (2008). Inquiring minds want to know: Inquiry-based learning in the general education and inclusion science classroom. In: *Proceedings of the seventh annual college of education research conference: Urban and international education section* (pp. 129–137). Florida International University.

Ruttenberg-Rozen, R., Hynes, K., Habibi, S., Cardoza, S., & Muchmaker, J. (2021). *Towards a community of care: Counterspaces for women in sTem education*. 2021 IEEE international symposium on technology and society (ISTAS), pp. 1–4. https://doi.org/10.1109/ISTAS52410.2021.9629141

Ryan, R. M., & Deci, E. L. (2000). Intrinsic and extrinsic motivations: Classic definitions and new directions. *Contemporary Educational Psychology, 25*(1), 54–67. https://doi.org/10.1006/ceps.1999.1020

Saini, A. (2019). *Superior: The return of race science*. Boston : Beacon Press.
Sanchez, A. (2018). Canon fire: Decolonising the curriculum. *The Cambridge Journal of Anthropology, 36*(2), 1–6. https://doi.org/10.17863/CAM.33840
Sardar, D. K., Mayo, M. L., & Glickman, R. D. (2001). Optical characterization of melanin. *Journal of Biomedical Optics, 6*(4), 404. https://doi.org/10.1117/1.1411978
Sartin, J. S. (2004). J. Marion Sims, the father of gynecology: Hero or villain? *Southern Medical Journal, 97*(5), 500–505. https://doi.org/10.1097/00007611-200405000-00017
Schreffler, J., Vasquez, E., III, Chini, J., & James, W. (2019). Universal Design for Learning in postsecondary STEM education for students with disabilities: A systematic literature review. *International Journal of STEM Education, 6*(1), 8. https://doi.org/10.1186/s40594-019-0161-8
Seydel, C. (2021). The missing sex. *Nature Biotechnology, 39*(3), 260–265. https://doi.org/10.1038/s41587-021-00844-4
Shahjahan, R. A., Estera, A. L., Surla, K. L., & Edwards, K. T. (2022). "Decolonizing" curriculum and pedagogy: A comparative review across disciplines and global higher education contexts. *Review of Educational Research, 92*(1), 73–113. https://doi.org/10.3102/00346543211042423
Silko, L. M. (2012). *Storyteller*. Penguin Books.
Simpson, A. (2014). *Mohawk interruptus: Political life across the borders of settler states*. Duke University Press.
Simpson, D. Y., Beatty, A. E., & Ballen, C. J. (2021). Teaching between the lines: Representation in science textbooks. *Trends in Ecology and Evolution, 36*(1), 4–8. https://doi.org/10.1016/j.tree.2020.10.010
Sium, A., Desai, C., & Ritskes, E. (2012). Towards the 'tangible unknown': Decolonization and the Indigenous future. *Decolonization: Indigeneity, Education and Society, 1*(1), 1–13.
Skopec, M., Fyfe, M., Issa, H., Ippolito, K., Anderson, M., & Harris, M. (2021). Decolonization in a higher education STEMM institution—Is 'epistemic fragility' a barrier? *London Review of Education, 19*(1), 1–21. https://doi.org/10.14324/LRE.19.1.18
Smedley, A., & Smedley, B. D. (2005). Race as biology is fiction, racism as a social problem is real: Anthropological and historical perspectives on the social construction of race. *American Psychologist, 60*(1), 16–26. https://doi.org/10.1037/0003-066X.60.1.16
Spates, K. (2012). "The missing link": The exclusion of Black women in psychological research and the implications for Black women's mental health. *Sage Open, 2*(3), 2158244012455179. https://doi.org/10.1177/2158244012455179
Spivak, G. C. (1988). Can the subaltern speak? *Die Philosophin, 14*(27), 42–58.
Spivak, G. C. (1999). *A critique of postcolonial reason: Toward a history of the vanishing present*. Harvard University Press.
Stevens, K. R., Masters, K. S., Imoukhuede, P. I., Haynes, K. A., Setton, L. A., Cosgriff-Hernandez, E., Lediju Bell, M. A., Rangamani, P., Sakiyama-Elbert, S. E., Finley, S. D., Willits, R. K., Koppes, A. N., Chesler, N. C., Christman, K. L., Allen, J. B., Wong, J. Y., El-Samad, H., Desai, T. A., & Eniola-Adefeso, O. (2021). Fund Black scientists. *Cell, 184*(3), 561–565. https://doi.org/10.1016/j.cell.2021.01.011
Talanquer, V. (2014). DBER and STEM education reform: Are we up to the challenge? *Journal of Research in Science Teaching, 51*(6), 809–819. https://doi.org/10.1002/tea.21162
Taylor, Z. W. (2021). Breaking the English routine: Opening polylingual avenues to postsecondary and higher education access. *Journal of Praxis in Higher Education, 3*(1), 21.
Thomas, V. G. (2004). The psychology of Black women: Studying women's lives in context. *Journal of Black Psychology, 30*(3), 286–306. https://doi.org/10.1177/0095798404266044
Thompson, K.-L., Reece, N., Robinson, N., Fisher, H.-J., Ban, N. C., & Picard, C. R. (2019). "We monitor by living here": Community-driven actualization of a social-ecological monitoring program based in the knowledge of Indigenous harvesters. *FACETS, 4*(1), 293–314. https://doi.org/10.1139/facets-2019-0006
Tuck, E., & Yang, K. W. (2012). Decolonization is not a metaphor. *Decolonization: Indigeneity, Education and Society, 1*(1), 1–40.
van Eijck, M., & Roth, W.-M. (2011). Cultural diversity in science education through Novelization: Against the epicization of science and cultural centralization. *Journal of Research in Science Teaching, 48*(7), 824–847. https://doi.org/10.1002/tea.20422

Vázquez, R. (2011). Translation as erasure: Thoughts on modernity's epistemic violence. *Journal of Historical Sociology, 24*(1), 27–44. https://doi.org/10.1111/j.1467-6443.2011.01387.x

Webb, E. K., Etter, J. A., & Kwasa, J. A. (2022). Addressing racial and phenotypic bias in human neuroscience methods. *Nature Neuroscience, 25*(4), 410–414. https://doi.org/10.1038/s41593-022-01046-0

Weissmann, G. S., Ibarra, R. A., Howland-Davis, M., & Lammey, M. V. (2019). The multi-context path to redefining how we access and think about diversity, equity, and inclusion in STEM. *Journal of Geoscience Education, 67*(4), 320–329. https://doi.org/10.1080/10899995.2019.1620527

Wendler, D., Kington, R., Madans, J., Wye, G. V., Christ-Schmidt, H., Pratt, L. A., Brawley, O. W., Gross, C. P., & Emanuel, E. (2006). Are racial and ethnic minorities less willing to participate in health research? *PLoS Medicine, 3*(2), e19. https://doi.org/10.1371/journal.pmed.0030019

Williams, C. L., Hirschi, Q., Hulleman, C. S., & Roksa, J. (2021). Belonging in STEM: Growth mindset as a filter of contextual cues. *International Journal of Community Well-Being, 4*(4), 467–503. https://doi.org/10.1007/s42413-021-00111-z

Wilson-Kennedy, Z. S., Payton-Stewart, F., & Winfield, L. L. (2020). Toward intentional diversity, equity, and respect in Chemistry research and practice. *Journal of Chemical Education, 97*(8), 2041–2044. https://doi.org/10.1021/acs.jchemed.0c00963

Winberg, C., Adendorff, H., Bozalek, V., Conana, H., Pallitt, N., Wolff, K., Olsson, T., & Roxå, T. (2019). Learning to teach STEM disciplines in higher education: A critical review of the literature. *Teaching in Higher Education, 24*(8), 930–947. https://doi.org/10.1080/13562517.2018.1517735

Wong, B., Copsey-Blake, M., & ElMorally, R. (2021). *The student experiences in STEM report: Minority ethnic students in higher education*. University of Reading.

Xie, Y., Fang, M., & Shauman, K. (2015). STEM Education. *Annual Review of Sociology, 41*(1), 331–357. https://doi.org/10.1146/annurev-soc-071312-145659

Xu, X., Zuo, X., Wang, X., & Han, S. (2009). Do you feel my pain? Racial group membership modulates empathic neural responses. *Journal of Neuroscience, 29*(26), 8525–8529. https://doi.org/10.1523/JNEUROSCI.2418-09.2009

Yudell, M., Roberts, D., Desalle, R., & Tishkoff, S. (2016). Taking race out of human genetics. *Science, 351*(6273), 564–565. https://doi.org/10.1515/9781478004363-026

Zuo, X., & Han, S. (2013). Cultural experiences reduce racial bias in neural responses to others' suffering. *Culture and Brain, 1*(1), 34–46. https://doi.org/10.1007/s40167-013-0002-4

Chapter 9
(Un)Diluting the Diaspora: Acknowledging Cultural Identity Intersections in Postsecondary STEM Education

Takeshia Pierre and Rose Pringle

Abstract With the increasing efforts to broaden participation in STEM, race and racism have become a topic now regularly discussed within STEM education and the subject of many research ventures. We posit that at this juncture, the use of a theoretical framework that effectively considers the Black identity, placement, and experience of Black undergraduate students pursuing STEM education is warranted. A review of the literature indicates that too often these students are essentialized into one singular [deficit] identity, therefore, diminishing the range of experiences of populations within the Black Diaspora. This chapter discusses challenges and barriers to diversifying STEM participation in the larger culture of STEM education and (re)define constructs and theoretical frameworks that have the potential to contribute to the greater STEM community. Specifically, we highlight Black immigrant populations, who for decades have occupied careers in STEM, and have matriculated through respective postsecondary programs that have prepared them for these careers. We discuss two key areas: (1) it is important to learn from the experiences of Black first- and second-generation immigrants in postsecondary spaces, and (2) essentializing the experiences of first- and second-generation immigrants in STEM education is problematic, stressing the importance of the Critical Race Theory tenet of intersectionality to nuance their experiences.

Keywords STEM · Black immigrants · Intersectionality · Culture · Postsecondary education

T. Pierre (✉)
Tufts University, Medford, MA, USA
e-mail: takeshia.pierre@tufts.edu

R. Pringle
University of Florida, Gainesville, FL, USA
e-mail: rpringle@coe.ufl.edu

G. L. Cochran, J. D. Adams (eds.), *Equity-Oriented, Discipline-Based STEM Education Research*, Critical Studies of Education 19,
https://doi.org/10.1007/978-981-96-9136-4_9

The literature surrounding Science, Technology, Engineering, and Mathematics (STEM) has invested a significant amount of time discussing the importance of and the need to increase diversity within the twenty-first century STEM workforce (Briggs, 2017; National Science Foundation, National Center for Science and Engineering Statistics [NSF, NCSES], 2021; PEW Research Center [PEWRC], 2021b). In response, funding has been allocated to support interventions to recruit and retain diverse populations through preparation efforts and studies across K-12 learning environments (Chapman & Feldman, 2017; Garvin-Hudson & Jackson, 2018; National Aeronautics and Space Administration, 2022; National Science Foundation, National Center for Science and Engineering Statistics, 2021). Additionally, efforts have been made to target Black undergraduate students in order to broaden participation in STEM and improve the statistics surrounding their under-representation (Burt et al., 2017; Corneille et al., 2020; Gasman et al., 2017; Maton et al., 2012). With the increasing efforts to broaden participation, race and racism have become a topic now regularly discussed within STEM education (Fries-Britt, 2017; National Academies of Sciences, Engineering, and Medicine [NASEM], 2020; Watkins & Mensah, 2019) and the subject of many research ventures (Burt et al., 2018, 2020; McGee, 2020; Morton et al., 2019; Nasir & Vakil, 2017). We posit that at this juncture, the use of a theoretical framework that effectively considers the Black identity, placement, and experience of Black undergraduate students pursuing STEM education is warranted. A review of the literature indicates that too often these students are essentialized into one singular [deficit] identity, therefore, diminishing the range of experiences of populations within the Black diaspora. This chapter will discuss challenges and barriers to diversifying STEM participation in the larger culture of STEM education and (re)define constructs and theoretical frameworks that have the potential to contribute to the greater STEM community. Specifically, we highlight Black immigrant populations, who for decades have occupied careers in STEM, and have matriculated through respective postsecondary programs that have prepared them for these careers. We discuss two key areas: (1) it is important to learn from the experiences of Black first- and second-generation immigrants in postsecondary spaces, and (2) essentializing the experiences of first- and second-generation immigrants in STEM education is problematic, stressing the importance of the Critical Race Theory tenet of intersectionality to nuance their experiences.

9.1 First- and Second-Generation Black Immigrants in the United States

Black immigrants make up roughly 10% of the U.S. Black population, showcasing the rapid growth seen in the Black immigrant population over the past two decades (PEWRC, 2021a). Similarly, the immigration of Black foreigners has increased by more than 2000% in the past six decades (Hamilton, 2020). Roughly 88% of Black immigrants were born in African or Caribbean nations and in 2016, 9% of Black people were second-generation Americans—where they were born in the United

States but had at least one foreign-born parent (PEWRC, 2022). It is projected that by the year 2060, the U.S. Black population will be one-third foreign-born, accounting for more than double the current level (2022). Second-generation Black immigrants combined with the 12% of Black people who were born directly in foreign countries (first-generation Black immigrants) currently account for 21% of the Black population (PEWRC, 2022). Given these demographic changes, we posit that research should continue to highlight and disaggregate data relevant to Black populations to display the richness of diversity within the Black Diaspora, trouble the notion that Blackness is a monolith, and showcase the educational implications that are seen in the experiences of various Black people in America.

9.2 Black Immigrants in Postsecondary Education

Over the past two decades, an increasing number of Black immigrants hold at least a bachelor's degree in the United States (PEWRC, 2022). The increase in the number of Black immigrant bachelor's degree holders is faster than an array of racial/ethnic groups in the United States, for instance, the growth is faster than that of the Black U.S.-born population (8 percentage points), the entire U.S.-born population (9 points), and the overall immigration population (9 points) over the same period (2022). Black immigrants from Africa have some of the highest educational attainment levels in the country, with almost 40% having at least a bachelor's degree, 16% having an advanced degree, and a greater likelihood compared to the U.S.-born population to hold a master's, professional, or doctorate level degree (New American Economy Research Fund [NAERF], 2018). Keller and Tillman (2008) reported that the odds of first-generation Black immigrants attending college were 27.8% higher than U.S.-born Whites.

More than 12% of all Black undergraduate students enrolled in U.S. colleges and universities are foreign-born and first- and second-generation immigrants. Also, while first- and second-generation Black immigrants account for only 13% of the 18- to 19-year-old Black population, Black immigrants represent 27% of Black students at selective colleges and 41% of Black students at Ivy League institutions (Kent, 2007; Massey et al., 2007). In a study conducted about Black undergraduates attending Harvard, the author found that 55% of the Black students attending Harvard had at least one parent or grandparent that was foreign-born and 41% of Black students at Harvard were second-generation immigrants (Haynie, 2002).

While the data surrounding Black immigrants showcases postsecondary successes and representation in STEM and health fields, Black immigrants continue to experience disparities mainly attributed to racism and an additional layer of discrimination due to holding immigrant identities (Burt et al., 2017; Berhane et al., 2020; Fries-Britt, 2014a; Sparks et al., 2019). Furthermore, it is reported that Black immigrants from non-English speaking countries face linguistic discrimination due to their English-speaking proficiency and bilingual status (Kalmijn, 1996; Kim, 2014; Thomas, 2012; Thomas & Lonobile, 2021) and even in cases where Black immigrants speak English fluently, for instance, from those who do hail from English-speaking countries, some face ostracization related to their accents (Akomolafe, 2020; Showers, 2015). Researchers in the United States have found that even high-achieving Black immigrant students are deemed intellectually inferior if they are English-language learners (George Mwangi & Fries-Britt, 2015) further exacerbating the often lowered expectations from teachers toward Black students (Gershenson et al., 2016). Despite these challenges, the data indicate that Black immigrants continue to attain levels of success in postsecondary educational settings positioning them to enter a variety of STEM-related fields (NAERF, 2018). In 2015, about three out of every ten employed African immigrant worked in the broader healthcare and social services industry, meaning they were more than twice as likely as the U.S. working population overall, to hold jobs in this field. Of this group, roughly 24% of African immigrant workers worked in nursing, psychiatry, and home health aides, with an additional 19% working as registered nurses and 5% as physicians or surgeons (NAERF, 2018).

9.3 Why Culture Is Capital: The Importance of Considering the Intersectionality of Black Immigrants in STEM

9.3.1 Significance of Black Immigration in STEM

Foreign-born workers account for 30% of science and engineering careers (National Science Board, National Science Foundation [NSB, NSF], 2020). Sub-Saharan African immigrants have higher levels of educational attainment than the U.S. population as a whole and are more likely to have earned their degree in a STEM field (NAERF, 2018). African immigrants account for about one in three, or 33.4%, of STEM degrees in the United States (2018).

In 2020, more than one-fifth of all STEM-related doctorates earned by Black students at U.S. universities went to foreign students (National Center for Science and Engineering Statistics, 2021). With Black immigrants attaining STEM degrees at greater rates than American-born Black people, and in some cases White populations (NAERF, 2018; PEWRC, 2022), it is now important to investigate the areas that have contributed to the success of Black immigrants. Consistent with researchers who have studied education among Black immigrants (Burt et al., 2017; Berhane et al., 2020; Fries-Britt et al., 2014a, 2014b; Hernandez & Murray-Johnson, 2015), we seek to further highlight the cultural components of their identity that have contributed to their successes. Subsequently, we assess how these populations have navigated these educational spaces to successfully matriculate in STEM.

A review of the literature indicates that immigrants that hold careers in STEM are likely to raise children who will also enter these fields (Thomas & Lonobile, 2021), leveraging the ability to diversify these careers. Thomas and Lonobile (2021) further discuss that the children of STEM graduates are less likely to fall behind in school than the children of graduates in Business, Arts/Humanities, and other fields of study, with this relative STEM advantage being stronger in immigrant families over the U.S born families. This finding highlights an academic edge held by immigrants that should be explored further in research. Additionally, the children of immigrant STEM graduates have more favorable outcomes, such as a low risk of falling behind in school, than the children of U.S.-born STEM graduates; however, the favorable outcomes of the former are more consistent for children whose parents have U.S. rather than foreign STEM degrees (2021).

Other researchers indicate that family and cultural expectations may contribute to Black immigrants' college matriculation, identifying their collectivist worldview as a motivating force to succeed in education (George Mwangi, 2018; George Mwangi et al., 2017; Griffin et al., 2012). This finding supports the importance of exploring how culture affects the STEM matriculation of Black immigrants.

9.3.2 *Black Immigrants' Experiences in STEM: Unearthing Forms of Cultural Capital*

Discussions surrounding Black Immigrants' STEM participation and engagement have increased in recent years, but their success stories are far between (Fries-Britt et al., 2014a, 2014b; Onuma et al., 2020). These discussions have been spurred on due to the recognition and the national call to broaden participation in STEM. A diverse and capable workforce is vital to maintaining the nation's standard of excellence in STEM-related fields thus maintaining America's global competitiveness (Hrabowski III, 2012). However, the challenges faced by Black Immigrants are often unique and have been well-documented. These include unpacking their new minority status in America and feelings of being othered in STEM spaces due to their ethnic identity and/or immigrant status (Burt et al., 2017; Berhane et al., 2020; Fries-Britt, 2014a; Sparks et al., 2019).

Literature has described additional takeaways specific to Black immigrants in STEM. Researchers detail findings that highlight aspects of culture that promote perseverance in Black immigrants (King Miller, 2017; Onuma et al., 2020; Sparks et al., 2019), dealings with racialized experiences in STEM spaces (Burt et al., 2017; Berhane et al., 2020; Fries-Britt et al., 2014a, 2014b; King Miller, 2017; Sparks et al., 2019), the self-actualization, discourse, or contesting of their identity intersections within or due to STEM environments in the U.S. context (Fries-Britt et al., 2014a, 2014b; King Miller, 2017; Sparks et al., 2019), and intra-group differences among Black people navigating STEM spaces (Berhane, 2017; Berhane et al., 2017; Leggett-Robinson et al., 2017). Subsequently, these discussions have revealed areas of potential cultural capital, as well as areas to further nuance through critical theories. While articles discuss an additional layer of oppression in relation to nativity status (Burt et al., 2017; King Miller, 2017; Sparks et al., 2019), literature discussing Black immigrant experiences in STEM reveals various forms of cultural capital that contribute to their matriculation in STEM and these forms of capital may help to explain their perseverance in these majors (Fries-Britt et al., 2014a; Onuma et al., 2020).

In 2005, Tara Yosso described these forms of cultural capital as follows: (1) aspirational, the ability to maintain hopes and dreams for the future, even in the face of real and perceived barriers, (2) navigational, the skills of maneuvering through social institutions; historically, this infers the ability to maneuver through institutions not created with Communities of Color in mind, (3) social, can be understood as networks of people and community resources; these peer and other social contacts can provide both instrumental and emotional support to navigate through society's institutions, (4) linguistic, includes the intellectual and social skills attained through communication experiences in more than one language and/or style (see Faulstich Orellana, 2003), (5) familial to those cultural knowledges nurtured among familia (kin) that carry a sense of community history, memory, and cultural intuition, and (6) resistant capital, knowledges, and skills fostered through oppositional behavior that challenges inequality.

While each form of cultural capital can be applied to the experiences of Black immigrants, we focus on forms of capital most prevalent in the literature describing Black immigrant experiences in STEM. As aforementioned, Black immigrants tend to hold collectivist views that encourage their college matriculation (George Mwangi, 2018; George Mwangi et al., 2017; Griffin et al., 2012). These cultural values can be defined as familial capital, where researchers discussed parental sacrifice and strategy motivated first- and second-generation Black immigrants to "pay it forward" by performing well in school and matriculating specifically in STEM majors (Onuma et al., 2020; King Miller, 2017; Sparks et al., 2019). The documentation from these researchers reveals that past and present Black immigrant students expressed that their parents worked blue-collar jobs to serve as "steppingstones" for better positions and to afford the best education for their children. Black immigrant students see attending college as part of a family investment (George Mwangi, 2018; George Mwangi et al., 2017) and described educational achievement as cultural norms embedded in their ethnic and/or immigrant identities (Griffin et al., 2012; Griffin & George Mwangi, 2020).

In addition to familial capital, Black immigrants showcase aspirational capital in their STEM education pursuits. Researchers note how first-generation Black immigrants reportedly leave their home countries to continue in STEM fields (Burt et al., 2020; Onuma et al., 2020; Sparks et al., 2019) sacrificing the comfort of their homeland and proximity to their family for educational attainment. Resistant capital can also be showcased in Black immigrants' plight to attain their STEM degrees. Despite the discrimination Black immigrants face for holding immigrant identities (Burt et al., 2017; King Miller, 2017; Sparks et al., 2019), they continue to make their mark in STEM careers (NAERF, 2018; PEWRC, 2022). While we celebrate the representation of Black immigrants in these spaces, we also trouble the racist structures in STEM education that make their journey to success difficult.

9.4 Eu[ro]thanization of Blackness in STEM Education

9.4.1 Deidentifying the Black Immigrant: From Culture to Color

The climate surrounding STEM education in recent years has gained considerable traction, particularly in reference to the hostility and unwelcoming nature it provides for minoritized populations such as women and Black people (NASEM, 2020). The image of STEM spaces has been perpetuated as an individualistic and largely exclusive space occupied by the stereotypical elite White male scientist (Johnson et al., 2011; Kachchaf et al., 2015; Ong et al., 2018). While studies have worked to shed light on the salience of hegemony and anti-Black, anti-female properties of white science within these spaces, cultural and nativity identity are largely left out of the conversation. For example, in a study focused on cross-cultural dynamics of Black native and foreign-born collegians in STEM, George Mwangi et al. (2016) contend that researchers must acknowledge the intraracial diversity of Black students, as those who are foreign-born detail that their nativity status ultimately shapes their experiences in academic programs. Authors state "understanding diversity among students based on nativity status will contribute to the existing discourse on diverse learning environments" (p. 790). We argue that the essentializing of the Black first- and second-generation immigrant experience, silencing their cultural intersections, further reifies White supremacy in STEM education and STEM-related workspaces.

Essentializing Black immigrants gives the notion that within American culture, and arguably science spaces, Blackness and Black identity must be palpable and conceptualizable by dominant groups. Categorizing intra-racial groups into one singular identity silences the experiences, cultures, customs, and personhood of Black immigrants in order to appease the standards set within the United States. In other words, the erasure of Black immigrants' cultural identity furthers society's perverse need to conquer Black bodies and force conformity to white standards that are considered "acceptable." Science and STEM spaces serve as an exclusive space that if Black people occupy, must uphold what the late Jamaican philosopher Charles Mills describes as the racial contract (1997). The racial contract refers to an invisible contractual agreement in society that requires the minoritized to be considered subordinate, and the spaces they occupy are not acknowledged until they are occupied by Whites—and even then, Black "presence is a form of absence" (p. 51). Spaces, under contractual agreement, must be normed and raced, and at the micro level, spaces refer to the non-White body. Black bodies are not held to a standard of value in white spaces, so the cultural identity of Black immigrants is also inconsequential. Mills goes on to share what Kenyan writer Ngugi wa Thiong'o describes as a "cultural bomb." wa Thiong'o (1986) states:

> …a cultural bomb is to annihilate a people's belief in their names, in their languages, in their environment, in their heritage of struggle, in their unity, in their capabilities and ultimately in themselves. It makes them see their past as one wasteland of non-achievement and it makes them want to distance themselves from that wasteland (pp. 3–5).

Although literature mentions that Black immigrant students are at first resistant to acknowledging racialized incidences experienced in STEM environments, the "cultural bomb," is eventually detonated, and the pressures of consistent racialized experiences in STEM education force them to consider their racial standing (subordination) within the United States. (Berhane et al., 2020; Fries-Britt et al., 2014a). Black immigrant student identity is often met with much resistance, where definitions of race and identity are enforced by a Black-White dichotomy, and emphasis on cultural identity may result in further subordination (Fries-Britt et al., 2014a; Sparks et al., 2019). This action removes the ability for Black immigrants to properly name and identify themselves, which should be a right afforded to all within STEM spaces (Rivera Maulucci & Mensah, 2015).

9.4.2 Epistemological Ethnocentrism: Discrediting the Origin of Knowledge Production

In addition, an essentialized view of Blackness stabilizes the notion that White, European science is the only science that exists. Literature has interrogated the ways in which White supremacy has been manifested in the teaching of "white science" (Dunac & Demir, 2017), but it is important to further unpack what this messaging can and has conveyed to Black immigrants. There is an additional layer of subordination associated with the Black immigrant that purports a negative narrative of the "underprivileged immigrant" from low socioeconomic backgrounds (Sparks et al., 2019)—though some studies refute this narrative (Onuma et al., 2020; NAERF, 2018)—automatically deeming their preparation for the rigor of an American education as hopeless. Naturally through the deficit-oriented perceptions of capability in STEM education already ascribed to Blackness (NASAM, 2020), an education outside of America is assumed to be subordinate, because they, in theory, have not learned "white science." Black people are not afforded the privilege (or skillset) to produce scientific knowledge and must learn to ascribe to the culture and knowledge production of white science.

Mills (1997) explains this further in reference to the "norming" dimensions of space within the racial contract, stating:

> The epistemological dimension is the consequence of the preemptive restriction of knowledge to European cognizers, which implies that in certain spaces real knowledge (knowledge of science, universals) is not possible. Significant cultural achievement, intellectual progress, is thus denied to those spaces, which are deemed (failing European intervention) to be permanently locked into a cognitive state of superstition and ignorance… "epistemological ethnocentrism"… [countervailing evidence] may be explained away as resulting from the intervention of outsiders, for example from a previously unknown contact with whites: since Africans would produce nothing of value; the technique of Yoruba statuary must have come from Egyptians; Benin art must be a Portuguese creation; the architectural achievement of Zimbabwe was due to Arab technicians, and Hausa and Buganda statecraft were inventions of white invaders (p. 44).

Historically science outside of European origin faces erasure and stories of the empires, wealth, and knowledge production from countries and continents of color are not taught in the American education system (Rosa & Mensah, 2021). The United States has a way of stripping the culture of Black people and repeating this method of coloniality whenever a Black immigrant enters the STEM space. Norming, racing, and establishing social order through the Black-White, superior–Subordinate dichotomy must be enforced in the newly occupied space. Leonardo (2004) emphasizes this "cultural imperialism" where those who are othered, are constructed, controlled, and written into inferiority, and must refer to the White, dominant frame of reference for civilization, moral development, and rationality (p. 146). Discrediting the origin of knowledge production is not alien to the euro-centric standard of education, and the perpetuation of racism continues when those who are "newly othered" in STEM spaces, where Black immigrants mention the hypervisibility of race as a new experience (Fries-Britt et al., 2014a; Sparks et al., 2019), requires the particular attention that can most effectively be achieved through a lens of intersectionality.

9.5 Where Race and Culture Converge: Broadening Conversations of Intersectionality

9.5.1 Origins of Intersectionality and Cultural Introductions

In 1989, Kimberle Crenshaw made a groundbreaking contribution to Critical Race Theory with the introduction of a theory called intersectionality. Serving as an additional tenet of CRT, intersectionality highlighted the dual marginalization of Black women, arguing that the combined effect of these identity categories (race and gender), is difficult to disaggregate when considering discriminatory actions or the societal placement of Black women. Crenshaw offered legal case proceedings to substantiate claims that prevented Black women from delineating the discriminatory motivations of their employers were solely determined by their race or gender. Crenshaw expanded this work in 1991, further expounding upon the violence brought upon Women of Color and alluded to other areas of intersectionality such as culture and nativity status. In this work, Crenshaw describes the cultural barriers immigrant Women of Color face in terms of their standing in the United States when experiencing marital violence, and the inability to advocate for themselves for fear of deportation and displacement due to their lack of citizenship. She also describes the many areas of intersectionality that incorporate social class, ability, and sexuality.

In 2013, authors expressed how intersectionality has since been embraced globally, where international discourses have been taken up beyond the North American context, highlighting the convergence of cultural components of identity newly incorporated within this theory (Carbado et al., 2021). Literature has continued to evolve beyond legal studies, utilizing intersectionality across other disciplines, as well as assessing marginalization beyond race and gender, with the inclusion of culture. Examples came in the form of advocating for Asian immigrant women advocates (Chun et al., 2013), addressing Asian high school male students' language and gender roles (Qin & Li, 2020), assessing the experiences of Afro-Caribbean immigrant women navigating higher education (Esnard and Cobb-Roberts), and eventually making its way into STEM education through the experiences of Women of Color (Blosser, 2020; Jones, 2019; Ong et al., 2018; Morton & Parsons, 2018). While some have used intersectionality as a theoretical framework, other authors have adapted the theory to describe the experiences of Black students in STEM or used intersectionality as an "approach," to unpack the plight of those minoritized in these same spaces (O'Brien et al., 2015; Sparks et al., 2019). For example, O'Brien and colleagues (2015) conducted a series of quantitative studies observing ethnic variation in gender-related stereotypes and STEM participation within U.S. university contexts. The intersectional approach accounted for the comparisons made between STEM majors that were identified as African American women and European women.

As of recent, the incorporation of intersectionality has rightfully made its way into STEM education to explain the dual marginalization of Women of Color, while also considering the underrepresentation of women and People of Color within STEM fields (Blosser, 2020; Jones, 2019; Ong et al., 2018; Morton & Parsons, 2018). Though we champion the continued use of intersectionality to explain the race and gender bias within STEM environments, we propose further utilization of culture as an intersection, to help explicate the unique cultural experiences of first- and second-generation immigrants maneuvering postsecondary STEM education, regardless of gender identity.

Culture has also been discussed in STEM, but mainly to explain aspects of communalism within Historically Black Colleges and Universities (HBCU) culture (Gasman et al., 2017; Jett, 2013; Lockett et al., 2018), and how STEM has created a "culture of exclusion" by ways of racism and sexism (Blosser, 2020; Ong et al., 2018; Jones, 2019). In addition, cultural experiences have been minorly discussed in STEM spaces, but usually not in terms of the Black immigrant, with literature largely highlighting the unique cultural experiences and immigration status of Hispanic/Latinx students in higher education and STEM spaces (Locke & Gonzalez, 2020; Pietri et al., 2019). Discussions of anti-essentialism have also been introduced into STEM, specifically through problematizing STEM policy's habit of grouping underrepresented minoritized (URM) students as one collective of "minorities" rather than expressing their diverse and unique concerns (Basile & Lopez, 2015). While this interrogation of essentialism is important, the intra-racial Black student experiences are not effectively highlighted. The incorporation of culture as an intersection within STEM education seems promising, but a deeper analysis of the experiences of Black immigrant first- and second-generation students has been largely understudied.

9.6 Future Implications

9.6.1 Importance of Intersectionality

Exploring the intersectionality of Black immigrants invites different voices and experiences that are not otherwise considered in society as well as STEM environments. Discussing Black cultural differences through a lens of intersectionality shows signs of inclusivity within STEM education and refutes and resists the deficit-based perceptions of immigration, acknowledges male marginalization in STEM environments, and rejects the essentialism of Black populations.

It has been reported that by the third generation, immigrants are educationally and financially less successful than their first- and second-generation predecessors (NASEM, 2015; Waters, 1994). Devoting further research to explore the implication of race and racism of Black immigrants can help to understand more about what causes the third generation decline and promote possible interventions within educational settings. These discussions also open up possibilities for the future of education across the Black Diaspora. There are examples outside of white science that allow us to look at race and Blackness to not only include Black students, but to offer them a counter-narrative to a euro-centric science. Considering culture within STEM and science history of Black people can allow them to see themselves in STEM education and potentially as scientists.

Essentializing the experiences of Black people prevents researchers from understanding important factors relative to cultural identity that are not as clear when theoretical approaches that center race and culture in STEM education are absent. For example, familial impact may look different for Black people pursuing STEM, whereas, for example, first-generation Black immigrants may persist in STEM because they left their country of origin to pursue a lucrative career in hopes of securing financial stability for themselves and their families (Burt et al., 2020; Sparks et al., 2019), and in contrast, American-born Black students may persist in STEM education because they were inspired by a parent that entered a career in STEM (Fries-Britt, 2017; McGee & Spencer, 2015; Morton & Parsons, 2018; Ortiz et al., 2019; Strayhorn, 2015). While family is a motivating factor in both cases, the cultural nuances are important. The key to determining factors that are unique to the ethnicity, cultural values, and/or traditions of Black people that may contribute to their success in STEM careers, is where culture and race intersect and as emphasized in Critical Race Theory (CRT). The CRT tenet of intersectionality allows us to understand the predicament of those who live at the intersection of two or more categories.

Bearing this in mind, research has supported the importance of understanding the cultural diversity of Black immigrant students and connecting life experiences in the classroom through culturally responsive practices (Kumi-Yeboah et al., 2021). STEM curriculum is traditionally euro-centric (Dunac & Demir, 2017; Rosa & Mensah, 2021) and the omission of contributions of Black people showcases a narrative that their impact on science does not exist. This begs the question of if Black people have not supposedly contributed to scientific knowledge, what is their value to society? Learning about culturally tailored science topics, and cultures outside of the United States, and allowing students to learn that there was science before slavery can do much for American-born Black students across the Black Diaspora (Dunac & Demir, 2017; Garvin-Hudson & Jackson, 2018; Rosa & Mensah, 2021), improving students' self-image and belief that they can too, engage in science.

The utilization of intersectionality is a great construct to unpack the racial and cultural identities of Black first- and second-generation immigrants. Intersectionality allows for nuances relative to ethnicity and culture to be unearthed in research devoted to Black immigrants' experiences of success in STEM. We must consider as Scholars the impact of culture and race on individuals who come to this country and have the aspiration and potential to pursue an education in STEM. When entering the United States, Whiteness ingrained in the education system can mold and essentialize students' Black identity, as they have entered the United States already having a sense of identity and self (Berhane et al., 2020; Hernandez & Murray-Johnson, 2015). A piece of their lives is lost when essentialized and silenced, and as seen from literature, from generation to generation they repeat the cycle of conforming and assimilating to what the U.S. prescribes as Black identity (Stepick et al., 2003; Waters, 1994). This cycle of conforming and assimilating to a Black identity that is conceptualizable to a hegemonic society further supports the importance of guarding against the essentialism of Blackness within the U.S. and STEM spaces. There is an overall assumption that has been debunked, where the belief has been that a U.S. education is superior—this notion is not proving to be true considering that Black immigrants, who arrive in the United States during their higher education studies, are matriculating into STEM careers. Exploring the intersectionality of Black immigrants invites different voices and experiences that are not otherwise considered in society as well as STEM environments.

Research should be conducted to understand the importance of exploring how race and ethnicity have contributed to Black undergraduate STEM matriculation, rather than othering those identity intersections. Significant data has supported that Black immigration has had economic benefits to the United States, as Black immigrants see much financial success in their careers, but also how Black immigrants enter STEM careers in large numbers (NAERF, 2018).

A review of the literature reveals that an area that is largely understudied is the matriculation of second-generation immigrants into STEM careers. Second-generation Black immigrants have experienced an American education system their whole lives, being more at risk to experience the racism prevalent in the United States, and hold similar values and upbringings to their parents' culture (Stepick et al., 2003; Waters, 1994). Considering the profound impact discrimination has had on student learning (McGee & Bentley, 2017; Nasir et al., 2017; Ross et al., 2021), research must also be expanded to consider the implications of racial tensions and the emotional and mental health of second-generation Black immigrants. This is especially important because second-generation Black immigrants are largely represented in the Black population attending postsecondary institutions and STEM majors (Haynie, 2002; Kent, 2007; Massey et al., 2007; Thomas & Lonobile, 2021) and racism and discrimination experienced in STEM spaces attribute to the attrition of minoritized students (Ong et al., 2018; McGee & Bentley, 2017; Ross et al., 2021).

Discussing Black cultural differences through a lens of intersectionality shows signs of inclusivity within STEM education, refutes and resists the deficit-based perceptions of immigration, and rejects essentialism. This theoretical framework effectively considers the Black identity, placement, and experiences of Black immigrants pursuing STEM education. As shown through this chapter, future research could benefit from employing CRT to center race and culture, redefining Blackness outside of the hegemonic structures of education within the STEM fields.

References

Akomolafe, O. (2020). African-born scholars and accent discrimination: The last acceptable form of prejudice. In S. O. Abidde (Ed.), *African scholars and intellectuals in North American Academies* (pp. 199–226). Routledge. https://doi.org/10.4324/9780429202537-17

Basile, V., & Lopez, E. (2015). And still I see no changes: Enduring views of students of color in science and mathematics education policy reports. *Science Education, 99*(3), 519–548. https://doi.org/10.1002/sce.21156

Berhane, B. T. (2017). Networking 101: Exploring within-group differences between high achieving Black African and Black American engineering community college student peer groups. *Journal of African American Males in Education, 8*, 67–91.

Berhane, B., Onuma, F., & Secules, S. (2017). *Rethinking engineering pathways: An exploration of the diverse K–12 school experiences of six Black engineering undergraduates*. Paper presented at the 2017 ASEE Annual Conference and Exposition: American Society for Engineering Education, Columbus, OH.

Berhane, B., Secules, S., & Onuma, F. (2020). Learning while Black: Identity formation and experience for five Black men who transferred into engineering undergraduate programs. *Journal of Women and Minorities in Science and Engineering, 26*(2), 93–124. https://doi.org/10.1615/JWomenMinorScienEng.2020024994

Blosser, E. (2020). An examination of Black women's experiences in undergraduate engineering on a primarily white campus: Considering institutional strategies for change. *Journal of Engineering Education, 109*(1), 52–71. https://doi.org/10.1002/jee.20304

Briggs, C. (2017). THE POLICY OF STEM DIVERSITY: Diversifying STEM programs in higher education. *Journal of STEM Education, 17*(4), 5–7. Laboratory for Innovative Technology in Engineering Education (LITEE).

Burt, B. A., Knight, A., & Robeson, J. (2017). Racializing experiences of foreign-born and ethnically diverse Black male engineering graduate students: Implications for student affairs practice, policy, and research. *Journal of International Students, 7*(4), 925–943. https://doi.org/10.5281/zenodo.1029487

Burt, B. A., Williams, K. L., & Smith, W. A. (2018). Into the storm: Ecological and sociological impediments to Black males' persistence in engineering graduate programs. *American Educational Research Journal, 55*(5), 965–1006. https://doi.org/10.3102/0002831218763587

Burt, B. A., Roberson, J. J., Johnson, J. T., & Bonanno, A. (2020). Black men in engineering graduate programs: A theoretical model of the motivation to persist. *Teachers College Record, 122*, 58.

Carbado, D. W., Williams Crenshaw, K., Mays, V. M., & Tomlinson, B. (2021). INTERSECTIONALITY mapping the movements of a theory. *Du Bois Review, 10*(2), 303–312. https://doi.org/10.1017/S1742058X13000349

Chapman, A., & Feldman, A. (2017). Cultivation of science identity through authentic science in an urban high school classroom. *Cultural Studies of Science Education, 12*(2), 469–491. https://doi.org/10.1007/s11422-015-9723-3

Chun, J. J., Lipsitz, G., & Shin, Y. (2013). Intersectionality as a social movement strategy: Asian immigrant women advocates. *Journal of Women in Culture and Society, 38*(4), 917–940. https://doi.org/10.1086/669575

Corneille, M., Lee, A., Harris, K. N., Jackson, K. T., & Covington, M. (2020). Developing culturally and structurally responsive approaches to STEM education to advance education equity. *Journal of Negro Education, 89*(1), 48–57. https://doi.org/10.7709/JNEGROEDUCATION.89.1.0048

Crenshaw, K. (1989). *Demarginalizing the intersection of race and sex: A Black feminist critique of antidiscrimination doctrine, feminist theory and antiracist politics*. University of Chicago Legal Forum. Retrieved from http://chicagounbound.uchicago.edu/uclf; http://chicagounbound.uchicago.edu/uclf/vol1989/iss1/8

Crenshaw, K. (1991). *Stanford law review mapping the margins: Intersectionality, identity politics, and violence against women of. Source: Stanford Law Review* (Vol. 43). Retrieved from https://about.jstor.org/terms

Dunac, P. S., & Demir, K. (2017). Negotiating white science in a racially and ethnically diverse United States. *Educational Review, 69*(1), 25–50. https://doi.org/10.1080/00131911.2016.1150255

Faulstich Orellana, M. (2003). *In other words: en otras palabras: Learning from bilingual kids' translating/interpreting experiences*. School of Education and Social Policy, Northwestern University.

Fries-Britt, S. (2017). It takes more than academic preparation: A nuanced look at Black male success in STEM. *Journal of African American Males in Education (JAAME), 8*(1), 6–22.

Fries-Britt, S., George Mwangi, C. A., & Peralta, A. M. (2014a). Learning race in a U.S. context: An emergent framework on the perceptions of race among foreign-born students of color. *Journal of Diversity in Higher Education, 7*(1), 1–13. https://doi.org/10.1037/a0035636

Fries-Britt, S., Mwangi, C. A. G., & Peralta, A. M. (2014b). The acculturation experiences of foreign-born students of color in physics. *Journal of Student Affairs Research and Practice, 51*(4), 459–471. https://doi.org/10.1515/jsarp-2014-0045

Garvin-Hudson, B., & Jackson, T. O. (2018). A case for culturally relevant science education in the summer for African American youth. *International Journal of Qualitative Studies in Education, 31*(8), 708–725. https://doi.org/10.1080/09518398.2018.1478156

Gasman, M., Nguyen, T. H., Lundberg, T., Conrad, C. F., & Commodore, F. (2017). Black male success in STEM: A case study of morehouse college. *Journal of Diversity in Higher Education, 10*(2), 181–200. https://doi.org/10.1037/dhe0000013

George Mwangi, C. A. (2018). A family affair: Examining the college-going strategies of sub-Saharan African immigrants in the U.S. through funds of knowledge. In J. Kiyama & C. Rios-Aguilar (Eds.), *Funds of knowledge in higher education: Honoring students' cultural experiences and resources as strength* (pp. 10–124). Routledge.

George Mwangi, C. A., & Fries-Britt, S. (2015). Black within Black: The perceptions of Black immigrant collegians and their U.S. *College Experience. About Campus: Enriching the Student Learning Experience, 20*(2), 16–23. https://doi.org/10.1002/abc.21187

George Mwangi, C. A., Fries-Britt, S., Peralta, A. M., & Daoud, N. (2016). Examining intraracial dynamics and engagement between native-born and foreign-born Black collegians in STEM. *Journal of Black Studies, 47*(7), 773–794. https://doi.org/10.1177/0021934716658860

George Mwangi, C. A., Daoud, N., English, S., & Griffin, K. A. (2017). Me and my family: Ethnic differences and familial influences on academic motivations of Black collegians. *Journal of Negro Education, 86*(4), 479–493.

Gershenson, S., Holt, S. B., & Papageorge, N. W. (2016). *Who believes in me? The effect of student-teacher demographic match on teacher expectations*. IZA Discussion Papers, 9202. Retrieved from https://www.econstor.eu/handle/10419/114079

Griffin, K., & George Mwangi, C. A. (2020). Similar, but not the same: Considering the intersections of race, ethnicity, and immigrant status in the lives of Black students. In R. T. Teranishi, B. M. D. Nguyen, C. M. Alcantar, & E. R. Curammeng (Eds.), *Measuring race: Why disaggregating data matters for addressing educational inequality* (pp. 103–118). Teachers College Press.

Griffin, K., del Pilar, W., McIntosh, K., & Griffin, A. (2012). "Oh, of course I'm going to go to college": Understanding how habitus shapes the college choice process of Black immigrant students. *Journal of Diversity in Higher Education, 5*(2), 96–111.

Hamilton, T. G. (2020). Black immigrants and the changing portrait of Black America. *Annual Review of Sociology, 46*(1), 295–313. https://doi.org/10.1146/annurev-soc-121919-054728

Haynie, A. C. (2002). Not 'just Black' policy considerations: The influence of ethnicity on pathways to academic success amongst Black undergraduates at Harvard University. *Journal of Public and International Affairs, 13*, 40–62.

Hernandez, K.-A. C., & Murray-Johnson, K. K. (2015). Towards a different construction of blackness: Black immigrant scholars on racial identity development in the United States. *International Journal of Multicultural Education, 17*(2), 53–72.

Hrabowski, F. A., III. (2012). Broadening participation in the American STEM workforce. *BioScience, 62*(4), 325–326.

Jett, C. C. (2013). HBCUs Propel African American male mathematics majors. *Journal of African American Studies, 17*(2), 189–205. https://doi.org/10.1007/s12111-011-9194-x

Johnson, A., Brown, J., Carlone, H., & Cuevas, A. K. (2011). Authoring identity amidst the treacherous terrain of science: A multiracial feminist examination of the journeys of three women of color in science. *Journal of Research in Science Teaching, 48*(4), 339–366. https://doi.org/10.1002/tea.20411

Jones, T. C. (2019). Creating a World for Me: Students of Color Navigating STEM identity. *Journal of Negro Education, 88*(3), 358–378.

Kachchaf, R., Ko, L., Hodari, A., & Ong, M. (2015). Career-life balance for women of color: Experiences in science and engineering academia. *Journal of Diversity in Higher Education, 8*(3), 175–191. https://doi.org/10.1037/a0039068

Kalmijn, M. (1996). The socioeconomic assimilation of Caribbean American Blacks. *Social Forces, 74*(3), 911–930. Retrieved from https://about.jstor.org/terms

Keller, U., & Tillman, K. H. (2008). Post-secondary educational attainment of immigrant and native youth. *Social Forces, 87*(1), 121–152.

Kent, M. M. (2007). Immigration and America's Black population. *Population Bulletin, 62*(4), 3–16.

Kim, E. (2014). Bicultural socialization experiences of Black immigrant students at a predominantly white institution. *The Journal of Negro Education, 83*(4), 580–594. https://doi.org/10.7709/jnegroeducation.83.4.0580

King Miller, B. A. (2017). Navigating STEM: Afro Caribbean women overcoming barriers of gender and race. *SAGE Open, 7*(4), 215824401774268. https://doi.org/10.1177/2158244017742689

Kumi-Yeboah, A., Onyewuenyi, A. C., & Smith, P. (2021). Teaching black immigrant students in urban schools: Teacher and peer relationships and academic performances. *Urban Review, 53*(2), 218–242. https://doi.org/10.1007/s11256-020-00570-2

Leggett-Robinson, P. M., Villa, B., & Davis, N. C. (2017, June). *Native-born and foreign-born Black students in STEM: Addressing STEM identity and belonging barriers and their effects on STEM retention and persistence at the two-year college*. Paper presented at the 2017 ASEE Annual Conference and Exposition: American Society for Engineering Education, Columbus, OH.

Leonardo, Z. (2004). The color of supremacy: Beyond the discourse of 'white privilege.'. *Educational Philosophy and Theory, 36*(2), 137–152. https://doi.org/10.1111/j.1469-5812.2004.00057.x

Locke, L. A., & Gonzalez, C. (2020). Latina dreamers, the current political era, and what educational leaders need to know. *Peace and Conflict, 26*(4), 365–378. https://doi.org/10.1037/pac0000435

Lockett, A. W., Gasman, M., & Nguyen, T.-H. (2018). Senior level administrators and HBCUs: The role of support for Black Women's success in STEM. *Education Sciences, 8*(2), 12. https://doi.org/10.3390/educsci8020048
Massey, D. S., Mooney, M., Torres, K. C., & Charles, C. Z. (2007). Black immigrants and Black natives attending selective colleges and universities in the United States. *American Journal of Education, 113*(2), 243–271.
Maton, K. I., Pollard, S. A., McDougall Weise, T. V., & Hrabowski, F. A. (2012). Meyerhoff scholars program: A strengths-based, institution-wide approach to increasing diversity in science, technology, engineering, and mathematics. *Mount Sinai Journal of Medicine: A Journal of Translational and Personalized Medicine, 79*(5), 610–623. https://doi.org/10.1002/msj.21341
McGee, E. O. (2020). Interrogating structural racism in STEM higher education. *Educational Researcher, 49*(9), 633–644. https://doi.org/10.3102/0013189X20972718
McGee, E., & Bentley, L. (2017). The equity ethic: Black and Latinx college students reengineering their STEM careers toward justice. *American Journal of Education, 124*(1), 1–36. https://doi.org/10.1086/693954
McGee, E., & Spencer, M. B. (2015). Black parents as advocates, motivators, and teachers of mathematics. *Journal of Negro Education, 84*(3), 473–490. https://doi.org/10.7709/jnegroeducation.84.3.0473
Mills, C. W. (1997). *The racial contract*. Cornell University Press.
Morton, T. R., & Parsons, E. C. (2018). #BlackGirlMagic: The identity conceptualization of Black women in undergraduate STEM education. *Science Education, 102*(6), 1363–1393. https://doi.org/10.1002/sce.21477
Morton, T. R., Gee, D. S., & Woodson, A. N. (2019). Being vs. becoming: Transcending STEM identity development through afropessimism, moving toward a Black X consciousness in STEM. *The Journal of Negro Education, 88*(3), 327–342.
Nasir, N. S., & Vakil, S. (2017). STEM-focused academies in urban schools: Tensions and possibilities. *Journal of the Learning Sciences, 26*(3), 376–406. https://doi.org/10.1080/10508406.2017.1314215
Nasir, N. S., McKinney de Royston, M., O'Connor, K., & Wischnia, S. (2017). Knowing about racial stereotypes versus believing them. *Urban Education, 52*(4), 491–524. https://doi.org/10.1177/0042085916672290
National Academies of Sciences, Engineering, and Medicine. (2015). *The integration of immigrants into American Society*. The National Academies Press. https://doi.org/10.17226/21746
National Academies of Sciences, Engineering, and Medicine. (2020). *The impacts of racism and bias on Black people pursuing careers in science, engineering, and medicine: Proceedings of a workshop*. Washington, DC: The National Academies Press. https://doi.org/10.17226/25849
National Aeronautics and Space Administration. (2022). *Space grant K-12 inclusiveness and diversity in STEM (SG KIDS)*. Retrieved from https://www.grantsoffice.com/GrantDetails.aspx?gid=60722
National Center for Science and Engineering Statistics (NCSES). (2021). *Doctorate recipients from U.S. universities: 2020. NSF 22-300*. National Science Foundation. Retrieved from https://ncses.nsf.gov/pubs/nsf22300
National Science Board, National Science Foundation. (2020). *Science and engineering indicators 2020: The State of U.S. science and engineering. NSB-2020-1*. Retrieved from https://ncses.nsf.gov/pubs/nsb20201/
National Science Foundation, National Center for Science and Engineering Statistics. (2021). Women, minorities, and persons with disabilities in science and engineering: 2021. Special Report NSF 21-321. Alexandria, VA. Available at https://www.nsf.gov/statistics/wmpd
New American Economy Research Fund. (2018). *Power of the purse: How Sub-Saharan Africans contribute to the U.S. economy*. Retrieved from http://research.newamericaneconomy.org/wpcontent/uploads/sites/2/2018/01/NAE_African_V6.pdf
O'Brien, L. T., Blodorn, A., Adams, G., Garcia, D. M., & Hammer, E. (2015). Ethnic variation in gender-STEM stereotypes and STEM participation: An intersectional approach. *Cultural Diversity and Ethnic Minority Psychology, 21*(2), 169–180. https://doi.org/10.1037/a0037944

Ong, M., Smith, J. M., & Ko, L. T. (2018). Counterspaces for women of color in STEM higher education: Marginal and central spaces for persistence and success. *Journal of Research in Science Teaching, 55*(2), 206–245. https://doi.org/10.1002/tea.21417

Onuma, F. J., Berhane, B., & Fries-Britt, S. L. V. (2020). "I've always been in private school": The role of familial norms and supports in Black immigrant students' preparation for STEM majors. *Journal of Diversity in Higher Education*. https://doi.org/10.1037/dhe0000285

Ortiz, N. A., Morton, T. R., & Roby, R. S. (2019). What about us? Exploring the challenges and sources of support influencing Black students' STEM identity development in postsecondary education. *The Journal of Negro Education, 88*(3), 311–326.

Pew Research Center. (2021a). *Key findings about Black America*. Retrieved from https://www.pewresearch.org/short-reads/2023/02/10/key-facts-about-black-americans/

Pew Research Center. (2021b). *STEM jobs see uneven progress in increasing gender, racial and ethnic diversity*. Retrieved from https://www.pewresearch.org/science/2021/04/01/stem-jobs-seeuneven-progress-in-increasing-gender-racial-and-ethnic-diversity/

Pew Research Center. (2022). *Key findings about Black immigrants in the U.S.* Retrieved from https://www.pewresearch.org/short-reads/2022/01/27/key-findings-about-blackmmigrants-in-the-u-s/

Pietri, E. S., Drawbaugh, M. L., Lewis, A. N., & Johnson, I. R. (2019). Who encourages Latina women to feel a sense of identity-safety in STEM environments? *Journal of Experimental Social Psychology, 84*, 103827. https://doi.org/10.1016/j.jesp.2019.103827

Qin, K., & Li, G. (2020). Understanding Immigrant Youths' negotiation of racialized masculinities in One U.S. High School: An intersectionality lens on race, gender, and language. *Sexuality and Culture, 24*(4), 1046–1063. https://doi.org/10.1007/s12119-020-09751-3

Rivera Maulucci, M. S., & Mensah, F. M. (2015). Naming ourselves and others. *Journal of Research in Science Teaching, 52*(1), 1–5. https://doi.org/10.1002/tea.21196

Rosa, K., & Mensah, F. M. (2021). Decoloniality in STEM research: (re)framing success. *Cultural Studies of Science Education, 16*(2), 501–508. https://doi.org/10.1007/s11422-020-10008-6

Ross, M. S., Huff, J. L., & Godwin, A. (2021). Resilient engineering identity development critical to prolonged engagement of Black women in engineering. *Journal of Engineering Education, 110*(1), 92–113. https://doi.org/10.1002/jee.20374

Showers, F. (2015). Being black, foreign and woman: African immigrant identities in the United States. *Ethnic and Racial Studies, 38*(10), 1815–1830. https://doi.org/10.1080/01419870.2015.1036763

Sparks, D. M., Nandakumar, V., & Libii, J. N. (2019). "We are shaped by our experiences" intersectionality and the African international STEM student. *Research in the Schools, 26*(2), 17–31.

Stepick, A., Grenier, G., Castro, M., & Dunn, M. (2003). *This land is our land: Immigrants and power in Miami*. University of California Press.

Strayhorn, T. L. (2015). Factors influencing Black males' preparation for college and success. *Western Journal of Black Studies, 39*(1), 45. Spring.

Thomas, K. J. A. (2012). *A demographic profile of Black Caribbean immigrants in the United States*. Retrieved from www.migrationpolicy.org

Thomas, K. J. A., & Lonobile, C. (2021). Parental STEM credentials and children's schooling progress in immigrant and U.S. born families. *Social Science Research, 100*, 102601. https://doi.org/10.1016/j.ssresearch.2021.102601

wa Thiong'o, N. (1986). *Decolonising the mind: The politics of language in African literature*. Heinemann.

Waters, M. C. (1994). Ethnic and racial identities of second-generation Black immigrants in New York City. *The International Migration Review, 28*(4), 795–820. Winter.

Watkins, S. E., & Mensah, F. M. (2019). Peer support and STEM success for one African American female engineer. *Journal of Negro Education, 88*(2), 181–193. https://doi.org/10.7709/jnegroeducation.88.2.0181

Yosso, T. J. (2005). Whose culture has capital? A critical race theory discussion of community cultural wealth. *Race Ethnicity and Education, 8*(1), 69–91. https://doi.org/10.1080/1361332052000341006

Chapter 10
It's in the Way We Talk: Exploring the Use of Everyday Language in Teaching Technical Content STEM Learning Environments

Nicole Pitterson and Benjamin Goldschneider

Abstract Language is an important aspect of everyday life. It is through language that we build connections with each other and learn about the world. In terms of scientific knowledge, language is the bridge that connects what we know about our everyday context to the technical realm of science. For STEM (science, technology, engineering, and mathematics) education, language allows students to build complex understandings from everyday common-sense meanings and unpack technical knowledge into accessible forms. The role of language in learning cannot be understated. When students are first exposed to content in formal learning environments, they are expected to learn the relationships between various concepts, how to apply several problem-solving strategies and to transfer this knowledge to other complex and advanced courses. However, in engineering and STEM more broadly, more research is needed to explore how the language used in formal classrooms and course documentation, such as lecture notes/videos, syllabi, worksheets and assessment documents, shapes how students come to make meaning and develop disciplinary expertise of the content. Further, in any given classroom there are students with varied backgrounds and experiences. Consequently, language plays a pivotal role in fostering equitable and inclusive learning environments by providing the means by which to legitimize cultural and disciplinary norms, promote access and highlight linguistic barriers where they exist. Understanding the best ways to reach diverse groups of students using common everyday language is becoming more important in the context of STEM learning environments, especially as the student

---

N. Pitterson (✉)
Department of Engineering Education, Virginia Tech, Blacksburg, VA, USA
e-mail: pitterson@vt.edu

B. Goldschneider
First Year Engineering Center, University of Virginia, Charlottesville, VA, USA
e-mail: bgoldsch@virginia.edu

G. L. Cochran, J. D. Adams (eds.), *Equity-Oriented, Discipline-Based STEM Education Research*, Critical Studies of Education 19,
https://doi.org/10.1007/978-981-96-9136-4_10

demographic continues to change due to robust recruitment and retention efforts. This chapter will seek to explore the role of language helping students learn while also creating equitable and inclusive STEM learning environments.

Keywords Equity in STEM · Role of language · Foundational STEM concepts · Inclusive pedagogy

10.1 Introduction

In any given classroom, and more specifically STEM classrooms, language is the medium through which students explore new ideas, engage with unknown concepts, and solidify their own understandings of the content being presented (McComas, 2013). Language serves as a powerful tool that shapes our understanding of the world, facilitates communication, and influences the way we learn and interact. In fact, language is particularly important in STEM because STEM disciplines are centered on discovery, problem-solving, and negotiation of meaning. The complex and technical nature of STEM subjects often present challenges for learners. As newcomers to STEM join their respective disciplines, they are expected to learn the technical content of their area of specialization while also learning how to communicate in a manner that is indicative of the discipline. For example, several researchers have attributed the challenges in learning fundamental STEM concepts to their technical, abstract, and in some cases informationally dense features (Bernhard & Carstensen, 2002; Biswas et al., 2001; Carstensen & Bernhard, 2007). By extension, when students are first exposed to fundamental disciplinary STEM concepts in formal learning environments, they are expected to learn the relationships between concepts, how to apply several problem-solving strategies and to transfer this knowledge to other complex and advanced courses (Georghiades, 2000; Johnstone, 1991; Moore et al., 2013). In STEM classrooms, language acts as a bridge that connects students to the vast expanse of information and ideas present in educational materials. According to Vygotsky and Cole's sociocultural theory, language serves as a tool for thinking and learning. When students engage with language-rich content, they can decode complex concepts and internalize abstract notions more effectively (Vygotsky & Cole, 1978). Similarly, other researchers have highlighted the significance of language proficiency in content area learning, emphasizing that a strong grasp of language leads to improved comprehension and higher academic achievement (Cummins, 1981; García & Wei, 2014; Larsen-Freeman & Long, 2014). However, there remains space for educators to explore and learn how the language used in formal classrooms and course documentation shapes how students come to make meaning and develop disciplinary expertise of the often highly abstract and complex technical content. Further, it is important to understand how this knowledge is communicated to students of various backgrounds and experiences since significant recruitment efforts have been undertaken to increase the diversity of student intake in STEM.

To ensure successful outcomes of these efforts, i.e., a diverse group of STEM graduates, it is critical that learning environments are equitable and inclusive so that students can connect the new content with their prior knowledge and experiences. There is a growing need for research that seeks to understand, document, and distribute across the STEM education landscape how we can better design learning experiences, assess learning, and communicate knowledge to help students become better learners and therefore better STEM professionals. Exploring how knowledge of these concepts is communicated and how students come to understand is critical for several reasons: (1) educators can better design learning experiences to meet students learning needs, (2) educators can use discipline-specific pedagogies to reach a diverse group of students therefore helping them integrate into the discipline, (3) provide students with a strong foundation in the discipline thereby enabling success in key courses later on in the curriculum and aiding retention efforts. As student demographics continue to change in STEM education, it is important for educators to understand how they can use language to connect with their students, thereby paving the way for them to learn and increase their overall engagement within their chosen discipline.

10.2 Background

Learning, from a constructivist perspective, involves interactive engagement among learners where knowledge is constructed "through a process of negotiation and consensus building" (Kittleson & Southerland, 2004, p. 270). As students learn new information, they integrate it with their existing knowledge and experiences, forming a structured mental framework. Language plays a crucial role in this process of constructing meaning. Through language, students organize their thoughts, link concepts, and create mental models that aid in long-term retention and application of knowledge (Johnson-Laird, 2010; Madaan & Yazdanbakhsh, 2022). Language also supports metacognitive processes, allowing students to reflect on their own thinking and learning strategies. Through self-assessment, self-regulation, and discussing their learning experiences, students develop a deeper understanding of how they learn best and how to improve their learning strategies. Additionally, the conceptual change theorists posit that learners bring their individual prior knowledge and experiences to the classroom and in order for learning to be meaningful (and lasting) instruction should make connections between what they already know and what they need to learn (Ambrose et al., 2010; Bernhard & Carstensen, 2002). However, in any classroom, there are several Discourses—ways of thinking, acting, interacting, knowing, believing, using tools, symbols, and objects—at play. Consequently, when students enter the classroom, they are often met with competing Discourses (Allie et al., 2009; Johnstone, 2018; Lee et al., 2007; Moje et al., 2001). In a typical STEM classroom, there are disciplinary, classroom/social, and students' everyday experiences Discourses (Moll et al., 1992). Understanding how these Discourses interact and the ways in which they individually influence students' learning is important to STEM teaching and learning. STEM educators can benefit from understanding how students experience the interaction of these Discourses as "studies of learning need to acknowledge both the knowledge and ways of knowing valued by the learners, the learning context, and the discipline" (Kittleson & Southerland, 2004, p. 270). These values are communicated using language that is both written and spoken.

Being able to acknowledge the role of language in the classroom is highly recommended in STEM settings. Language is not just a means of conveying information; it is also a means of fostering critical thinking skills (Manalo & Sheppard, 2016). Through language, students are encouraged to analyze, evaluate, and synthesize information, thus developing their cognitive abilities. Liang and Fung (2021) advocate for the use of language to frame and reformulate questions, facilitating deeper understanding and encouraging students to approach problems from multiple perspectives. They posit that as students engage in discussions with their peers and instructor, respond to questions, and complete written assignments, they refine their ability to think critically and communicate their thoughts effectively. Similarly, Halliday and colleagues (Halliday, 1993; Alaei & Ahangari, 2016; Urban, 1981) assert that language is the essential condition of learning, as language is the process by which experience becomes knowledge. It is through language we build connections with each other and learn about the world. In terms of scientific knowledge, language is the bridge that connects what we know about our everyday context to the technical realm of science (Doran, 2017; Lemke, 1990). Consequently, using everyday language to explain technical concepts can facilitate a deeper understanding among learners. By drawing parallels between abstract concepts and relatable experiences, educators can help students to break down and comprehend complex ideas. This approach encourages learners to build on their existing knowledge and transfer skills from their everyday lives to STEM contexts, promoting a more holistic and integrated understanding. In STEM, language allows professionals to build complex understandings from everyday common-sense meanings and unpack technical knowledge into accessible forms. For example, in mathematics, formulae are forms of language and we can see this use of language enacted in classroom dialogue, course materials, textbooks, assessments, students' work, etc. (Doran, 2017; Dreyfus, 1991; Schoenfeld, 2016; Schwartz & Moore, 1998).

Studies of the barriers to instructional change in STEM have concluded that the key reason for the lack of holistic reform can be attributed to the piecemeal approach to research that yields isolated findings that are difficult to implement (Seymour, 2002; Smith et al., 2004, 2005a). Currently, there is a dearth of studies that seek to understand how to improve STEM teaching and learning in a holistic manner. This demands extensive exploration of all the intervening factors in the classroom such as what is taught, how it is taught, and how it is assessed (Pitterson, 2015). Early curriculum reform efforts were aimed at changing classroom dynamics, how students were being taught and helping instructors develop an appreciation for the breadth of experiences and background knowledge students brought to the classroom (McDermott, 1993; Radcliffe, 2006; Vest, 2008). Since then, several studies have been conducted to explore the use of active learning and evidence-based instructional practices (Ambrose et al., 2010; Lord et al., 2012; Prince, 2004), the role of students' prior knowledge and experience (Borrego & Henderson, 2014; DeMonbrun et al., 2017; National Research Council, 2015), the design of learning environments and the STEM curriculum (Smith et al., 1993, 2005b; Stevens et al., 2014; Wankat, 1999), how to improve students' learning and conceptual understanding, and instructors' overall preparedness to teach STEM courses (Ambrose et al., 2010; Chickering & Gamson, 1987; Fink et al., 2005). These studies have resulted in several recommendations for course designers, educators, and stakeholders. Yet, the curriculum and changes to instructional practice still appear to lag research findings (Gabriele, 2005). In 2005, Gabriele (Gabriele, 2005) found there is often a 15-year lag between what research recommends and how soon these recommendations are adapted in the curriculum. Instead, we need a holistic view of what happens in the classroom especially as it relates to how language use can foster learning because instructors must consider all of these factors simultaneously. This is of particular importance since language barriers can not only impede learning but equitable access to STEM education as well and ultimately perpetuate disparities in achievement. For example, students who do not speak English as their first language may struggle to grasp complex STEM concepts due to language-related obstacles. To address this issue, educators should provide scaffolded instruction that gradually bridges the language gap. Strategies such as incorporating translanguaging, a pedagogical approach that encourages students to draw on their native languages to make meaning, have shown promising results (García & Kleifgen, 2010). This approach not only facilitates content comprehension but also nurtures a positive linguistic identity by validating students' multilingual capabilities. By recognizing and embracing linguistic diversity, educators can create an environment where students' language skills are seen as assets rather than obstacles.

10.3 Interacting Factors in STEM Learning Environments

In any given classroom, there are several factors at play. Instructors have philosophical beliefs about their courses, the overall curriculum, and how best to teach and reach their students (Hill et al., 2008; Luft et al., 2011; Magnusson et al., 1999), students do not enter the classroom *tabula rasa*, instead they bring background experiences that help them to make sense of the formal instruction (Ainsworth, 2008; Reiner et al., 2000; Slotta & Chi, 2006), and language plays an important role in connecting instructors' philosophical beliefs and students' everyday experiences (Lee et al., 2007; Luykx & Lee, 2007; Moje et al., 2001). Particularly, emphasis on the role of language in learning STEM concepts includes how language (1) can be used an indicator of conceptual understanding, (2) manifests mastery of STEM practices, beliefs, and mindsets, and (3) how meaning is encoded in course content through text, spoken word, and images/equations.

10.3.1 Instructors' Beliefs About Teaching and Learning

Detailed research evidence now exists on the ways in which instructor's actions are deeply influenced by their beliefs on teaching and learning. Many of these beliefs tend to be formed during their own experiences in the classroom when they were students (National Research Council, 2000). Basing teaching on their own experiences is often cited as the main reason for the continuation of relatively traditional teaching practices; instructors tend to teach in the same ways that they were taught. Thus, there is a significant field of work focused on professional development related to teaching, which aims to shift instructors' beliefs about teaching and learning. At the same time, as found in many studies (Borrego et al., 2013; DeMonbrun et al., 2017; Lee et al., 2007), the relationship between beliefs and practices is not straightforward. There is significant evidence that espoused beliefs do not match the practices that are observed in the classroom. Several studies have sought to design means to capture how instructors' beliefs about teaching and learning influence their classroom practices (Luft & Roehrig, 2007; Magnusson et al., 1999). For example, one study explored how teachers' epistemological beliefs are enacted in their practice and found constructive beliefs often lead to the design of constructive learning environments (Luft & Roehrig, 2007). Researchers have also identified potential gaps between instructors' beliefs and practice (Lavoie & Roth, 2005). Consequently, there are significant benefits to understanding how instructors use their own knowledge and background to design and deliver content in a manner their students can learn. In particular, when instructors use their knowledge of the content and understanding of how to teach to design learning experiences for their students, they communicate this knowledge using language with which they are familiar.

However, the language used by the instructors may not always be easily understood by the students because the examples used are outside of the students' frame of reference. Traditional teaching methods tend to rely on jargon-laden language. This can inadvertently create barriers to effective learning. This also creates an equity issue in that the language being used in the classroom might only reach students of a particular demographic or may be seen to privilege one set of experiences over another. For example, instructors may use analogies of everyday events to help students make connections between what they know and what the instructor hopes they learn. Research has indicated analogies are useful learning tools for several reasons (Joshua & Dupin, 1987; Sfard, 1998; Treagust et al., 1992). Firstly, learning itself is inherently comparative, meaning whether overtly or subtly we are always comparing one thing to another to see what similarities or differences exist between the two and based on these comparisons we update our own mental registers of what the item of exploration is. Incorporating everyday language can also enhance learner engagement. Real-world examples and analogies resonate more strongly with learners, making the content feel relevant and applicable. This approach captivates learners' attention and encourages active participation, creating a positive feedback loop between comprehension and motivation. Secondly, from a conceptual understanding and constructivist perspective, we have preconceptions about the world and knowledge as we develop through childhood and all the stages of human development through engaging with the world. Since students do not come to the classroom as blank slates it is critical that pedagogical practices take into consideration the breadth of differences students bring as an asset to the learning space. One approach to highlight students' cultural backgrounds and experiences as a strength in the classroom is through the use of culturally responsive pedagogy. Culturally responsive pedagogy recognizes and values the diverse cultural backgrounds of students, integrating these perspectives into the teaching and learning process. Language is a central component of culture, influencing communication patterns, worldview, and cognitive processes. By incorporating students' native languages and cultural experiences, educators can make STEM content more relatable and accessible. Research by Ladson-Billings (2021) underscores that culturally responsive pedagogy enhances students' engagement and learning outcomes. Inclusive language practices that validate students' linguistic diversity build their confidence and self-efficacy in STEM subjects. Moreover, when educators incorporate diverse examples and real-world applications in their lessons, students from underrepresented backgrounds can better connect with the subject matter.

10.3.2 Using Students' Everyday Experiences to Improve Learning

Research suggests when instruction makes specific links to students' prior knowledge and experience, they are better able to make sense of the unknown (Barton & Tan, 2009). The disconnect perceived by students between their everyday knowledge and the scientific content they are exposed to in formal classrooms is a major challenge for learning (Allie et al., 2009; Kittleson & Southerland, 2004). Studies advocate that while scientific knowledge may appear to be very separate from the everyday world and experiences students possess, students often rely on these experiences when they are expected to construct new meaning and understandings in the classroom (National Research Council, 2015). Primarily for technical and abstract content, it is important that instructors use "instructional approaches that conceptualize science learning in terms of students' everyday experiences" (Kervinen et al., 2020, p. 2). By making these connections, instructors can help students channel the rich experiences they bring to the classroom (Gee, 2000; Lee & Fradd, 1998; McCabe, 2017). As discussed before, analogies and metaphors are often the most used vehicles to make these connections (Lakoff & Johnson, 1980a, 1980b, 1980c) and their role is a function of how language works.

Analogies can help to motivate students desire to learn by provoking their interest in a particular body of knowledge thus helping them to restructure their underlying knowledge frameworks by making the unknown more relatable and, by extension, more understandable. While their power in making the unknown, known is a major advantage of analogies their use comes with a bit of caution. In fact, "before using an analogy in the classroom, the strength and usefulness must be considered and must be based on how accurately it relates one concept to another and the prior knowledge and experiences of the students for whom it is intended" (Dagher, 1995, p. 297). Accordingly, some researchers posit "the use of analogies should be constantly assessed and based upon students' background knowledge, their experiences, and their needs. Good analogies must be personally and culturally relevant to students" (McComas, 2013, p. 6). For example, if an instructor uses a sport analogy in their classroom with which students are unfamiliar, this may cause confusion or invoke disinterest among the students who feel the example used is too far outside the scope of their experiences. Consider how often the phrase "it's like riding a bike" comes up in regular conversation or even a classroom. To the student who has never ridden a bike or calls a bike something else there is a space for confusion or misunderstanding. In this regard, that use of the analogy would be ineffective.

The use of everyday language aligns with principles of inclusive pedagogy by accommodating diverse learning styles and backgrounds, for example, culturally responsive pedagogy discussed earlier. Learners from non-STEM disciplines or with limited exposure to technical language can benefit from a teaching approach that bridges the gap between their existing knowledge and new STEM content. This inclusivity not only supports a broader range of learners but also promotes equity in STEM education. Educators can employ various strategies to incorporate everyday language into their teaching methods. These strategies include utilizing metaphors, analogies, visual aids, and storytelling to present technical content in relatable terms.

10.3.3 Role of Language in Bridging Everyday Experiences and Learning

In any context, language is a systematic resource for expressing meaning (Halliday, 1993; Lemke, 1990, 1997). Studies of the role of language in context is often focused on how people construct and exchange meanings (Bartlett, 2017; Gendlin, 1997; Glynn, 2008; Halliday, 2009). This view of language as a system for meaning making implies that language is not only a well-defined system that involves a set of grammatical sentences, but that language exists in contexts such as professional settings, classrooms, and texts and must be studied in these contexts. In short, "when we talk science, we are helping to create, or re-create a community of people who share certain beliefs and values" (Lemke, 1990, p. x). Therefore, the influence of language in teaching and learning should not be understated. "The foregrounding of language in science teaching and learning demands that teachers have a better understanding of the central role language plays in shaping experience/reality and hence in learning, so that they can become more proactive and effective in apprenticing students to the discourse community of science" (Halliday, 2009, p. 345). Employing everyday language is particularly effective in bridging the gap between novice and expert learners. As experts tend to take their specialized knowledge for granted, they might unintentionally overlook the challenges beginners face in grasping fundamental concepts. The use of relatable language forces experts to reevaluate their explanations, leading to clearer and more concise communication.

Language plays a central role in creating equitable and inclusive STEM learning environments. Broadly, language influences how individuals perceive themselves and others, shapes classroom dynamics, and impacts students' confidence, participation, and sense of belonging. Inclusive language ensures that all students, regardless of their backgrounds, feel seen and acknowledged. As expressed earlier, using diverse examples, case studies, and role models from various ethnicities, genders, and backgrounds can help students relate to the content and see themselves as potential contributors to STEM fields. Additionally, using culturally sensitive language demonstrates respect for students' diverse backgrounds and experiences. This avoids unintentional exclusion or alienation of certain groups and promotes a welcoming environment where all students feel valued. Further, STEM fields have traditionally been male-dominated, and using gender-inclusive language helps create an environment where all genders feel welcome and respected. This includes avoiding gender stereotypes and using terms that encompass all gender identities.

Across the literature, several studies have been conducted in international contexts to showcase the multiple, varied yet crucially important roles that language serves in STEM learning environments. Some of these roles include; acquisition of knowledge, communication, critical thinking, reflection and metacognition, constructing meaning, and highlighting cultural backgrounds and differences in meaning making. Below, we provide an annotated discussion along with the full reference citation of a sample of these works.

Savaria, M., & Monteiro, K. (2017). A critical discourse analysis of engineering course syllabi and recommendations for increasing engagement among women in STEM. Journal of STEM Education: Innovations and Research, 18(1), 92.

This short article describes the use of critical discourse analysis on a pair of First-Year Engineering syllabi to examine what messages the language used can send to students in the class. The authors argue that the lack of student learning outcomes, connections to information outside of engineering, encouragement of student-Peer and student-Faculty relationships, explorations of personal and societal impacts, and acknowledgments of underrepresented groups—particularly women—in STEM all carry negative messages that can affect student participation and persistence in STEM. The centrality of the syllabus, in addition to its early introduction in courses, makes it an important tool for making first impressions and communicating core values in a classroom.

Jahan, I., & Hamid, M. O. (2019). English as a medium of instruction and the discursive construction of elite identity. Journal of Sociolinguistics, 23(4), 386–408. https://doi.org/10.1111/josl.12360

In this in-depth exploration of the Bangladeshi educational context, this article focuses on exploring the experiences of Bangla- and English-medium students, the quality of each educational system, and the opportunities they provide. It examines how languages are "placed on a social hierarchy" and the influence that this has on students' identity relative to their national peers. While this paper does not focus on STEM settings directly, its findings and recommendations are still quite relevant in demonstrating the power of language use in international contexts.

Hatmaker, D. M. (2012). Practicing engineers: Professional identity construction through role configuration. Engineering Studies, 4(2), 121–144. https://doi.org/10.1080/19378629.2012.683793

Geared toward professional engineers, but potentially reflected in the classroom, this journal paper explores how male and female engineers describe the process of partaking in, juggling in, and succeeding at six distinct engineering roles in the workplace. The results indicate that women are often more likely to discuss the challenges of juggling multiple roles and are much more likely to talk about their work in coordinating projects and communication—a skill that several participants attribute to their gender. While the connection to this chapter is somewhat tangentially, this work is an interesting example about how students and professionals who identify in different ways use language in their self-descriptions of gender-specific challenges they face in the classroom and at work.

Zepke, N., & Leach, L. (2010). Beyond hard outcomes: "Soft" outcomes and engagement as student success. Teaching in Higher Education, 15(6), 661–673. https://doi.org/10.1080/13562517.2010.522084

The authors of this journal article address the language surrounding how "success" is perceived, arguing that "hard" measures like GPA, employment, and graduation rates do not accurately reflect student achievement. They instead propose alternative "soft" outcomes that are more abstract but measure students' perceptions of their own progression toward university and personal objectives. In conjunction with outcomes stemming from student engagement literature, it is proposed that student progress can be better recognized and uplifted in terms of the soft objectives. They measure student success with respect to these soft objectives via a matrix with individual effort and collaborative effort on one axis and institution support and non-institution support on the other.

Kumpulainen, K., & Rajala, A. (2017). Dialogic teaching and students' discursive identity negotiation in the learning of science. Learning and Instruction, 48, 23–31. https://doi.org/10.1016/j.learninstruc.2016.05.002

This journal article is a foray into the impacts of dialogic teaching approaches that emphasize open communication and the recognition of diverse student ideas and beliefs in the classroom environment. The background for this study is elementary science education, so while it may be outside the scope of undergraduate STEM education, its focus on dialogic teaching—at least among other language-focused pedagogical approaches is quite pertinent. This paper focuses on how dialogic teaching affects the negotiation of various identities as a science learner.

Wylie, C. D. (2019). Socialization through stories of disaster in engineering laboratories. Social Studies of Science, 49(6), 817–838. https://doi.org/10.1177/0306312719880266

This paper looks at how senior members of an environment—in this case, graduate students and faculty in a research lab—can impart confidence and knowledge to their junior colleagues via stories of their own experiences with failure. Given the difficulty and technical nature of lab work, the authors believe this could carry over to the technical content that is new to early-career STEM students. The article also explores how disaster stories can be interpreted when coming from different groups, as well as how their impact can change depending on how they are delivered.

Brown, B. A. (2004). Discursive identity: Assimilation into the culture of science and its implications for minority students. Journal of Research in Science Teaching, 41(8), 810–834. https://doi.org/10.1002/tea.20228

This paper explores how students acquire and subsequently choose to use scientific language. The key point of the paper boils down to the existence of four groups: those that have the knowledge and can articulate it, those who have the knowledge and are attempting to figure out how to use it, those who reject the scientific discourse and language despite demonstrated proficiency in it, and those who reject it altogether. This relates to how scientific discourse and language find their place in the classroom. The work is based in the middle school context but still holds significant implications beyond this context.

Allie, S., Armien, M. N., Burgoyne, N., Case, J. M., Collier-Reed, B. I., Craig, T. S., Deacon, A., Fraser, D. M., Geyer, Z., Jacobs, C., Jawitz, J., Kloot, B., Kotta, L., Langdon, G., le Roux, K., Marshall, D., Mogashana, D., Shaw, C., Sheridan, G., & Wolmarans, N. (2009). Learning as acquiring a discursive identity through participation in a community: Improving student learning in engineering education. European Journal of Engineering Education, 34(4), 359–367. https://doi.org/10.1080/03043790902989457

Framing success as the ability to understand and appropriately use a discipline's specific discourse in a learning or work environment, this paper argues that language is the very essence of understanding in education. To acquire understanding of this discourse, students must participate in the learning process through interactions with peers and instructors. Instructors must also be aware of the fact that their implicit knowledge of the discourse will influence the way they present it, so they must try to recognize what their students know and do not know in order to most effectively present the discourse to them. Their understanding may be deeper than students, leading them to skim over important descriptions or details because they already know it when their students do not.

Clements, H., Benedict, B., Godwin, A., Rohde, J., & Chen, S. (2020). "Adversary or Ally": Undergraduate Engineering Students' Perceptions of Faculty. https://doi.org/10.18260/1-2%2D%2D33966

The key takeaways from this conference paper are centered around how language can be used to develop positive relationships between students and their instructors. While its focus is not specifically about conveying complex material, it does highlight the importance of communicating course content at an understandable level for students. It also touches on language that encourages feedback and participation, giving students a greater feeling of involvement in the classroom, leaving them more likely to view their instructors as allies.

Dall'Alba, G. (2009). Learning professional ways of being: Ambiguities of becoming. Educational Philosophy and Theory, 41(1), 34–45. https://doi.org/10.1111/j.1469-5812.2008.00475.x

When in a classroom, students are exposed to language and beliefs from their instructor, more often about the course content, but sometimes on other topics as well. This paper argues that as students (or learners in general) are exposed to their professors (the experts/senior members of a field), they learn how to "be" by adapting what they are exposed to. They begin to think and act as the experts do, while also beginning to develop the ability to make judgements about what they are exposed to and decision about what they wish to incorporate into their own actions and belief systems. This is geared toward professional training, but links into some of the other papers here with slight reframing.

N. M. Zain, Rohandi, and A. Jusoh, "Instructional congruence and changing students' attitudes and interests toward science in 'low performing' secondary schools," Procedia—Soc. Behav. Sci., vol. 2, no. 2, pp. 1260–1265, 2010.

A short article detailing the effects of the implementation of greater instructional congruence on student perceptions of and future aspirations in science. These impacts are almost universally positive, with some not being statistically significant. It also provides a useful definition of instructional congruence in terms of recognizing, using, and incorporating students' everyday discourse in collaboration with scientific discourse. This makes the content more relatable and approachable for many students.

N. M.Zain, "the Effect of Instructional Congruence on Students' Interest Towards Learning Science," Eur. J Phys. Educ., vol. 1, no. ISBN 1309-7202, pp. 20–27, 2010.

Seemingly an expanded version of the paper directly above (same primary author), this one goes a bit more into detail about how the teachers were trained and the various components of instructional congruence that they were trying to capture in the study.

L. P. Rivard and A. Levesque, "Three francophone teachers' use of language-based activities in science classrooms," Can. Mod. Lang. Rev., vol. 67, no. 3, pp. 323–350, 2011.

Exploring the use of language-based activities like group discussion, oral presentations, and active question and answer periods, this article sought to understand how these language-focused teaching strategies could provide rich opportunities to minority language students—as in the case of French speakers in Canada. In this study, teachers were willing to learn about and implement these strategies with proper support in terms of both materials and training.

O. Lee and S. H. Fradd, "Instructional congruence to promote science learning and literacy development for linguistically diverse students," in Models of Science

Teacher Preparation: Theory into Practice, D. R. Lavoie and W.-M. Roth, Eds. New York/Boston/Dordrecht/London/Moscow: Kluwer Academic Publishers, 2002, pp. 109–126.

Coming from some of the big names behind the theory, this chapter delves into the importance of developing the scientific literacy of students and the challenges associated with varied levels of comfort with the language used in the classroom. In a melting pot like a college classroom—though the example in this chapter is an elementary school environment with Spanish, Haitian Creole, and English being represented—this is a valuable consideration. Learning English is a challenge of its own, learning science in a language you are not comfortable with further builds upon this challenge. This in turn forces instructors to consider how they can frame their teaching with experiences and terminology that are accessible by all students.

10.4 Instructional Congruence

Instructional congruence is the process of mediating the nature of academic content with students' language and culture experiences to make content accessible, meaningful, and relevant for a diverse group of students (Fradd & Lee, 1999; Lee & Fradd, 1998). In particular, instructional congruence is of utmost importance in the classroom to ensure that the range of diverse backgrounds and experiences are taken into consideration when designing learning experiences. Students bring their whole selves into the learning space. This includes values associated with their upbringing, their culture, their ways of viewing the world which is shaped by their personal experiences and familial background. Hence, it is important that as instructors are designing their instruction, they are aware of how the prior background and cultural experiences of their students are framing how they learn and how much they learn. According to Lee and Fradd, "the emerging literature on multicultural science education indicates that students from diverse languages and cultures display alternative ways of knowing that are sometimes incompatible with the nature of science or the way science is taught in school" (Lee & Fradd, 2005, p. 110). Hence, it is important that the designers of learning environments are cognizant of practices they can apply in their settings to reach a diverse group of students. Instructional congruence is a way to achieve this. Below we present a model of instructional congruence that emerged from years of work and a series of research projects that is aimed at "integrating science learning and literacy development for linguistically diverse students" (Lee & Fradd, 2005, p. 112). Even though this model was designed based on English Language Learners (ELL), its principles can still be applied to STEM learning environments since learning content in a new discipline can be likened to learning a new language.

The model of instructional congruence as put forth by Lee and Fradd (2005) consists of four defining features: These are:

1. What instructors do through classroom instruction and assessment is directly targeted to meeting the needs of a diverse group of students. This feature governs the knowledge, experiences, and practices of instructors that show up in the classroom that they can attune to their students. When instructors and students have shared cultural experiences and use language in similar ways this can lead to improved and increased participation and engagement with the students. Instructors should attempt to learn as much about their student to the extent that they are able so that when they present content to the students, they can do so in a way that integrates students backgrounds and experiences into the content being taught.
2. There are subject-specific pedagogies that can engage a diverse set of students. Much work has been done on culturally relevant pedagogies and researchers have demonstrated how important it is for classrooms to be culturally compatible. Instructors have full and complete grasps on the content to be taught. It is thus very important that they are able to present this content in a way that takes into consideration that various ways in which the content is applicable to students' everyday lives, language, and culture.
3. Classroom instruction should promote students' understanding of the content but also provide them with the ability to be conversant about the content through written, verbal, and other illustrative forms. Thus, learning environments should be hands on while also promoting the multiple ways in which the content can be represented. The language used in the classroom should be such that it encourages students to move from concrete to abstract thinking, as well as providing opportunities for them to engage in dialogue so that they can describe, reflect, integrate, and apply their understandings in multiple contexts.
4. Students should be provided with the opportunity to construct knowledge of the target content and approaches to understanding said content through the lens of their own cultural background and understanding of language. At the heart of instructional congruence is a constructivist approach to learning which advocates that students should be given the space and opportunity to construct their own understanding of the content through the design of intentional instruction. Often, the language and experience of students are not taken into consideration in the formal classroom. However, instructional congruence seeks to build on these unique individual features of the students as extremely valuable resources within the classroom.

Overall instructional congruence is important in STEM learning environments because it is a way to promote, enact, and improve learning for a diverse set of students that communicates to them that what they bring to the classroom is welcomed and valuable.

10.5 Recommendations

- Models of instructional congruence are useful pedagogical principles that help students acquire discipline-specific understandings so that they can engage in the language and practices of their chosen discipline in a more meaningful way. There is a need for a model designed specifically for STEM disciplines that explores the influence of instructors' beliefs about teaching and learning, using students' everyday experiences to improve learning and the role of language in bridging everyday experiences and learning. The model should also highlight steps that can be taken to achieve instructional congruence in STEM learning environments. Researchers and educators can then be drawn into conversation about how the model can be applied to STEM and non-STEM disciplines with the express goal of demonstration its dynamism and flexibility in application and adaptability.
- Course coordinators and designers may consider designing open source, open access course design clinics that share models of instructional congruence with specific, actionable insights that instructors can modify to use in their particular settings.
- Instructors and educators should take into consideration how their own backgrounds, experiences, and beliefs about teaching and learning influence how they design their courses, teach, and assess their students. To create equitable and inclusive STEM learning environments, it is important that the experiences and ways of being of all students in the classroom are not only recognized but it integrated in classroom activities and instruction. For example, when presenting a complex or difficult concept, the instructor should consider not only presenting examples or analogies that have worked to help them understand or conceptualize the content. Instead, they should open up the discussion to the class and soliciting from the students' examples of how they have conceived of these topics or what experiences in their backgrounds can they call upon to help them access the information being presented.
- Reward systems should be modified to take into consideration the ways in which instructors are designing their courses and classrooms to acknowledge the variety of backgrounds and experiences that students bring into the learning space. It is not enough to continue to recruit diverse groups of students to STEM when the activities that take place within the classroom do not acknowledge and bring into the mainstream conversation the power of students' varied experiences.
- Funding agencies should also continue to support researchers who are focused on classroom teaching and assessment especially those who study how the values and experiences students bring to the classroom may not always be aligned with those of their instructors. Using research to find ways to bring both perspectives into alignment is beneficial to all concerned.

Inclusive STEM education is not just about providing access; it is about fostering an environment where all students can thrive and contribute meaningfully. Language is a fundamental element that influences how students learn, communicate, and perceive their place in the STEM field. By implementing culturally responsive pedagogy, promoting active engagement, addressing linguistic barriers, and adopting inclusive assessment practices, educators can create equitable and inclusive STEM learning environments. As society continues to diversify, embracing the role of language in education is essential for shaping the next generation of innovative thinkers and problem solvers.

References

Alaei, M., & Ahangari, S. (2016). A study of ideational metafunction in Joseph Conrad's "Heart of darkness": A critical discourse analysis. *English Language Teaching, 9*(4), 203–213.

Allie, S., Armien, M. N., Burgoyne, N., Case, J. M., Collier-Reed, B. I., Craig, T. S., et al. (2009). Learning as acquiring a discursive identity through participation in a community: Improving student learning in engineering education. *European Journal of Engineering Education, 34*(4), 359–367.

Ambrose, S. A., Bridges, M. W., DiPietro, M., Lovett, M. C., & Norman, M. K. (2010). *How learning works: 7 research-based principles for smart teaching*. Jossey-Bass.

Bartlett, T. (2017). Context in systemic functional linguistics: Towards scalar supervenience? In *The Routledge handbook of systemic functional linguistics* (pp. 375–390). Routledge. https://doi.org/10.4324/9781315413891

Barton, A. C., & Tan, E. (2009). Funds of knowledge and discourses and hybrid space. *Journal of Research in Science Teaching: The Official Journal of the National Association for Research in Science Teaching, 46*(1), 50–73.

Bernhard, J., & Carstensen, A.-K. (2002). *Learning and teaching electrical circuit theory*. PTEE 2002: Physics Teaching in Engineering Education, pp. 163–178.

Biswas, G., Schwartz, D. L., Bhuva, B., Bransford, J., Holton, D., Verma, A., & Pfaffman, J. (2001). Assessing problem solving skills in understanding and troubleshooting AC circuits. *Standard Form, 298*(0704), 1–39.

Borrego, M., Cutler, S., Prince, M., Henderson, C., & Froyd, J. E. (2013). Fidelity of implementation of research-based instructional strategies (RBIS) in engineering science courses. *Jornal of Engineering Education, 102*(3), 394–425.

Borrego, M., & Henderson, C. (2014). Increasing the use of evidence-based teaching in STEM higher education: A comparison of eight change strategies. *Journal of Engineering Education, 103*(2), 220–252. https://doi.org/10.1002/jee.20040

Carstensen, A.-K., & Bernhard, J. (2007). Threshold concepts and keys to the portal of understanding: Some examples from electrical engineering. In R. Land, J. H. Meyer, & J. Smith (Eds.), *Threshold concepts within the disciplines* (pp. 1–12). Sense Publishers.

Chickering, A. W., & Gamson, Z. F. (1987). Seven principles for good practice in undergraduate education. *AAHE Bulletin, 1*, 3–7. https://doi.org/10.1016/0307-4412(89)90094-0

Cummins, J. (1981). The role of primary language development in promoting educational success for language minority students. In California State Department of Education (Ed.), *Schooling and language minority students: A theoretical framework* (pp. 3–49). California State Department of Education.

Dagher, Z. R. (1995). Review of studies on the effectiveness of instructional analogies in science education. *Science Education, 79*(3), 295–312.

DeMonbrun, M., Finelli, C. J., Prince, M., Borrego, M., Shekhar, P., Henderson, C., & Waters, C. (2017). Creating an instrument to measure student response to instructional practices. *Journal of Engineering Education, 106*(2), 273–298.

Doran, Y. J. (2017). *The discourse of physics: Building knowledge through language, mathematics and image*. Routledge.
Dreyfus, T. (1991). Advanced mathematical thinking processes. In *Advanced mathematical thinking* (pp. 25–41). Springer Netherlands.
Fink, L. D., Ambrose, S., & Wheeler, D. (2005). Becoming a professional engineering educator: A new role for a new era. *Journal of Engineering Education, 94*(January), 185–194. https://doi.org/10.1002/j.2168-9830.2005.tb00837.x
Fradd, S. H., & Lee, O. (1999). Teachers' roles in promoting science inquiry with students from diverse language backgrounds. *Educational Researcher, 28*(6), 14–42. https://doi.org/10.3102/0013189X028006014
Gabriele, G. A. (2005). Advancing engineering education in a flattened world. *Journal of Engineering Education, 94*(3), 285–286. https://doi.org/10.1002/j.2168-9830.2005.tb00852.x
García, O., & Kleifgen, J. (2010). Bilingualism for equity and excellence in minority education: The United States. In *Equity and excellence in education* (pp. 178–201). Routledge.
García, O., & Wei, L. (2014). Language, bilingualism and education. In *Translanguaging: Language, bilingualism and education*. Palgrave Pivot. https://doi.org/10.1057/9781137385765_4
Gee, J. P. (2000). *Identity as an analytic lens for research in education. Department of Curriculum and Instruction. Madison, WI, University of Wisconsin-Madison*. Mimeo.
Gendlin, E. (1997). *Experiencing and the creation of meaning: A philosophical and psychological approach to the subjective*. Northwestern University Press.
Georghiades, P. (2000). Beyond conceptual change learning in science education: focusing on transfer, durability and metacognition. *Educational Research, 42*(2), 119–139. https://doi.org/10.1080/001318800363773
Glynn, S. M. (2008). Making science concepts meaningful to students: Teaching with analogies. In *Four decades of research in science education: From curriculum development to quality improvement* (pp. 113–125). Waxmann. http://books.google.com/books?hl=en&lr=&id=l3GfZTYb9fcC&oi=fnd&pg=PA113&dq=Making+science+concepts+meaningful+to+students+:+teaching+with+analogies&ots=406FaAAIXT&sig=YgP3-6O0a08MW7oZYRMcINbllNg
Halliday, M. A. (1993). Towards a language-based theory of learning. *Linguistics and Education, 5*(2), 93–116.
Halliday, M. A. K. (2009). Methods—Techniques—Problems. In M. A. K. Halliday & J. J. Webster (Eds.), *Bloomsbury companion to systemic functional linguistics* (pp. 59–86). Continuum International Publishing Group.
Hill, H. C., Ball, D. L., & Schilling, S. G. (2008). Unpacking pedagogical content knowledge: Conceptualizing and measuring teachers' topic-specific knowledge of students. *Journal for Research in Mathematics Education, 39*(4), 372–400. https://doi.org/Article
Jarman, M., Lamp, S., Mitchell, D., Nepveux, D., Nowell, N., & Snyder, S. (2002). Theorising disability as political subjectivity: Work by the UIC disability collective on political subjectivities. *Disability & Society, 17*(5), 555–569.
Johnson-Laird, P. N. (2010). Mental models and human reasoning. *Proceedings of the National Academy of Sciences, 107*(43), 18243–18250.
Johnstone, A. H. (1991). Why is science difficult to learn? Things are seldom what they seem. *Journal of Computer Assisted Learning, 7*, 75–83.
Johnstone, B. (2018). Discourse Analysis (3rd ed.). Wiley Blackwell.
Joshua, S., & Dupin, J. J. (1987). Taking into account student conceptions in instructional strategy: An example in physics. *Cognition and Instruction, 4*(2), 117–135.
Kervinen, A., Roth, W. M., Juuti, K., & Uitto, A. (2020). The resurgence of everyday experiences in school science learning activities. *Cultural Studies of Science Education, 15*, 1019–1045.
Kittleson, J. M., & Southerland, S. A. (2004). The role of discourse in group knowledge construction: A case study of engineering students. *Journal of Research in Science Teaching, 41*(3), 267–293.
Ladson-Billings, G. (2021). *Culturally relevant pedagogy: Asking a different question*. Teachers College Press.

Lakoff, G., & Johnson, M. (1980a). Conceptual metaphor in everyday language. *Journal of Philosophy, 77*(8), 453–486.

Lakoff, G., & Johnson, M. (1980b). The metaphorical structure of the human conceptual system. *Cognitive Science, 4*, 195–208. http://onlinelibrary.wiley.com/doi/10.1207/s15516709cog0402_4/abstract

Lakoff, G., & Johnson, M. C. (1980c). *Metaphors we live by*. University of Chicago Press.

Larsen-Freeman, D., & Long, M. H. (2014). *An introduction to second language acquisition research*. Routledge.

Lavoie, D. R., & Roth, W. M. (Eds.). (2005). *Models of science teacher preparation: Theory into practice* (Vol. 13). Springer Science & Business Media.

Lee, O., & Fradd, S. H. (1998). Science for all, including students from non-English-Language backgrounds. *Educational Researcher, 27*(4), 12–21. https://doi.org/10.3102/0013189X027004012

Lee, O., & Fradd, S. H. (2005). Instructional congruence to promote science learning and literacy development for linguistically diverse students. In *Models of Science Teacher Preparation* (pp. 109–126). Kluwer Academic Publishers. https://doi.org/10.1007/0-306-47230-9_7

Lee, O., Luykx, A., Buxton, C., & Shaver, A. (2007). The challenge of altering elementary school teachers' beliefs and practices regarding linguistic and cultural diversity in science instruction. *Journal of Research in Science Teaching, 44*(9), 1269–1291.

Lemke, J. L. (1990). *Talking science: Language, learning, and values*. Ablex Publishing. (hardback: ISBN-0-89391-565-3; paperback: ISBN-0-89391-566-1).

Lemke, J. L. (1997). Cognition, context, and learning: A social semiotic perspective. In D. Kirshner (Ed.), Situated Cognition: Social, Semiotic, and Psychological Perspectives (pp. 37–55). Erlbaum. http://books.google.com/books?id=ivqlQw9Dy-8C&pgis=1

Liang, W., & Fung, D. (2021). Fostering critical thinking in English-as-a-second-language classrooms: Challenges and opportunities. *Thinking Skills and Creativity, 39*, 100769.

Lord, S. M., Prince, M. J., Stefanou, C. R., Stolk, J. D., & Chen, J. C. (2012). The effect of different active learning environments on student outcomes related to lifelong learning. *International Journal of Engineering Education, 28*(3), 606–620. http://www.scopus.com/inward/record.url?eid=2-s2.0-84861629726&partnerID=tZOtx3y1

Luft, J. A., & Roehrig, G. H. (2007). Capturing science teachers' epistemological beliefs: The development of the teacher beliefs interview. 11(2), pp. 38–63. *The Electronic Journal for Research in Science and Mathematics Education*.

Luft, J. A., Firestone, J. B., Wong, S. S., Ortega, I., Adams, K., & Bang, E. (2011). Beginning secondary science teacher induction: A two-year mixed methods study. *Journal of Research in Science Teaching, 48*(10), 1199–1224.

Luykx, A., & Lee, O. (2007). Measuring instructional congruence in elementary science classrooms: Pedagogical and methodological components of a theoretical framework. *Journal of Research in Science Teaching: The Official Journal of the National Association for Research in Science Teaching, 44*(3), 424–447.

Madaan, A., & Yazdanbakhsh, A. (2022). *Text and patterns: For effective chain of thought, it takes two to tango*. arXiv preprint arXiv:2209.07686. https://doi.org/10.48550/arXiv.2209.07686

Magnusson, S., Krajcik, J., & Borko, H. (1999). Nature, sources, and development of pedagogical content knowledge for science teaching. In *Examining pedagogical content knowledge: The construct and its implications for science education* (pp. 95–132). Springer Netherlands.

Manalo, E., & Sheppard, C. (2016). How might language affect critical thinking performance? *Thinking Skills and Creativity, 21*, 41–49.

McCabe, A. (2017). Systemic functional linguistics and language teaching. In *The Routledge handbook of systemic functional linguistics* (pp. 615–628). Routledge.

McComas, W. F. (Ed.). (2013). *The language of science education: An expanded glossary of key terms and concepts in science teaching and learning*. Springer Science & Business Media.

McDermott, L. C. (1993). Guest comment: How we teach and how students learn—A mismatch? *American Journal of Physics, 61*(4), 295. https://doi.org/10.1119/1.17258

Moje, E. B., Collazo, T., Carrillo, R., & Marx, R. W. (2001). "Maestro, what is 'quality'?": Language, literacy, and discourse in project-based science. *Journal of Research in Science Teaching: The Official Journal of the National Association for Research in Science Teaching, 38*(4), 469–498.

Moll, L. C., Amanti, C., Neff, D., & Gonzalez, N. (1992). Funds of knowledge for teaching: Using a qualitative approach to connect homes and classrooms. *Theory into Practice, 31*(2), 132–141. https://doi.org/10.1080/00405849209543534

Moore, T. J., Miller, R. L., Lesh, R., Stohlmann, M. S., & Kim, Y. R. (2013). Modeling in engineering: The role of representational fluency in students' conceptual understanding. *Journal of Engineering Education, 102*(1), 141–178. https://doi.org/10.1002/jee.20004

National Research Council. (2000). How People Learn: Brain, Mind, Experience, and School. Expanded Edition (J. D. Bransford, A. L. Brown, & R. R. Cocking, Eds.). National Academies Press. https://doi.org/10.1016/0885-2014(91)90049-J

National Research Council. (2015). *Reaching students: What research says about effective instruction in undergraduate science and engineering*. https://doi.org/10.1017/CBO9781107415324.004

Pitterson, N. P. (2015). *Undergraduate engineering students' understanding of complex circuit concepts; An investigation of the intersection of students' prior knowledge, design of learning environments and the nature of the content*. Purdue University.

Prince, M. (2004). Does active learning work? A review of the research. *Journal of Engineering Education, 93*(3), 223–231.

Radcliffe, D. F. (2006). Shaping the discipline of engineering education. *Journal of Engineering Education, 95*(4), 263. Retrieved from http://www.jee.org/2005/july/3.pdf

Reiner, M., Slotta, J. D., Chi, M. T., & Resnick, L. B. (2000). Naive physics reasoning: A commitment to substance-based conceptions. *Cognition and Instruction, 18*(1), 1–34.

Schoenfeld, A. H. (2016). Learning to think mathematically: Problem solving, metacognition, and sense making in mathematics (Reprint). *Journal of Education, 196*(2), 1–38.

Schwartz, D. L., & Moore, J. L. (1998). On the role of mathematics in explaining the material world: Mental models for proportional reasoning. *Cognitive Science, 22*(4), 471–516.

Seymour, E. (2002). Tracking the processes of change in US undergraduate education in science, mathematics, engineering, and technology. *Science Education, 86*(1), 79–105. https://doi.org/10.1002/sce.1044

Sfard, A. (1998). On two metaphors for learning and the dangers of choosing just one. *Educational Researcher, 27*(2), 4–13.

Slotta, J. D., & Chi, M. T. (2006). Helping students understand challenging topics in science through ontology training. *Cognition and Instruction, 24*(2), 261–289.

Smith, E. L., Blakeslee, T. D., & Anderson, C. W. (1993). Teaching strategies associated with conceptual change learning in Science. *Journal of Research in Science Teaching, 30*(2), 111–126. https://doi.org/10.1002/tea.3660300202

Smith, K. A., Linse, A., Turns, J., & Atman, C. (2004). *Engineering change*. American Society for Engineering Education Annual Conference and Exposition, 1997, pp. 1–18.

Smith, K. A., Sheppard, S. D., Johnson, D. W., & Johnson, R. T. (2005a). Pedagogies of engagement: Classroom-based practices. *Journal of Engineering Education, 94*(1), 87–101. https://doi.org/10.1002/bmb.20204

Smith, K. A., Sheppard, S. D., Johnson, D. W., & Johnson, R. T. (2005b). Pedagogies of engagement: Classroom-based practices. *Journal of Engineering Education, 94*(1), 87–101. https://doi.org/10.1002/j.2168-9830.2005.tb00831.x

Stevens, R., Johri, A., & Connor, K. O. (2014). Professional engineering work. In *Cambridge handbook of engineering education research* (pp. 119–137). Cambridge University Press. https://doi.org/10.1017/CBO9781139013451.010

Treagust, D. F., Duit, R., Joslin, P., & Lindauer, I. (1992). Science teachers' use of analogies: Observations from classroom practice. *International Journal of science education, 14*(4), 413–422.

Urban, G. (1981). Reviewed work(s): Language as social semiotic: The social interpretation of language and meaning by M.A.K Halliday. *American Anthropologist, 83*(3), 659–661.
Vest, C. M. (2008). Special guest editorial: Context and challenge for twenty-first century engineering education. *Journal of Engineering Education, 97*, 235–237. https://doi.org/10.1002/j.2168-9830.2008.tb00973.x
Vygotsky, L. S., & Cole, M. (1978). *Mind in society: Development of higher psychological processes*. Harvard University Press.
Wankat, P. C. (1999). An analysis of the articles in the journal of engineering education. *Journal of Engineering Education, 88*(1), 37–42. Retrieved from http://www.scopus.com/inward/record.url?eid=2-s2.0-19544389328&partnerID=40&md5=2cc8b12120bf2681a788c2e3d5cccd32

Chapter 11
From Here to Equity: Promoting STEM Inclusion and Equity by Engaging in Authentic Field-Based Ecology Experiences

Christina Colon and Gillian U. Bayne

Abstract This chapter describes a field-based, hands-on ecology course designed to promote greater engagement with and enthusiasm for science among biology majors at a community college. The authors describe a series of weekly inquiry-based research activities and field trips on campus, to a local beach and to local informal science institutions. All activities are linked to or building up to citizen science and local field research. These field labs are designed to be student driven and collaborative as a means to shift student perceptions of themselves from passive learners to future scientists. The long-term objective is to promote academic success and retention in STEM fields among traditionally underserved and marginalized demographics. The lead author couches this narrative within a context of her own academic journeys as a woman of color, while the second author, also a woman of color, provides supporting literature documenting the ongoing barriers to success in STEM involvement, including those in STEM careers experienced by marginalized demographics The activities are juxtaposed to studies that demonstrate best practices and their outcomes of effective and accessible inclusive science education.

Keywords Inquiry-based · Field-based · Hands-on · Student-driven · Equitable · Field ecology · Community college · Citizen science · Civic engagement · Undergraduate research

---

C. Colon (✉)
Department of Biological Sciences, Kingsborough Community College, City University of New York, Brooklyn, NY, USA
e-mail: christina.colon@kbcc.cuny.edu

G. U. Bayne
Department of Middle and High School Education, Lehman College, City University of New York, Bronx, NY, USA

G. L. Cochran, J. D. Adams (eds.), *Equity-Oriented, Discipline-Based STEM Education Research*, Critical Studies of Education 19,
https://doi.org/10.1007/978-981-96-9136-4_11

11.1 Overview

That it took me so long to finally teach a fully field-based, hands-on, inquiry-based ecology class is perhaps a testament to the overwhelming logistical, administrative, and fiscal barriers this approach takes. Yet, for all these barriers, I hope to never teach my Ecology elective for bio majors any other way. Although I have been using this approach throughout my career, I was not able to have a fully field-based course until COVID-19. After 2 years of online teaching due to lockdown, these field labs were an attempt to return to an in-person format while safely avoiding the post-COVID-19 constraints imposed on indoor learning. Trained as an ecologist and a zoo educator, teaching outdoors is my preferred modality, and the overwhelmingly favorable student response informed me that they too preferred this approach. This course builds on the students' foundation of concepts and skills in life sciences through successful completion of a pre-requisite two semester biology course for biology majors. My primary goal of this course is to provide tangible scientific research experiences and foster a sense of accomplishment and inclusion as future scientists, per the findings of Elmesky and Tobin (2005). In order to ascertain whether I was meeting this goal, I embedded open-ended questions about their experience in a post semester reflection piece.

11.1.1 Positionality

There are two authors for this current study. The primary author's teaching pedagogy is informed by her decades of experience as an informal science educator and training as a biology professor at an urban community college which serves a diverse audience of underserved and economically challenged students. She is a PhD Ecologist and a veteran zoo educator with a master's degree from New York University in Environmental Conservation Education. She grew up poor in an urban landscape, but was privileged to have a private Quaker education. In pursuit of nature, she sought the solace of quiet nooks such as playing fields, pocket parks, backyards, and cemeteries. Unable to afford other options, her hippie mother took her brother and her camping for a few days every summer. Her gravitation toward nature was both recreational and spiritual. Ever since Quakerism taught her to worship as she saw fit, she has seen the divine in nature and she pursued ecology with the fervor of a missionary. Sharing her love of nature is therefore more than an academic pursuit.

As the daughter of an absent Puerto Rican father and a Slavic and Danish mother, she was raised among the Ukrainian diaspora of Philadelphia. She grew up seeing the world through the eyes of a White family; they were all White, and thus to her mind, so was she. Her Baba (grandmother) frequently reminded her that she was Hutsul, of proud Ukrainian ancestry; this refusal to acknowledge her Puerto Rican side was her first exposure to racism. Today she self identifies as a Pseudo-Rican: considered Latina among non-Hispanics, and decidedly NOT Latina among Hispanics. This intersectionality of racial and socioeconomic extremes allows her the ability to empathize with and relate to a particularly wide demographic. The chapter is written from her perspective and experience.

The second author of this chapter is a Black second-generation Guyanese American science education scholar, who has over 30-years' experience in the field, both in the United States and abroad. Her work as a secondary science teacher, adult basic educator, urban science educator, researcher, and scholar has centered on diverse perspectives, lived experiences, and the individual and collective needs of people who have been marginalized—all as a means to move increasingly toward equity and justice-oriented practices in science education. Engaging in community social action, both inside and outside of classrooms, while engaging young people in transformative educational experiences from trial and error, represents only a part of her commitment to social justice. In her research, scholarly work and teaching, she argues passionately for culturally and linguistically embedded pedagogy—pedagogy that firmly and deeply incorporates the individual and collective lifeworlds, lived experiences, community wisdom, and oral and written histories of diverse people. These important facets of lived experiences have been used creatively as primary and positive contexts for teaching, learning, learning to teach science, and as components of an important foundation that informs her research. Within this chapter, she has contributed to the grounding of the theoretical framework within this study.

11.1.2 Limitations and Threats to Validity

Potential threats to an autoethnographic narrative such as this include selective memory and other forms of bias resulting from my own experience and unique personality type. My teaching style has been shaped by a reflection of what has worked and not worked for me educationally; however, I remain cognizant that all learners are unique and strive to deliver content that will resonate with all learners in a way that they as individuals can absorb. Both positive and negative feedback from students and colleagues is considered a welcome opportunity to continually assess and adjust my pedagogical approach. I recognize this approach lacks broad educational applicability and cannot be disseminated universally; I am confident that elements can be adapted and applied in many other circumstances. Likewise, it is acknowledged that while the authors here have attempted to framed this narrative in a theoretical context, there is much in the literature that does not support the assertions made here. Finally, this pilot was only replicated once to date, and both took place within the broader context of a once in a lifetime global pandemic. Therefore, all conclusions made here are to be considered preliminary and subjective.

11.2 Existing Barriers to Authentic Field-Based Learning

Higher education continues to pose barriers to underrepresented groups in all areas of Science, Technology, Engineering, and Math (STEM) (NCSES, 2021; Pew Research Center, 2021), including ecology and environmental sciences (Chaudhury & Colla, 2021). As a STEM educator in an urban, minority serving community college, I have a tremendous opportunity for inclusive practices in my ecology elective due to small class size, maturity of the students, support of the administration, and flexibility in the curriculum. Since many biology classes cater to students educated in the New York public school system, many students come to expect predictable and static learning experiences in their lab work. In contrast, I have embraced outdoor field experiences, which provide dynamic, open-ended, active-learning opportunities that change with each season and vary from year to year. These explorations provide a crucial link between traditional learning and real-life engagement (Rennie, 2014). I recall that my experience as a biology student was an endless series of instructions for snipping, pinning, and focusing on inanimate, foul smelling, oddly colored objects with only a passing resemblance to the once living organisms they were. I certainly did not feel more equipped to succeed outside of the lab having dissected a cat or a frog or a worm. The emergence of a civic engagement graduation requirement allowed me to justify getting students out of the lab and into the real world. I interpreted the mandate to have students engage in solving real-world problems by involving them in citizen science, where they could contribute to open access research as volunteer scientists. In this chapter, I describe some of the challenges and benefits of field-based learning. I also draw on the expertise of the second author, an expert in science education to frame my narrative within a matrix of relevant pedagogy.

Through my years of running teen internships at the Bronx Zoo in New York City, I became aware of many barriers to learning. I was surprised when several urban youth informally shared that they are actively discouraged from entering city parks due to alleged nefarious activities therein. Activities like bird watching, hiking, and botanizing are not encouraged, and outdoor learning is a foreign concept to many of my students who admit they had never taken a nature hike before they did field work with me. I recall my Puerto Rican grandpa was mortified that my mother would take my brother and me to the museum for the whole day. His reluctance to join us, or to bring our cousins left me baffled; I was raised to see learning as fun and assumed that was universal, but I now saw a different perspective. My mother and her mother both enjoyed teaching and learning and sharing knowledge; their example no doubt influenced me to embrace education as fun and empowering. I believe that expanding opportunities for learning beyond the classroom provides students with opportunities to access science education in different, fun, and engaging ways that may spark their interest as lifelong learners.

Another challenge is the stigma associated with community service, which is looked upon by some as punishment for criminal acts rather than as an opportunity to contribute or give back. Therefore, beach cleanups or volunteering in a park is often not viewed as connecting with nature, but as unpaid work. Students lose out on the potential benefits that come with volunteer activities, such as the strengthening of identity, increased analytical skills, stress relief, or improved mental health and wellbeing (Mann et al., 2021). E.O. Wilson's 1993 theory of *Biophilia*, and Richard Louv's (2010) book *Last Child in the Woods*, both assert that exposure to nature promotes deeper understanding, curiosity, compassion, inspiration as well as emotional and physical resilience. It is therefore important for students to develop connections to nature so they can engender a sense of caring for, and valuing urban nature (see Bellino & Adams, 2017).

In an urban landscape, cultural institutions like museums and zoos are designed to help expand people's knowledge about nature and natural sciences. However, these institutions are all too often seen by urban youth from underserved communities as destinations for tourists or wealthy residents, moms with strollers, or a place you went to once in grade school. Free days or free community tickets are rarely well publicized, and students lack the needed information (or chutzpah) to take advantage of these opportunities (Adams, 2007). To facilitate an understanding that these are dynamic world-class research institutions, I require students take a focused and purpose-driven examination or survey of artifacts, exhibits, and specimens which fosters an appreciation of the research and data produced therein.

In fact, many museums with collections that exceed their curatorial and scientific capacity forge alliances with the public to help organize, digitize, transcribe, and process historical and archival data sets online (see Hill et al., 2012). Scientist's reliance on the public to assist with data collection or processing has been formalized throughout the emergence of citizen science. Volunteer scientific opportunities such as the Audubon Christmas bird count, Zooniverse, and iNaturalist enlist the help of citizens to collect and process data. Experiences such as these can be leveraged to empower students to become producers of knowledge, solvers of real-world problems, and viable agents of change. Engaging with researchers and academics on large-scale, long-term scientific monitoring projects as trained research volunteers (aka citizen scientist) can lead to career opportunities and awareness through networking and resume building, and can hone student decision-making processes toward a fulfilling career.

Experiences such as field trips and field research provide students the opportunity to consider STEM more broadly. I learned that the notion of open-ended, low stakes exploration is often a new concept to students, some of whom had never been on a hike or nature walk prior to taking my class. My world was profoundly changed through a mandatory 11th grade month-long internship. My experience as a volunteer at the Elmwood Park Zoo in New Jersey (filled with enormous real-life responsibilities and hands-on interactions with numerous wild animals), inspired me to pursue a career in zoo education and ultimately a PhD. Caring for wild animals was my first taste of being seen by others as fully autonomous and responsible, which in turn made me believe in myself and my potential for doing more. I now see that it takes far less to inspire a different career path; just becoming someone who takes hikes or nature walks is in itself a major life choice with profound consequences.

Educators know that students can hone critical thinking, math and science skills by being asked to problem solve as opposed to simply read about how others solved real-world problems. Freeman et al. (2014) demonstrate that such active learning (e.g., problem formulation, examination, and attempted solving) increases student performance when compared to passive learning (e.g., being instructed on how to solve a problem). Theobald et al. (2020) and Haak et al. (2011) both found that active learning reduces the achievement gap among underrepresented undergraduates in STEM. Student learning is not a passive process and requires students to be active participants. Eddy and Hogan (2014) also found an increase in course performance, satisfaction, engagement, and participation among first generation and black students through structured active learning. Student-centered and evidence-based learning can also create a culture of self-efficacy and a sense of belonging among students according to Ballen et al. (2017). A major goal of all my curricula is to enhance student interest, enthusiasm, and confidence through inquiry-based, active/outdoor labs, and citizen science, where every student is a collaborator, a contributor, and a colleague. Promoting this sense of belonging is a vital component of retention and achievement among those who have not been actively welcomed into the realm of scientific inquiry and engagement, or who have not been encouraged, coached, or primed to believe in their ability to prevail and succeed in scientific tasks.

Racial and gender inclusivity in environmental fields continue to lag behind other STEM pathways (Taylor, 2014). This is despite the fact that modern ecology has benefitted from a diversity of non-traditional researchers who challenged many assumptions such as competition, conflict, and male hierarchy. Famed primatologists Dian Fossey and Jane Goodall replaced the conflict-driven image of great apes with gentler and more gender equitable understandings (Eckholm, 1984). Gay researchers at The British Museum unearthed long buried and actively ignored evidence of same sex chick rearing and pair bonding among many animal species including Adélie penguins (see Russell et al., 2012). Suzanne Simard up-ended the global forestry industry by replacing decades of prevailing views about the primacy of competition and parasitism, with a paradigm of cooperation, collaboration, and mutualism now recognized in forests worldwide (see Simard, 2021). Museums such as The American Museum of Natural History (AMNH) and The Smithsonian National Museum of Natural History are recontextualizing their exhibits to now be curated by and fully accessible to members of the ethnic groups whose artifacts are being displayed and/or repatriated. While not an ideal solution to the inherent racial bias of museums, these steps represent an important shift away from an attitude of colonization toward one of inclusivity and equity for future generations of curators and academics. Only when we actively include researchers with different experiences and diverse thinking and learning processes, can we access and benefit from this untapped intellectual and scientific resource (see Medin & Bang, 2014).

11.3 Backdrop of My Academic Institution

The context of this field course is through a backdrop of an urban, minority-serving community college with an agenda of inclusiveness, accessibility, and affordability. This course was also aligned with mandates for civic engagement (aka service learning requiring that students explore solving societal problems), open access (aka free textbooks and lab manuals), and reading/writing intensive learning with a particular focus on students who are English language learners, and students who need remediation. The community college is a part of one of the largest urban college and university systems in the United States, with a large demographics of students who have been, for example, underserved, marginalized as a result of systemic oppression and those who have been unjustly served because of their immigrant or minority background. As of Fall 2022, the student population represented 142 nationalities, speaking 73 different languages. More than 38% of the student body are African-American, over 18% are LatinX, close to 16% are Asian/Pacific Islander, and over 27% are White. Most students (75%) come from families with annual incomes under $40,000 per year, and over 52% of students are first-generation immigrants and the first to go to college. Many are returning to school after a hiatus which often involved raising a family and/or serving in the military. The Department of Biological Sciences attracts students interested in allied health, medicine, biotech, and biology.

Stern et al. (2022) have found racial and cultural alignment in discerning the impacts that instructors of color can have on youth of color as they relate to interest and involvement in scientific field trips. Data collected at a 2-year community college from Black's research (2020) evidenced statistically significant positive relationships between a teacher's race-ethnicity and students' teaching expectations. Teacher gender was also significantly associated with student teaching expectations. While this demonstrates a positive connection, more research is needed to more clearly understand the nuances of the alignment. However, in this research, the fact that I am part Latina therefore might play an important role in positively influencing student attitudes, enthusiasm, and involvement in the work at hand because these characteristics are integral to my approach to my work as an educator.

11.4 Cutting My Teeth as an Informal Educator

As a multiracial Latina educated in urban White-dominated private Quaker schools, I understand some of the challenges to feeling welcome and valued in STEM. My entry into the natural sciences was through informal science institutions, such as the Elmwood Park Zoo and Academy of Natural Sciences in Philadelphia. Free days at the Philadelphia Zoo and Philadelphia Art Museum allowed my single White mother to gain access to these cultural institutions for her children. Camping, gardening, and birdwatching were affordable alternatives to summer camps or theme parks.

These free-range childhood experiences drew me to STEM and formed the lens through which I now teach. Adams and Gupta (2017) not only underscore the increased value of informal science institutions in enriching student science experiences, but also emphasize that they are extremely important sites where learner-centered teacher identity development is commonly strengthened.

My goal is to dispel the myth that scientists are born different, or smarter than everybody else. I still smile when I hear students comment that the friendly guy who came with us on the beach to tag crabs has the same name as the scientist who wrote all those papers on horseshoe crab research. My goal is met if I can blur the distinction between student and expert, teacher and learner and make clear to my students that they each have much to teach as well as much to learn. My own college field trips to the Drew campus pond, Woods Hole, or to the Hofstra Marine Lab in Jamaica remain embedded in my memory not only as a learning experience but as opportunities to "be" a scientist for a week. I got to snorkel on coral reefs at night, survey scallops aboard a NOAA research vessel, collect, handle, and examine live brittle stars and sea biscuits, and walk among nests in a black backed gull breeding colony. What I learned in the field I will never forget. Thompson and Jensen-Ryan (2018) emphasized that a person's belief that they are indeed a scientist is one feature commonly used to define science identity. The perception by others that this indeed is the case also contributes to a person's science identity, as Carlone and Johnson (2007) elucidated when studying Women of Color in science. It is not an identity that is static, but is one that is dynamic, developed over time, and influenced by a variety of social structures. While out in the field doing science I not only saw myself as a scientist, but I saw my professor, Dr. Pollock view his students as emerging scientists. This profoundly influenced my teaching philosophy and as a result, I design all my courses to allow students to step into the shoes (or rubber boots, or dive flippers) of a scientist at some point during the semester.

The stark difference between feeling like a static, confounded, and incompetent student in lab, and a dynamic, observant, passionate researcher in their field is a metamorphosis that I hope that all my STEM students can experience as I did. I was never a passionate lab scientist, thus I am not a passionate lab professor; but in the field where I do my research, my passion is palpable and contagious. As an educator, I am expected to transmit not only content knowledge but to also the tools, skills, insight, and drive to become scientifically literate. Yet just as the progression of science is ongoing, so too is the process of scientific literacy. Therefore, as a scientist and a science educator, I feel it is vital that I reveal myself as a lifelong learner. To hide this progression would deny my students critical insights into my own journey, trials, triumphs, and setbacks. Unfortunately, not much has changed since Malcolm, Hall, and Brown's seminal report, *The Double Bind: The Price of Being a Minority Woman in Science* (1976), which shed light on the underrepresentation of Women of Color in the sciences, and the causes of this underrepresentation including biases, stereotypes, and limited access to resources. To portray myself as unflawed, unwavering, and impenetrable would only perpetuate the stereotype of scientists being born, not wrought through hard work and ongoing struggles—especially as it pertains to Women of Color. That such gritty vulnerability is more readily visible in the field when I am doing science, rather than in the lab or lecture hall when I am teaching science, makes it all the more important to get students out into nature, addressing real-life STEM challenges as part of my research team.

Another STEM strength borne out of weakness is my penchant for collaborations. Ecology lends itself particularly well to collaborations since teamwork is critical to success, and building on an existing effort is invariably easier than starting something new. My horseshoe crab research dovetailed with the work of several collaborating institutions, including Fordham University, the International Union for the Conservation of Nature (IUCN), and NYC Audubon (now NYC Bird Alliance). Learning the ropes and identifying opportunities is much easier when volunteering, then later on slotting into an existing project. Collaborations are also fortuitous when community college participation or inclusion of underserved students ticks a box on their grant applications. Involving students in research affords new insights not only into the interpretation of select artifacts and resources, it helps to inform educators and provide new insights into the art of teaching and learning (Bayne, 2013). Including my students in research such as horseshoe crab surveys and horseshoe crab tagging blitzes through Cornell, allows them to add renowned institutions on their resumes and name drop to potential employers and admissions officers.

Despite my challenges, I am keenly aware of my own privilege and understand that the educational journey of many of my students is far more impressive. Most of my students are from low income households, half work full-time or part-time, and one in five support children, siblings, or elders. The imperative to do better for these deserving individuals drives my enthusiasm for my research and my scholarship of teaching and learning. The idea that I can allow them to bring a sibling, child, or parent to join us on many of our field trips, helps me to extend the reach of my educational goals while enhancing the learning opportunity for many students who juggle multiple priorities.

11.5 The Evolution of an Ecology Elective

I adapted an existing Ecology lab course to leverage Kingsborough's coastal setting and campus green space. I gradually developed inquiry-based, hands-on labs to replace the existing active-learning labs that had a game-like appeal, but were disconnected from the organisms and interactions they attempted to represent. My active-learning labs require students to directly observe organisms, generate and analyze original data, draw conclusions, and link their findings to the scientific literature. In other words, students were required to behave like ecologists by doing ecology and thereby came to see themselves as future scientists. This comported with the findings of Ballen et al. (2017) who found that active-learning increased as a result of increasing self-efficacy in underrepresented Students of Color. Students who can see themselves as scientists become more active in their learning of science. With this in mind, I ditched the lab manual and its fuzzy diagrams of lifecycles and predictable dissections. Instead we explored ecological concepts by observing common species such as competition among shorebirds, prey preference among gulls, tree phenology, plant diversity, etc. Students also modeled scientists by specializing on one taxa per lab, sharing their knowledge with peers, then merging data while testing different hypotheses.

A campus-wide civic engagement mandate that requires students help solve real-world problems allowed me to engage all students in volunteer-based citizen science. My research on horseshoe crabs at a nearby beach is a long-standing collaboration with NYC Bird Alliance, who tag and monitor spawning crabs to protect this vital species and educate the public. Not only does blood from these crabs play a vital role in the biomedical field, but the crab population is declining due to habitat loss and overharvest. Citizen science allows regular folks to play a role in large-scale, long-term scientific research that would not be possible otherwise. It effectively merges volunteer work, community outreach, and data collection to raise awareness and protect biodiversity. Turns out it is also a wonderful learning opportunity.

Reframing students as volunteer scientists and future peers allowed me to involve them all in my research, certify them for CE credits, model scientific behavior, and create a pipeline for future interns. Students can now redefine themselves as practitioners and envision a path forward in STEM. Indeed, Habig and Gupta (2021) found that underrepresented youth's participation in authentic informal research promotes mastery of skills requisite for science practices and improved interest in scientific research. In 2017, student feedback on campus field trips was positive, and visits to our campus farm, beach, or marina proved good practice for the annual capstone field trips to Plumb Beach to study horseshoe crabs not as students, but as scientists.

11.6 Ecology in the Age of COVID

COVID-19 lockdown presented new challenges and new opportunities when all my fieldwork was replaced with online labs. Thankfully, I was already using an online open access database called iNaturalist, where folks can upload photos of plants, animals, or fungi they observe, thereby contributing to an ever-growing global database of wildlife sightings. Previously I had students use it to help ID specimens, but I could now enlist students to become contributors by surveying the biodiversity in their neighborhood, map the population density of whales, or find the peak season for mushroom sightings. Students could also use the map feature to find the nearest observations of a slime mold, sand shark, or ginkgo in their own borough. Students invariably marvel at the richness of local ecosystems and the scientific power of this online tool/phone app. Student feedback was overwhelmingly positive toward websites that allowed them to engage with nature, science, and a community of other citizen scientists.

Another remote learning tool that became invaluable during lockdown was Zooniverse; a clearing-house website where researchers from around the world solicit help to process mountains of remote data such as trail-cam photos. An intuitive series of questions and pictograms about size, color, and shape of critters captures on camera lead novice viewers to correct identifications of species, making this tool ideal for learners at all levels. I assigned students to process 100 images from a wildlife surveys in L.A., then compare these with an existing Excel data set of coyote sightings in New York from Gotham Coyote, a remote sensing collaboration through AMNH. Other citizen-science dataset including Journey North and National Phenology Network are among many open access resources I have yet to fully explore.

11.7 Planning Post COVID Trips to NYC Informal Science Institutions

Returning to campus was fraught with logistical, safety, and financial challenges, so I was grateful to be teaching Ecology as a hybrid with lectures online. Social distancing and contact-free labs eliminated group work, dissections and even microscopy. With so many barriers (literal and figurative), I seized the opportunity to get back outdoors and use in-person field trips to local cultural institutions to enrich student learning.

After running a several Ecology Club field trips in the early 20 teens to the Brooklyn Museum, Shark Discovery at Time Square and whale watching aboard the American Princess, I became increasingly alarmed at the lack of engagement with cultural institutions I saw among my club members. As a result, back in 2018, I took my entire ecology class to the Coney Island Aquarium which opened their eyes to the spectacular local diversity of species just below the water's surface. Students were surprised to learn sharks, stingrays, and whales are considered local. They were amazed at what they saw and were unaware that these institutions were accessible for free to local residents. Inspired by this success, I decided to add additional field trips to local cultural institutions starting in 2021. The rest so to speak, is [natural] history.

In-person field trips were planned for after spring break, and included Brooklyn Botanic Garden, Prospect Park Zoo, Coney Island Aquarium, and AMNH. Each trip targeted specific galleries where students selected a few exhibits to make observations and collect data. This allowed students to understand the exhibits as scientific resources and to see themselves as curators and creators of content in these otherwise exclusionary spaces. In year two, I wrangled curator mini talks and tours of their labs which allowed students to peer behind the public facade and see scientists in action.

When I broke my foot near the start of the spring 2022 semester, I was forced to become the guide on the side, not the sage on the stage, as I was taught thorough ongoing professional development (see King, 1993). I sat near an entrance and sent the students off to collect their data, finally able to resist the temptation to scamper from group-to-group prodding or encouraging them, finally seeing that the more responsibility that students are given, the more they accept. I've watched exhausted faculty juggling tasks while student look on with blank stares. I have done this myself and lived to regret training them to be passive recipients. Now my students know that the sooner they figure out and complete their task the sooner they can sign out and go home. They also learned that if confused, I was on hand to help. A quick review of their datasheets allowed me to assess their work and correct it if needed.

11.8 Reflecting Back on a Pilot

With grades in, and commencement over, I was finally able to reflect on the semester. In full disclosure, this was the only class I taught that spring, and the student roster was unusually small with only 18 students as opposed to the usual 24. All students completed the course, and none dropped out after week one. These variables alone could dramatically alter the outcomes of any class; I am a far better professor when teaching a small group of students 1 day per week. Similarly, that the grades were unusually high with nobody falling below a "C-" can be attributed to the fact that students were able to resubmit work, and students missing major assignments were given an incomplete, allowing them to finish later.

During the first 4 weeks, labs were held on campus (see attached syllabus). A computer lab was procured for the first 2 weeks in order to process camera trap data from Gotham Coyote and Zooniverse. Students reported that they liked these labs because they were able to learn about local wildlife and about the many citizen science projects across the world. Jumping into processing data rather than working up to it was strategic in terms of letting them "learn by doing." The following week they processed a photo set of horseshoe crabs and their epibionts (aka hitchhikers) I collected during lockdown. Moving from terrestrial animals to marine species provided a contrast between the big and visible creatures they saw in the camera trap data, with the small aquatic and less visible taxa offshore. Introducing them to horseshoe crabs also established a baseline of information that the students would need later in the semester when we studied the crabs on the beach.

After 2 weeks in the computer lab, students were tasked with collecting data on campus. A survey of predation by gulls required that they collect shells from crustaceans and mollusks dropped by gulls around campus. A study of competition among birds required that they observe and record aggressive behaviors on an ethogram (aka data sheet for recording behaviors). Finally, a phenology study of campus trees allowed them to observe and compare the timing of budbreak and leaf flush among the plants they otherwise barely notice. Studying vertebrates, invertebrates, terrestrial, aquatic, plants, and animals built on and reinforced the taxonomy learned in their prerequisite.

Student enthusiasm and high attendance increased when we began field trips off campus, even despite the higher logistical challenges. Getting to the zoo or museum in the midst of a pandemic was a perfect excuse for absences, although students would have to make up the lab through a solo trip sometime that week. If a student was unable to go for quarantine or mobility reasons, an equivalent assignment was devised using remote data collection. For example, a student unable to come to Brooklyn Botanic Garden had to do a biodiversity index of trees in the Prospect Park using data from iNaturalist.

Allowing time for travel to and from campus required that trips be shortened by about 45 minutes depending on the distance to the destination. Arrival times for each student varied since some came from campus, others from home. Therefore, I designed each lab to allow students to complete specific tasks individually or in pairs. This allowed me to remain at the entrance and greet students as they arrived. All worksheets and instructions were posted online but I would provide copies and a sign in sheet to keep track of each student's arrival and departure time. If a given student came late, they could simply remain longer to complete their tasks. If a student needed to leave early, they knew to arrive early.

My field trip protocol was to be at the entrance to the institution specified in their pre-lab instructions. I would usher students in as they arrived and dole out tickets, clip-boards, pens, and data sheets as needed. Having a few one-on-one minutes with students was pivotal in building relationships. Once a critical mass of students was already inside, I would make my way to the first of two designated galleries to check on students. Movement to a second gallery was allowed only once a student's worksheet was approved. There was usually a compare and contrast between the galleries that helped prevent students from dividing up the work. Each student was assigned a subset of species or specific dioramas which prevented them from copying one another's work. Questions like "Describe the feel of this exhibit to an alien from another planet" prompted creative narratives and made plagiarism or AI nearly impossible. Students were free to leave once they showed me their work and signed out. The occasional late student would be required to dedicate a minimum of two hours on site. While I was at liberty to leave, I often preferred to stay and accompany the stragglers, again as an opportunity to connect and engage. While I presumed this would be a deterrent, some students preferred the extra attention.

One complaint from students was that these informal science institutions were too crowded and frenetic. Flexibility on which exhibits to review and in which order helped students to mitigate the inevitable crowds, while exhibits with a single entrance allowed me to greet, monitor, and aid students as needed. Text messages from students allowed me to guide them if they got lost or ran into problems, which is a frequent issue with many students who are often unfamiliar with NYC cultural institutions.

The first two field trips (AMNH and Coney Island Aquarium) were indoors and thus easier to manage. The second two trips (Brooklyn Botanic Garden and Prospect Park Zoo) were small enough for students to backtrack or find me as needed. The Bronx Zoo and New York Botanic Garden are large and difficult to access and navigate, so given the choice, students overwhelmingly opted for smaller and closer venues despite one having a $4 entrance fee.

The final 2 weeks were the culmination of the semester. This was when the students participated in my decade-long research on the American horseshoe crab

(*Limulus polyphemus*) whose blood is a biomedical blockbuster product used to test surgical equipment for bacterial contamination. This protected species is also a critical food supply for endangered migratory shorebirds so my research on their spawning activity and my role on the IUCN SSC Horseshoe Crab Specialist Group is part of an ongoing effort by the conservation community to monitor this vital species. The horseshoe crabs were on their predictable march toward the shores, and our task was to go out to their spawning beach and tag as many as we could. In preparation, students read and summarized one of my recent peer reviewed articles based on a decade of my research on the same beach with the same species. Students were also already familiar with the crabs from our week three study of their epibionts attached to their shell such as algae, barnacles, and bryozoans. They also watched video clips of my work from a local TV show. As the end of the semester approached, term papers and final exams loomed. In the past, I would have ramped up content delivery to get it all in under the wire, but now I have learned to skip the paperwork and instead let their performance in the field serve as a lab practical for their final task. Since the final topic was Conservation Ecology, what better way to learn than to contribute to the conservation of the horseshoe crab?

Students were briefed on how to get to the beach by car and where to meet me if they required transportation via campus vehicle or Uber. I assigned students specific tasks to reduce chaos and broke them into two teams. The students took their tasks seriously and worked diligently; some scanned the beach in search of crabs; others measured their size, identified gender, age, and epibiont load, while others recorded data, took photos, or attached the tags. Those who arrived early were given leadership roles and became peer mentors. This was a spontaneous decision which has become entrenched in my teaching approach. I have found that these early students are often prepared and focused. Rather than repeat myself to the latecomers, I seized on the opportunity to let the early and eager students brief late students while I step aside and listen in, to verify they got it right or needed help. Their ability to convey the information also reassured me they were well prepared for a leadership role, allowing me to differentiate my teaching on the fly.

While in the field, the students were laser focused on their tasks. They lost track of time and quickly fell into a well-organized unit. In one case, the fear of touching a crab the size of a hubcap was too much; that student was instead charged with recording data. However, the following week the same student was spotted running up and down the tidal flats, hijab blowing in the breeze, happily collecting crabs, and apparently well over her initial fear. When dismissed, a chorus of indignant cries emerged about how lab was too short and they wanted to continue to tag. This was the highlights of my semester. This was the kind of learning that cannot happen in a lab simulation. The power of emotional learning is profound, and in my opinion, deeply undervalued in our education system.

The weather was less cooperative the following and final week. Attendance was the only requirement to get full credit. Students were again invited to bring friends, family, and associates. My only instructions were to swap jobs with someone else and train their own replacement. There was a celebratory vibe among the students; they were now teammates and colleagues. They were now taking charge of their own learning and leadership. Chilly weather and a much longer hike out to find the crabs did little to sway their enthusiasm at first, but soon the work began to take a toll (fieldwork is much more fun in balmy weather!) The tide was lower and the crabs further out on the flats; wet feet and chilly arms tired more quickly. The novelty had clearly worn off. While it is vital to see a field site at both its best and under challenging conditions, it was clear the students had learned enough for their taste. One group prematurely declared they were done, leaving others to finish up their work. Those who remained either owed me extra time, or were already hooked, and happy to stick it out.

Overall, the final field trips were a success. While a written reflection of their civic engagement experience was required, their narratives were unable to convey the full range of emotions and reflections possible in person. In past years, samples were collected in the field then processed in lab, which allowed for a deeper reflection on their field experience. In future years, I may require their term paper to hinge upon one of the crabs epibionts, thus allowing me to close the loop and take a deeper dive into fewer species, as opposed to a curriculum that is broad but lacks depth.

11.9 Student Reflections

Student reflections at the end of the semester provided feedback to the course overall, while weekly reflections helped me adjust expectations and assignments in real time. Because these writings were required, I could include student comments by simply requesting permission for work they had already done. All students were required to review, correct, and comment on each other's work, so all assignments benefitted from peer review. Their feedback provided excellent guidance and allows me to acknowledge them as contributors here. Participating in scientific research and contributing to scholarly works has been found to have a transformative effect on students to become further be involved with science (Carpi et al., 2017).

Written reflections focused on field work and civic engagement and revealed overwhelmingly positive feedback. Words such as "favorite," "loved" "great," "amazing" "awesome" and "best experience I ever had" were pervasive. Feeling like a scientist and seeing themselves as future scientists stood out as evidence of success. Several students commented on how unexpected and different this class was. Excerpts include:

> Horseshoe crab tagging was one of my favorite activities…here at KBCC!
>
> I did not know that regular everyday folks [without] much of a scientific background can contribute and help further science so much by collecting and posting data!
>
> This class made me inspired.
>
> The hands-on aspect of the class….allowed me to have an immersive understanding of what it's like to be in the field as a scientist.

Prof. Colon made a difference in my life…doing what I did in ecology felt like therapy.

The [crab tagging] lab played a particularly special role in my journey as a scientist.

Civic engagement showed me what it can be like after this course; a future where I might just be in the field doing who knows what.

The horseshoe crab tagging field lab was amazing; it's an experience I didn't know I would ever do in my life.

It's awesome that we are able to make a change with nature..[and].. help them thrive.

11.10 Moving Forward and Expanding the Model

In the future, I would like to add a trip to the Metropolitan Museum of Art to search for elements of endangered species on display in the art and artifacts (e.g., bone, ivory, tortoise, mahogany, amber, pearl, etc.). I would also like to expand on the reflection elements of the field trip destinations as a form of social exclusion or inclusion. I plan to ask whether the exhibition of live or stuffed endangered species is justified. I'd like to ask students how this course format impacted their perceptions of what it means to do science or to be a scientist. After all, being scientifically literate, active, and engaged is a form of civic engagement. Therefore, I plan to assign a pre and post semester reflection on their sense of inclusion in, and access to science as a practice and as a career. This may help address "imposter syndrome" which appears to disproportionately impact first-generation college students and Students of Color, and can lead to a decrease in equity and diversity among college students and young professionals (see Le, 2019; Chrousos & Mentis, 2020).

Field trips have always been a fun part of many curricula, yet have traditionally been removed from learning at the secondary or undergraduate level. I challenge this and recommend each college course include at least one field experience; be it a poetry reading, historic site, walking tour, guest speaker, museum, or hospital tour. I would also like to formalize my inclusion of fun as an integral part of the labs themselves. To this end, I also advocate for including family in the learning experience. While I cannot by law allow parents siblings or children onto my campus or into my teaching lab, field trips to public spaces or cultural institutions are a wonderful way to include the family. Some of my students are on parent duty, while others are watching their elders or siblings. I designed all my field trips to be fully accessible to anyone and everyone who wishes to participate, and I always encourage extra credit trips to be a family affair. One student brought her mother to Brooklyn Botanic Garden and while she was welcome to sit with me at the entrance, she opted to walk with her daughter to assist (or not) with her botanical search. Another brought her son who added both enthusiasm and assistance with reading signs and finding target species. Sending students off on self-guided assignments worked well for them to include others. The low or free admission options we selected also reduced the barriers of admission costs and in some cases, extra tickets could be passed on to members of the entourage. If my goal is to reduce barriers to STEM learning, then it makes sense to allow students to meet their family duties as well as extend the learning to their circle.

I would also like to expand the role of students in the teaching as well as the learning. Once students had mastered their preliminary role in the field work, they were happy to train their replacement the following week. Students at every stage of their academic career need to have leadership, public speaking, and content expertise "included" in their educational experience. Therefore, these elements are always included, both as part of the planning and opportunistically. For example, when a student shows up late after I finish explaining things, I will invite students who were present and attentive to explain things to the latecomer. I can listen to them explain as a way to assess how well my instructions were absorbed.

Another aspect of inclusive education I hope to expand is community. When I received a request to run a field trip or demonstration for a group of afterschool students, I was more than happy to include them in our horseshoe crab work. By now my students were already content experts so I was excited to let them hone their leadership skills. One student brought a colleague who was a vet tech seeking experience working with wild invertebrates; another student brought a friend, another brought her younger siblings, a third brought her son, and one fellow even brought a date! All of these additional people added value to the process since there was more than enough work to go around. I also included two film crews and a journalism student writing a feature on community outreach. Finally, I included research colleagues, who worked alongside of the students and rounded out this exceedingly diverse and very inclusive experience.

11.11 In Conclusion

Education is a progression from passive learning (e.g., note taking) to active leadership (e.g., knowledge creation). This is perhaps most pressing among students breaking out of a marginalized demographic. All students require the opportunity to make this transition, yet all too many are denied this vital rite of passage. Time, money, resources, and administrative flexibility are in short supply outside the elite academic institutions. These limitations become burdens borne by the consumers of affordable, mass education experiences. Many students do not learn well in a large passive learning experience and therefore drop out or give up before they are given the opportunity to take control of their learning. Many students with financial, emotional or familial stress, Post-Traumatic Stress Disorder (PTSD), or undiagnosed or untreated learning (dis)abilities are particularly vulnerable to premature attrition. My friend's brother suffered this fate despite his higher than average intelligence. His learning style was tactile, visual, and anything but passive, so consuming education in a traditional setting was untenable. Yet he was fully capable of consuming information faster and more voraciously than anyone I have ever met. He could fix or repair anything. He could read a book in a day. And he could spend countless hours transfixed in a museum gallery, obsessively reading and absorbing every sign, every fact, and every item on display. He did this alone, quietly and for his own edification. There were no grades, no tests, and no teachers, and perhaps if there were, he may well have opted out.

I help many students who falter in the classroom or lab, shine in the field; perhaps because I was among their ranks. In middle school, I somehow confused classes, lost my way or was late due to poor time management. In college, I fell asleep during class after waitressing till 2 am or studying all night. I would loose the thread during lecture and fly into a panic while scribbling half sentences and nonsensical fragments gleaned from the chalkboard. In lab, I was perpetually grossed out by dissections, and often got lost in the lab manual instructions. I often mismeasured or inadvertently skipped a step in chemistry lab, resulting in a failed experiment with no understanding of where I had gone wrong. In hindsight, I suspect I did not fall through the cracks due to the privledged education I received, which included a robust academic safety net. This luxury is denied all too many of my students, who are at risk of giving up or falling through the cracks.

I see active learning and authentic experience-based learning as a form of equity in education. Hands-on autonomous learning opportunities, free from a didactic lecturer is a way to create a more inclusive learning environment where multiple intelligences, problem-solving, logistical capabilities, and sheer passion can foster motivation to seek a deeper understanding of concepts that come easily for some, yet may require additional effort for others. This may be particularly true among demographics previously marginalized in STEM education pathways. As we look to the future and see a shift in the demographics of America toward one with a higher ratio of Black, Hispanic, and Asian youth, we must actively work to demystify what it means to be a scientist, and facilitate opportunities for them to engage with science both inside and outside of the classroom (see Bayne, 2021).

This approach is not without risks, and it is not for everyone. Some students see it as a dereliction of duty on the part of Professor Colon to sit by and dole out tasks, cheering on at the sideline. I have been accused of such in evaluations by students who felt cheated by this style. And while it is easy for me to write these comments off as a lack of vision or comprehension of my goals, the truth is they have a valid point on many levels. Turning over the reins to a student is not a guarantee of success, nor is it the cake walk it may appear. However, allowing students, colleagues, and administrators exposure to a wide diversity of teaching methodologies as part of their academic experience is perhaps a vital part of their educational journey. So while the risk of criticism is high, the rewards of providing an authentic and inclusive experience are invaluable. Fine tuning my perception of students who require or desire more traditional approaches is not only another form of inclusion, but also an essential element of inclusive teaching practice (see Rodgers, 2002). A truly inclusive and equitable educational experience is one where all needs are addressed, all learners are respected, and all approaches are "included."

Appendix: Course Syllabus Summary

Biology 5300: Ecology 2021
Lecture 1: Chapter 1—Introduction to Ecology
Lab 1: **GOTHAM COYOTE** Campus computer lab (citizen science)
Lecture 2: Chapter 5—Populations
Lab 2: **ZOONIVERSE** Computer lab on citizen science
Lecture 3: Chapter 7 Predation
Lab 3: **CAMPUS BEACH SURVEY** Evidence of gull predation

Lecture 4: Chapter 6 Competition
Lab 4: **BIRD BEHAVIOR** Evidence of bird competition with scan sampling

Lecture 5: Chapter 8 Parasitism and Mutualism **TEST 1**
Lab 5: EPIBIONTS Hitchhikers on oysters and horseshoe crabs

Lecture 6: Chapter 9 Communities
Lab 6: CAMPUS TREE PHENOLOGY Tree bud burst and flower flush

Lecture 7: Chapter 4.1 Terrestrial Ecosystems
Lab 7: AMERICAN MUSEUM OF NATURAL HISTORY Biomes. Meet at AMNH

Lecture 8: Chapter 4.4 Aquatic Ecosystems
Lab 8: CONEY ISLAND AQUARIUM Shark ethogram. Meet at aquarium

Lecture 9: Chapter 14 Conservation
Lab 9: **PROSPECT PARK ZOO** Survey of endangered species. Meet at zoo

Lecture 10: Chapter 10 Biodiversity **TEST 2**
Lab 10: BROOKLYN BOTANIC GARDEN Tree biodiversity index. Meet at BBG

Lecture 11 Field Ecology and Civic Engagement
Lab 11: PLUMB BEACH Horseshoe crab tagging. Meet on campus or at beach

Lecture 12: Horseshoe Crab Research and Civic Engagement reflections
Lab 12: **PLUMB BEACH** Horseshoe crab tagging. Meet on campus or at beach

References

Adams, J. D. (2007). The historical context of science and education at the American Museum of Natural History. *Cultural Studies of Science Education, 2*, 393–440. https://doi.org/10.1007/s11422-007-9059-8

Adams, J., & Gupta, P. (2017). Informal science institutions and learning to teach: An examination of identity, agency and affordances. *Journal of Research in Science Teaching, 54*(1), 121–138.

Ballen, C. J., Wieman, C., Salehi, S., Searle, J. B., & Zamudio, K. R. (2017). Enhancing diversity in undergraduate science: Self-efficacy drives performance gains with active learning. *CBE—Life Sciences Education, 16*(4), 1–6. https://doi.org/10.1187/cbe.16-12-0344

Bayne, G. U. (2013). Coteaching, peer tutoring and curriculum writing: Lasting effects of involving students in talking about science. *Pedagogies: International Journal of Education, 8*(4), 369–383.

Bayne, G. U. (2021). Global voices: Personal and professional lived experience of black and brown women's culture in science. In M. M. Atwater (Ed.), *International handbook of research on multicultural science education* (Springer international handbooks of education). Springer. https://doi.org/10.1007/978-3-030-37743-4_45-1

Bellino, M. E., & Adams, J. D. (2017). A critical urban environmental pedagogy: Relevant urban environmental education for and by youth. *The Journal of Environmental Education, 48*(4), 270–284. https://doi.org/10.1080/00958964.2017.1336976

Black, M. D. (2020). Exploring relationships between a teacher's race-ethnicity and gender and student teaching expectations. *Education Inquiry, 12*(2), 202–216. https://doi.org/10.1080/20004508.2020.1824343

Carlone, H. B., & Johnson, A. (2007). Understanding the science experiences of women of color: Science identity as an analytic lens. *Journal of Research in Science Teaching, 44*(8), 1187–1218.

Carpi, A., Ronan, D. M., Falconer, H. M., & Lents, N. H. (2017). Cultivating minority scientists: Undergraduate research increases self-efficacy and career ambitions for underrepresented students in STEM. *Journal of Research in Science Teaching, 54*, 169–194.

Chaudhury, A., & Colla, S. (2021). Next steps in dismantling discrimination: Lessons from ecology and conservation science. *Conservation Letters, 14*, e12774.

Chrousos, G., & Mentis, A. (2020). Imposter syndrome threatens diversity. *Science, 367*(6479), 749–750.

Eckholm, E. (1984). New View of Female Primates Assails Stereotypes. The New York Times.

Eddy, S. L., & Hogan, K. A. (2014). Getting under the hood: How and for whom does increasing course structure work? *CBE—Life Sciences Education, 13*(3), 453–468.

Elmesky, R., & Tobin, K. (2005). Expanding our understandings of urban science education by expanding the roles of students as researchers. *Journal of Research in Science Teaching, 42*, 807–828.

Freeman, S., Eddy, S. L., McDonough, M., Smith, M. K., Okoroafor, N., Jordt, H., & Wenderoth, M. P. (2014). Active learning boosts performance in science, engineering, and mathematics. *Proceedings of the National Academy of Sciences, 111*(23), 8410–8415. https://doi.org/10.1073/pnas.1319030111

Haak, D. C., HilleRis Lambers, J., Pitre, E., & Freeman, S. (2011). Increased structure and active learning reduce the achievement gap in introductory biology. *Science, 332*(6034), 1213–1216.

Habig, B., & Gupta, P. (2021). Authentic STEM research, practices of science, and interest development in an informal science education program. *International Journal of STEM Education, 8*. https://doi.org/10.1186/s40594-021-00314-y

Hill, A., Guralnick, R., Smith, A., Sallans, A., Gillespie, R., Denslow, M., Gross, J., Murrell, Z., Conyers, T., Oboyski, P., Ball, J., Thomer, A., Prys-Jones, R., de la Torre, J., Kociolek, P., & Fortson, L. (2012). The notes from nature tool for unlocking biodiversity records from museum records through citizen science. *ZooKeys, 209*, 219–233.

King, A. (1993). From sage on the stage to guide on the side. *College Teaching, 41*(1), 30–35. https://doi.org/10.1080/87567555.1993.9926781

Le, L. (2019). Unpacking the imposter syndrome and mental health as a person of color first generation college student within institutions of higher education. *McNair Research Journal SJSU, 15*, Article 5.

Louv, R. (2010). *Last child in the woods*. Atlantic Books.

Malcolm, S. M., Hall, P. Q., & Brown, J. W. (1976). *The double bind: The price of being a minority woman in science*. American Association for the Advancement of Science.

Mann, J., Gray, T., Truong, S., Sahlberg, P., Bentsen, P., Passy, R., Ho, S., Ward, K., & Cowper, R. (2021). A systematic review protocol to identify the key benefits and efficacy of nature-based learning in outdoor educational settings. *International Journal of Environmental Research and Public Health, 18*(3), 1199. https://doi.org/10.3390/ijerph18031199

Medin, D. L., & Bang, M. (2014). *Who's asking? Native science, western science, and science education*. MIT Press. https://doi.org/10.7551/mitpress/9755.001.0001

National Center for Science and Engineering Statistics [NCSES]. (2021). *Women, minorities, and persons with disabilities in science and engineering: 2021*. Special Report NSF 21-321. National Science Foundation. Retrieved from https://ncses.nsf.gov/wmpd

Pew Research Center. (2021). *STEM jobs see uneven progress in increasing gender, racial and ethnic diversity*. Retrieved from https://www.pewresearch.org/science/2021/04/01/stem-jobs-see-uneven-progress-in-increasing-gender-racial-and-ethnic-diversity/

Rennie, L. J. (2014). Learning science outside of school. In *Handbook of research on science education* (Vol. 2, pp. 134–158). Routledge.

Rodgers, C. (2002). Seeing student learning: Teacher change and the role of reflection. *Harvard Educational Review, 72*(2), 230–253.

Russell, D. G. D., Sladen, W. J. L., & Ainley, D. G. (2012). *Dr. George Murray Levick (1876–1956): Unpublished notes on the sexual habits of the Adélie penguin*. Published online by Cambridge University Press.

Simard, S. (2021). *Finding the mother tree: Discovering the wisdom of the forest*. Alfred A. Knopf.

Stern, M. J., Powell, R. B., & Frensley, B. T. (2022). Environmental education, age, race, and socioeconomic class: An exploration of differential impacts of field trips on adolescent youth in the United States. *Environmental Education Research, 28*(2), 197–215. https://doi.org/10.1080/13504622.2021.1990865

Taylor, D. E. (2014). *The state of diversity in environmental organizations*. Green 2.0.Working Group. Belgum.

Theobald, E. J., Hill, M. J., Tran, E., Agrawal, S., Arroyo, E. N., Behling, S., Chambwe, N., Cintrón, D. L., Cooper, J. D., Dunster, G., & Grummer, J. A. (2020). Active learning narrows achievement gaps for underrepresented students in undergraduate science, technology, engineering, and math. *Proceedings of the National Academy of Sciences, 117*(12), 6476–6483.

Thompson, J. J., & Jensen-Ryan, D. (2018). Becoming a "science person": Faculty recognition and the development of cultural capital in the context of undergraduate biology research. *CBE—Life Sciences Education, 17*, ar62.1–ar62.17.

Chapter 12
Promoting a Culture of Inclusion in Graduate Engineering Education: A Case Study

Walter C. Lee, Jeremi S. London, Adam S. Masters, Teirra K. Holloman, David B. Knight, and Eunsil Lee

Abstract It is well established that equitable participation in graduate programs is not always facilitated for students from all backgrounds, and there is a need to continue improving graduate STEM education practices and research cultures that are characterized as unwelcoming environments. In this chapter, the authors share insights from a 3-year effort to develop, implement, and refine a mandatory seminar course focused on promoting equity in STEM. Over the course of six semesters (i.e., Fall 2019–Spring 2022), the team taught the course to approximately 1500 graduate students from 12 engineering departments at a predominantly White institution, continuously evolving their pedagogical practices based on student experiences and feedback. Part I of this chapter summarizes insights on graduate education that informed the course design. Part II details the course design process. Part III overviews the assessment techniques used to evaluate course effectiveness. And Part IV presents lessons learned from the instructor team's group reflection on their experiences.

Keywords Multiculturalism · Higher education · Course design · Teach teaching · Doctoral institutions

W. C. Lee (✉) · A. S. Masters · T. K. Holloman · D. B. Knight
Department of Engineering Education, Virginia Tech, Blacksburg, VA, USA
e-mail: walterl@vt.edu; masters1@vt.edu; teirrah@vt.edu; dbknight@vt.edu

J. S. London
Office of the Provost, Vanderbilt University, Nashville, TN, USA
e-mail: Jeremi.London@vanderbilt.edu

E. Lee
Department of Engineering Education, University at Buffalo, Buffalo, NY, USA
e-mail: eunsille@buffalo.edu

G. L. Cochran, J. D. Adams (eds.), *Equity-Oriented, Discipline-Based STEM Education Research*, Critical Studies of Education 19,
https://doi.org/10.1007/978-981-96-9136-4_12

12.1 Introduction

According to the National Academies of Sciences, Engineering, and Medicine (2018), in an ideal STEM graduate education system, "Students from all backgrounds would fully participate and achieve their greatest potential during their educational experience through transparent institutional action to enhance diversity and promote inclusive and equitable learning environments" (p. 3). However, it is well established that full participation across graduate programs is not always facilitated, and there is a need to continue improving graduate STEM education practices and research cultures that are characterized as unwelcoming environments (e.g., Sanczyk et al., 2021; Burt et al., 2021; Limas et al., 2022; Moss & Mahmoudi, 2021).

Studies on graduate students' sense of belonging, which serves as a measure of a student's perceived inclusiveness within an academic unit, echo this claim. For example, Gardner et al. (2014) showed that engineering graduate students reported a relatively lower sense of belonging within their academic department compared to students in other disciplines. In a related study, O'Meara et al. (2017) pointed out that there are fewer facilitators (e.g., a critical mass of women, micro affirmations, etc.) for creating an inclusive culture in STEM graduate program environments than in non-STEM programs, supporting the need to develop new mechanisms for promoting an inclusive culture in STEM graduate education. Across these prior studies, the need to identify and disrupt the teaching, learning, and practice norms in STEM graduate education is quite evident.

In this chapter, we share insights from a 3-year effort at Virginia Tech (VT) to develop, implement, and refine a mandatory graduate seminar course focused on promoting equity in the College of Engineering (COE). Over the course of six semesters (i.e., Fall 2019–Spring 2022), we taught the course to approximately 1500 graduate students from 12 engineering departments, continuously evolving our pedagogical practices based on student experiences and feedback. Our insights from this experience are organized as follows. Part I summarizes insights on graduate education that informed our course design. Part II details our course design process. Part III summarizes the assessment techniques used to evaluate effectiveness. Part IV includes results from our group reflection on our experience as instructors. These four elements are consistent with Boyer's (1990) Scholarship of Teaching and Learning (SoTL). The insights in this chapter will help educators at other universities promote a culture of inclusion in graduate STEM education.

12.2 Part I: Knowledge About STEM Graduate Education

Each part of this chapter is shaped by a guiding question. The question guiding our approach to Part I is: *What existing scholarship is important for us to consider when aiming to improve STEM graduate education?* Improving STEM graduate education to be more inclusive requires an understanding of its many distinct cultural characteristics. For example, in many STEM disciplines, graduate student experiences frequently occur within team and lab settings. These collaborative spaces are where many graduate students develop their expertise and acquire knowledge, spending intensive time working on projects and tasks with other peers (e.g., Burt, 2017; Crede & Borrego, 2012).

In discipline-based STEM education research, the overwhelming majority of literature focuses on doctoral students and graduate students' research experiences. Researchers and practitioners often overlook the experience of graduate students in course-based master's programs, as well as all graduate students' non-research experiences (e.g., see National Academies of Sciences, Engineering, and Medicine, 2018)—experiences that have implications for teaching and learning. In light of this gap in STEM graduate education literature, our team has reflected here on our experience teaching a graduate engineering seminar course focused on equity to a heterogeneous student population. In Part I, we discuss the importance of (a) addressing salient equity issues in graduate STEM education by (b) more intentionally socializing incoming students to graduate engineering education.

12.2.1 Salient Equity Issues in Graduate STEM Education

There is no shortage of equity issues in graduate STEM education. For example, bullying, harassment, precarious contracts, and competitive research environments—which are all too common in STEM—can have a negative effect on a student's mental health (Limas et al., 2022) and academic outcomes. To document the nature and extent of academic bullying behaviors in STEM, Moss and Mahmoudi (2021) collected and analyzed survey responses from 2006 people and found that an overwhelming majority of their participants (84%) had experienced some form of bullying. In addition to these issues, STEM graduate students frequently encounter competitive and socially isolating settings (e.g., Rodriguez et al., 2022); unclear, unspoken, and inconsistently applied expectations (e.g., Bahnson et al., 2022); and unequal access to opportunities (e.g., opportunity hoarding, Matthew Effect), which can be associated with students' funding mechanisms (Grote et al., 2021). Beyond STEM, graduate education can also be stressful for an array of reasons, such as poor work/school–life balance, pressure from academic responsibilities, and professional isolation (El-Ghoroury et al., 2012). Gardner's (2009) study of graduate student attrition found that common reasons for students ending a program early include personal problems (e.g., family and physical/mental health issues), departmental issues (e.g., poor advising, department policies, and politics), and a poor personal fit with graduate school. Combined, these issues create a high-stakes environment ripe with competition and inequities, so much so that prior scholars have referred to graduate education as *an ongoing interview* where students are continually becoming a member of their respective professions (Peters, 1997).

In addition to understanding these issues, it is also important to understand how they manifest. Tuma et al. (2021), for instance, shed light on how the practices, values, and norms of STEM and academic research can negatively influence graduate experiences, contributing to the aforementioned issues being so prevalent in STEM. Through analyzing interviews with 40 life science doctoral students, they highlight the negative effect of (1) gatekeeping intended to "weed out" students; (2) an incentive structure that discourages quality mentorship; (3) power structures that create a hierarchical relationship between faculty and students; and (4) tension between the roles of being a student versus an employee (Tuma et al., 2021). These findings allow us to conclude that the equity issues students face in graduate school are not only the result of inequitable policies and practices, inequity also shows up in interpersonal interactions between peers and when interacting with people in positions of power (e.g., advisors). An awareness of these issues and how they manifest informed our approach to leveraging a graduate seminar course to intentionally socialize incoming students and suggest strategies for how to navigate these situations.

12.2.2 The Role of Socializing Students to Address Equity Issues

We adopted the concept of graduate student socialization (Weidman & Stein, 2003) in the course design. Graduate student socialization refers to the process of becoming a member of the academic community (Weidman & Stein, 2003); it is known to be significant for graduate student success and persistence (Fisher et al., 2019; O'Meara et al., 2017). According to Weidman's Graduate Student Socialization Model, socialization requires graduate students to interact with others in their unit (e.g., faculty and peers) to develop a sense of integration and learn knowledge and skills relevant to their field (Twale et al., 2016; Weidman & Stein, 2003). We consequently designed our seminar with the purpose of educating students about interpersonal interactions they should anticipate encountering and the associated expectations.

Student interactions with faculty and other students were the most salient in graduate STEM education literature. For example, regarding faculty-to-student interactions, it is not uncommon for an advisor–advisee relationship to be characterized as a supervisor–supervisee relationship, where interactions focus on academic performance and work obligation as opposed to psychological and social matters (Primé et al., 2015). Student progress and funding are also often closely tied to the advisor's research, further impacting these relationships. Prior research also highlights the tendency for marginalized STEM graduate students (e.g., women, Students of Color) to be neglected by their mentors (Gay, 2004; Sanczyk et al., 2021).

Regarding student-to-student interactions, it is important for students to be prepared to work collaboratively within team and lab settings. Most STEM graduate students develop their expertise and acquire knowledge through spending intensive time working on projects and tasks with peers in these collaborative spaces.

Consequently, peer relationships within lab-based STEM disciplines are a vital aspect of graduate training that is influenced by factors associated with lab-based settings, including the lab structure and lines of communication, and departmental culture (Golde et al., 2009; Jeong et al., 2019).

In light of the collaborative nature of STEM graduate education, a large part of being socialized entails learning how to successfully navigate interactions with faculty and peers. Additionally, it is important to ensure that students are socialized to not be neutral; in essence, they should not just avoid reinforcing negative peer-to-peer interactions. Consequently, our team taught students how to navigate existing norms, as well as how to combat negative ones.

12.3 Part II: Curriculum Development and Implementation

The question guiding our approach to Part II is: *How might we plan a graduate seminar to disrupt facets of exclusionary engineering graduate education culture?* At this point, we sought to determine how to take what we know about STEM graduate education more broadly and apply it to our specific context within the VT COE. Subsequently, our course design is best understood when situated in Lattuca and Stark's (2015) model of an Academic Plan in Sociocultural Context. The Academic Plan model provides a framework for viewing a curriculum as an academic plan or blueprint. The model assumes an intentional planning process in the development of the curriculum while acknowledging the role of internal and external influences; it suggests that all of the components of the model are included in every academic plan, whether consciously or unconsciously (Lattuca & Stark, 2015). Those components include purpose, content, sequence, learners, instructional resources, instructional processes, plus assessment and evaluation. This perspective highlights the various decisions that we made and the sociocultural context in which these decisions were situated, and Part II is situated in these components. More details about how we designed and implemented the course, as well as examples of course content can be found here (Holloman et al., 2021).

12.3.1 Internal Influences

Both institutional and unit-level influences shaped course design. We developed the seminar course based on the results of an assessment performed by the VT COE Dean's Office in partnership with the VT Graduate School around graduate students' experiences in the COE. As a result of this assessment, the adviser/advisee relationship was found to be a key area affecting student success. More specifically, three primary issues were identified: (1) misaligned goals and expectations between advisers and advisees; (2) poor communication between faculty and graduate students; and (3) the potential for race, gender, and culture to influence interactions between faculty and graduate students. In response to these issues, the COE sought ways to help faculty and students avoid these potential points of conflict, and enhance the graduate student mentoring culture within the COE.

Around the same time and occurring independently of that assessment effort, the VT Commission on Graduate Studies and Policies (CGSP) approved a resolution to incorporate an inclusion and diversity education component into the graduate education of *all* degree-seeking graduate students by Spring 2022 (Virginia Tech Graduate School, n.d.). Individual graduate programs were given the autonomy to determine the most appropriate way to enact that requirement with the one constraint that programs must be able to track student completion. In anticipation of the inclusion and diversity requirement going into effect, VT's COE recognized a unique opportunity to also strengthen the training provided to incoming students about navigating advisor–advisee relationships. COE administrators partnered with the VT Department of Engineering Education to create Graduate Student Success in Multicultural Environments (GSSME), a one-credit, pass/fail seminar course that could address some of the mentoring challenges identified by the COE assessment, while simultaneously meeting CGSP's new inclusion and diversity education requirement (Holloman et al., 2021). Departments from across the COE were invited to have their students meet the CGSP requirement through this course, and 12 of 13 departments independently signed on, resulting in the GSSME course having an annual enrollment of approximately 500 graduate students.

Strengthening the training of graduate students also required strengthening the training of faculty members who could lead such an effort. As a result, two members of our team participated in training offered by the Center for the Improvement of Mentored Experiences in Research (CIMER) focused on their "Entering Research" curriculum (Branchaw et al., 2010, 2018; Balster et al., 2010). CIMER is composed of researchers and practitioners who focus on improving research mentoring relationships among postsecondary researchers. We used the "Entering Research" curriculum as a guide when creating the first offering of this course. Over time, however, we adapted the content (topics, examples, activities, assignments) to better align with the needs of our college. The remainder of this chapter is based on our revised curriculum.

12.3.2 Course Purpose

According to Lattuca and Stark (2015), the purpose refers to the learning outcomes for the course, and the content is the subject matter used to achieve those outcomes. The GSSME course had a dual purpose: (1) to improve the research mentoring experiences for engineering graduate students, and (2) to meet a new requirement for graduate programs across the university to incorporate an inclusion and diversity component. Accordingly, we developed the following learning objectives, where successful completion of the course would result in a student's ability to:

- Recognize and understand essential relationships found in graduate education
- Assess the consequences of personal actions and language
- Understand and respect the needs, perspectives, and actions of others
- Recognize interpersonal and intrapersonal conflicts, and identify possible solutions and
- Reflect on one's role in the Virginia Tech community and engineering profession

These learning objectives were created through an iterative process. Initially, our objectives were unintentionally informed more by the COE's interest in improving mentoring than the Graduate School's interest in advancing diversity and inclusion (Holloman et al., 2021). Through iteration, we developed the final version of our learning objectives (listed above). These objectives are the result of considering how integral relationships are to graduate student success (e.g., advisor/advisee, peer/peer, student/staff, student/mentor, etc.), and how working relationships rely on respect and an awareness of how our differences might impact our communication. Combined, the course purpose and learning objectives were intended to promote equity by explicitly addressing the hidden curriculum of graduate engineering education and confronting inequitable norms that need to be addressed by the entire community.

12.3.3 Learners

Understanding the prior experiences and preparation of students (i.e., learners) coming into a course was essential in determining how to best present the content. Our course design considered prior knowledge and preparation in various ways, including disciplinary background, degree type (e.g., Ph.D. vs. MS), race/ethnicity, cultural differences, gender, sexual orientation, disability, and socioeconomic status. While many of these considerations translate across STEM contexts, we want to highlight two that are potentially more unique to engineering. First, students enroll in the seminar from a variety of engineering departments, such as Aerospace Engineering, Biomedical Engineering, Civil Engineering, Electrical Engineering, Industrial & Systems Engineering, and Mechanical Engineering. This variation was important for us to consider because each discipline has its own subculture (e.g., Knight et al., 2012), and these differences in norms translate to their own unique issues related to equity. Second, more than half of our graduate students were classified as international; these demographics are a large shift from U.S.-based undergraduate engineering programs, where the number of international students is significantly lower (American Society for Engineering Education, 2021). This was important for us to consider because we could not assume students had an understanding of equity in the context of the United States, similar socializations to engineering, or prior experiences in an environment with similar demographics of peers.

12.3.4 Course Content and Sequence

During the six-semester development process, we experimented with how we presented course topics. Table 12.1 lists the final sets of topics and how they were sequenced. Across these topics, we discussed the norms and expectations of graduate school and explored how the demographic realities of engineering disciplines shape engineering education practice. As noted in Table 12.1, each session focused on two distinct yet related topics. Whereas the topics themselves did not change significantly from one iteration to another, the sequencing and specific content we used to present the topics shifted as our understanding of the learners grew. We will offer one example of each.

In terms of sequencing, we initially covered *Addressing External Stressors* and *Managing Internal Stressors* (Parts 9 and 10) much earlier in the course, and concluded this course with *Ethical Behavior* and *Professionalism* (Parts 5 and 6). We changed this sequencing after concluding that (a) discussing how to address and manage stress was a much more effective way to conclude the course, and (b) complicating conventional ideas about ethics and professionalism sooner rather than later made it easier to then discuss the nuances of *Communicating Effectively* and *Dealing with Group Dynamics* (Parts 7 and 8).

In terms of content, we initially used a visual of the Gender Unicorn (i.e., a graphic illustration commonly used to introduce gender as a social construct) (Trans Student Educational Resources, 2015) to discuss how language and understanding of diversity topics evolve, but eventually shifted to using a published article (i.e., Castania, 2003). We made this change after receiving feedback from a student who felt the cartoonish nature of the Gender Unicorn graphic undermined the seriousness of the topic. Castania (2003), although flawed, acknowledges the ongoing nature of dialogues around how language about diversity is used.

We also experimented with forms of delivery that each met the university's minimum required contact hours for a 1-credit hour course. Initially, the standard course offering was structured as 3-h sessions that occurred once per week for 5 weeks and an "intensive" version of the course taught over two 8-h sessions on back-to-back days. The most notable difference between these offerings was that the 2-day

Table 12.1 Overview of course topics and sequence

Session	Topics
1	Part 1: Aligning Expectations
	Part 2: Establishing a Support Network
2	Part 3: Promoting Diversity, Equity, and Inclusion
	Part 4: Understanding Stereotypes and Bias
3	Part 5: Encouraging Ethical Behavior
	Part 6: Examining Professionalism
4	Part 7: Communicating Effectively
	Part 8: Dealing with Group Dynamics
5	Part 9: Addressing External Stressors
	Part 10: Managing Internal Stressors

"intensive" version made it possible for students to more singularly focus on the seminar, whereas the 5-week version provided time for reflection and task completion between each session. As a result of the COVID-19 pandemic, we also developed and offered online versions of both the standard and "intensive" offerings where synchronous class times were cut in half. We supplemented class time with multimedia assignments to prepare students to come to class ready to engage. When comparing the different modalities, we found in-person delivery made it easier for students to interact with one another, while online delivery lowered the barrier to participation as students could simply use the chat feature.

12.3.5 Instructional Resources and Processes

A group of two Engineering Education faculty members led the development of this course through the use of various instructional resources. Course development was aided by financial and administrative support from the COE, as well as evidence-based curricula developed by the Center for the Improvement of Mentored Experiences in Research (CIMER) (https://cimerproject.org/), which is housed at the University of Wisconsin-Madison. Additionally, two advanced graduate teaching assistants (GTAs) met weekly with the course instructors to plan the course curriculum, reflect on the previous class, and aid in classroom instruction. To ensure adequate resources were available during the 3-year course development period, the COE provided the financial resources necessary for two GTAs. Support like this from the COE was crucial to the success of this initiative because the department's normal assistantship allotment is one GTA per course.

Instructional processes are the activities that take place in the classroom (or through online media) to present the content to the learners as well as assessments of student learning. In the 5-week course format, sessions were discussion- and case-based. Each session included short lectures to introduce the main ideas, one or two activities for students to engage with the content, a discussion of inclusion considerations for each topic, and finally, a debriefing discussion. Every session ended with "Closing Tips" that drove home the key takeaways from the discussions.

12.3.6 Assessment of Learning

This seminar was offered on a pass/fail basis. Although attendance was mandatory and the completion of an end-of-course survey was required, student learning was only assessed based on two deliverables: critical reflections (CR) and a professional development plan (PDP). Each deliverable included four parts. The instructional team met and created a matrix to ensure that course content and assignments were aligned with the learning objectives. The CR assignment invited students to connect the topics from class to their experiences, career, and beliefs through reflections on each of the following topics: bias & stereotypes; diversity philosophies; perspectives on professionalism; and vulnerable populations. The PDP assignment prompted students to consider and apply what they learned throughout the course. In particular, students were tasked with establishing career goals, planning ahead, identifying ways to support their personal wellness, and aligning expectations with a mentor of their choice.

12.4 Part III: Evaluative Methods

The question guiding our approach to Part III is: *How can we systematically monitor our efforts as instructors?* Evaluation of course effectiveness is an essential part of the plan used to check whether the purpose was achieved, followed by making adjustments for the next implementation of the course (Lattuca & Stark, 2015). Our team used an array of evaluation methods to facilitate continuous improvement and monitor student reactions to adjustments in the curriculum or teaching practices. Table 12.2 summarizes the evaluation methods we used for each offering of the course. The timing is presented based on a 5-week version, but similar evaluation methods were used for other formats. To illustrate the usefulness of those methods, we will elaborate on the results from the end-of-course survey below.

Table 12.2 Summary of evaluation methods

Evaluation type (amount)	Purpose	Timing
Instructor team meetings (4)	Capture ongoing ideas for improvement of future seminar offerings	Weekly
Weekly surveys (4)	Identify student questions, comments, and concerns in preparation for the next section	Weekly in between class sessions
End-of-course survey (1)	Capture student feedback and suggestions immediately after the usefulness of the topics to inform following seminar offerings	After final session
Wrap-up meeting (1)	Capture overall ideas for improving the course design to improve following seminar offerings	End of semester
Student assignments (2)	Identifying misconceptions or common misunderstanding that suggested potential changes in the course content, delivery, or assignment instructions	End of semester
Teaching evaluation scores (1 per instructor)	Capture student feedback and suggestions at the end of the semester in which they took the course to inform following seminar offerings and evaluate our overall instruction	End of semester

Note: Evaluation types that occurred weekly were not implemented for the 2-day versions of the seminar

12.4.1 Student Feedback from End-of-Course Survey

Data were collected from students enrolled in the seminar course from Fall 2019 to Spring 2022. At the end of each seminar course, students completed a survey that included both closed- and open-ended questions on the students' perceived usefulness of the course content, challenges and barriers in various aspects of their graduate education experiences (e.g., advising relationships, faculty/peer interactions, etc.), and a demographic survey. Data analysis for this book chapter utilized the data from Fall 2020 to Spring 2022 considering that a pilot data collection was conducted prior to Fall 2020 with pilot survey questions.

Part IV discusses how we incorporated student feedback from these surveys. The remainder of this section focuses on students' perception of the course from years two to three. This data represents one way of gauging the effectiveness of the changes we made to the course. Table 12.3 shows descriptive statistics of students' satisfaction with the course and their perception of 13 topics covered in the course for the 2020–2021 (n = 407) and 2021–2022 (n = 395) academic years. This table also includes t-test results between the data collected in different implementations, shown to illustrate how it is possible to identify the impact of changes made from one year to another.

By conducting comparative analyses between the data from 2020–2021 to 2021–2022, our team found that students' responses on the course satisfaction and their perception of the course content suggest that the course successfully gained student buy-in. The mean value of the course satisfaction from 2020 to 2022 showed that students tended to be satisfied with the course (3.60 for 2020–2021 and 3.98 for 2021–2022; 3 = Neutral and 4 = Satisfied). Similarly, the mean value of the students' perceived usefulness of the course topics collectively ranged from 3.44 to 3.72 (3 = Somewhat useful and 4 = Very useful) for the 2020–2021 implementation, implying the overall positive reception of the course content. For the next implementation, results also showed enhanced students' perception of whether the course topics are valuable and helpful (ranged from 3.44 to 3.80), which collectively suggest an improvement made in the way of delivering the course throughout the iterations and adjustments.

12.5 Part IV: Lessons Learned

The question guiding our approach to Part IV is: *How might lessons from these experiences inform broader practices in STEM graduate education?* In addition to the aforementioned evaluation methods enabling us to improve the course, they also allowed us to identify high-level insights of value to the broader STEM community. The insights in this section are the direct results of the final two 1-h wrap-up

Table 12.3 Descriptive statistics and *t*-test results of end-of-course survey (2020–2021 & 2021–2022)

	2020–2021 ($n = 407$)		2021–2022 ($n = 395$)		
	Mean	Std. Dev.	Mean	Std. Dev.	*t*-test
Grad School Experience[a]	3.93	0.82	4.07	0.84	−2.55*
Course Satisfaction[a]	3.60	1.05	3.98	0.92	−5.53
Diversity, Equity, and Inclusion (DEI)[b]					
DEI and privilege	3.68	0.99	3.70	0.94	−0.32
Vulnerable populations	3.52	0.98	3.63	0.88	−1.68
Stereotypes, bias, and stereotype threat	3.72	0.96	3.72	0.92	0.00
Influence of D&I on language	3.67	0.95	3.65	0.90	0.30
Bystander intervention	3.48	0.99	3.44	0.98	0.54
Communication and Stress Management[b]					
Aligning expectations with advisors/supervisors/mentors	3.62	0.89	3.76	0.79	−2.32*
Effective communication/barriers to communication	3.72	0.93	3.78	0.90	−0.95
Establishing a support network	3.44	0.92	3.70	0.84	−4.25
Addressing external stressors	3.55	0.94	3.76	0.85	−3.31**
Managing internal stressors	3.59	0.94	3.79	0.88	−3.24**
Behavior[b]					
Constructive and destructive behaviors in groups	3.70	0.97	3.70	0.86	−0.50
Ethical behavior	3.70	0.93	3.68	0.89	0.25
Professional behavior	3.69	0.93	3.80	0.88	−1.69

[a]A response option for ***Course Satisfaction*** scale was arrayed on a five-point Likert scale from 1 (Very unsatisfied) to 5 (Very satisfied)

[b]Response options for **Diversity, Equity, and Inclusion (DEI)**, ***Communication and Stress Management***, and ***Behavior*** scales were arrayed on a five-point Likert scale from 1 (I do not remember covering this topic) to 5 (Extremely useful)

*$p < 0.05$; **$p < 0.001$

meetings among the instructional team. In preparation for writing this chapter, these meetings were recorded and analyzed more systematically than the meetings that occurred during the prior semesters. As a result, we identified three overarching strategies that summarized our engagement with the course. A summary of the strategies identified through this process is presented in Table 12.4, which we elaborate on in the subsequent sections. To facilitate others' implementation of these strategies, we also note specific ways (i.e., tactics) in which each strategy is relevant to course preparation, implementation, and improvement.

Table 12.4 Emergent strategies and tactics for expanding equity-oriented practices

	Tactics		
Strategy	Stage 1: Preparation	Stage 2: Implementation	Stage 3: Improvement
Embrace sincerity	Display intellectual humility	Engage in honest dialogue	Identify entry points
Foster accessibility	Know your audience	Leverage technology	Adapt to your students
Minimize harm	Establish productive norms	Disrupt problematic incidents	Implement feedback

12.5.1 Embracing Sincerity

The embrace sincerity strategy included displaying intellectual humility, engaging in honest dialogues, and identifying multiple entry points for reaching students. As we taught the course, we realized we needed to be transparent and vulnerable about our STEM experiences and beliefs. This honesty helped us balance making students feel like they belonged within the course and giving them tools to create a culture that exuded this sense of belonging throughout the entire COE. We navigated this fine balance by negotiating wanting students to feel welcomed, needing to prepare them for the fact that the culture is not welcoming everywhere, and equipping them with information to inspire change where possible.

12.5.1.1 Display Intellectual Humility During Course Preparation

The first tactic for embracing sincerity is beginning with an attitude of humility and recognizing that discussing topics through an equity lens requires drawing on various areas of expertise. We recommend treating this work as an academic endeavor that requires continuous learning while also being transparent about one's own imperfections and ongoing learning. We found that adopting this mindset was important for ensuring that we were sufficiently prepared for classroom conversations. By prepared, we mean having a degree of familiarity with each topic; being able to anticipate the range of questions, comments, and reactions that might come (both good and bad); and having a general plan for how we might respond, who might respond, and when. Team teaching was particularly useful for this tactic as we each had our own prior knowledge, experiences, and identities to draw upon and leverage when appropriate.

We collectively needed to be prepared to respond to any topic that arose during a session, being careful not to erode the sense of belonging and trust that we were developing with students when we addressed shortcomings and gaps in students' understanding. Preparation included updating our knowledge and awareness about the topics using both public and scholarly publications (e.g., blogs, news articles, journal articles, books), spending time developing the material we used in class (e.g., slides, handouts, 2-page facilitator guides), and meeting as an instructor team before class to discuss results from the weekly surveys to make sure we were prepared to facilitate the conversation.

12.5.1.2 Engage in Honest Dialogue During the Course Implementation

The second tactic for embracing sincerity is engaging in honest dialogue and being upfront about the current status of the university. We recommend being honest about the current situation at your institution, within your discipline(s), and about your current understanding (or lack thereof) regarding each topic. We found it important to acknowledge the shortcomings of the current system and not overpromise when it came to discussing university efforts. We made sure to explain why equity does not happen automatically and focused on providing students with tools for navigating and improving spaces they might personally occupy. We knew not to assume that delivering this seminar meant everything about the typical, exclusionary culture would change; all we could do was introduce the values and norms we wanted to see and prepare students for situations they might find themselves in. Finding balance here was central to how we implemented the seminar. Our hope was that the students in our seminar, regardless if they had experienced certain issues or not, would benefit from conversations about how to address these issues and potentially make them better allies in real-time for their peers.

It was important for us to be honest about our own experiences with these topics, our own biases and shortcomings, and when we did not know the best approach to addressing a situation. While we would share our opinions, we did not shy away from acknowledging when we were not sure if our personal approach would be the most appropriate or effective for all situations. We explained that everyone is a socialized human being with an imperfect brain (e.g., unconscious biases), so we must be proactive in managing how we show up in spaces. Lastly, in instances where student statements in class discussions were problematic, we aimed to use the situation as an opportunity to highlight an existing, problematic norm rather than rebuke the individual student, so as to not put anyone on trial in class.

12.5.1.3 Identify Entry Points During Course Improvement

The third tactic for embracing sincerity is identifying entry points to connect the topics to students' prior knowledge and experiences. We recommend constantly looking for ways to incorporate students' prior knowledge/experiences, and anticipate their questions and comments. After a few iterations of offering the course, we began to anticipate typical student responses and revised the seminar delivery accordingly. For example, we noticed that conversations about racism did not immediately resonate with some international students (e.g., those from countries with racially homogenous populations). One student suggested we also mention South Asia's caste system; we incorporated this point so we could more broadly discuss different forms of social stratification. Our mindset of intellectual humility led us to search for ways to more intentionally connect the topics to our students; openly welcoming feedback aided in this endeavor.

12.5.2 *Fostering Accessibility*

The second lesson we learned was the importance of fostering accessibility. This lesson included knowing our audience, leveraging technology, and adapting to our students. We wanted to make sure that there were minimal barriers to both engagement and understanding the content, while also accounting for the wide variety of students we were tasked with serving. We worked to achieve this goal while the course capacity increased (from ~300 to 500+ in 3 years), and we were forced to rethink what "accessible" meant during a global pandemic.

12.5.2.1 Know Your Audience During Course Preparation

The first tactic for fostering accessibility is ensuring you know your audience. We recommend looking for ways to explicitly connect course topics to students' disciplinary priorities and epistemological basis. We were careful in selecting which topics to focus on and how we framed them for an engineering audience. For example, because we were aware of the positivist/post-positivist paradigm influence in engineering, we incorporated quantitative data and research findings throughout our course material. We also intentionally sought out engineering or engineering-adjacent examples as much as possible, especially when it came to more social issues such as diversity, equity, and bias. In doing so, our goal was to clearly convey the relevance of the material, and minimize the likelihood of students thinking the material was, at best, tangential to engineering or, at worst, irrelevant to engineering. In practice, this meant that for every topic, we explicitly communicated the implications of the topic for engineering work or education. We created clear bridges between our topics and our context.

12.5.2.2 Leverage Technology During the Course Implementation

The second tactic for fostering accessibility is intentionally leveraging technology. We recommend thoughtfully using technology to remove barriers to engagement and participation. We found technology to be particularly helpful in three ways. First, we used survey administration software to collect information from students during the days leading up to each class session. Students who may have been reluctant to raise a question in front of their peers could submit them in advance. We regularly used student submissions to inform in-class conversations. Second, we used the Zoom chat feature to facilitate classroom discussions. Instead of having to communicate vocally, students could carefully craft their questions/comments before pressing send. We found Zoom lowered the barrier for people to ask honest questions, as students could send both public and direct-to-instructor messages. Providing numerous ways for students to engage helped make our class a much more inclusive and accessible space.

Lastly, Zoom provided the option for closed captions and made it possible for students to attend class if they were acutely ill. Some of the most complimentary pieces of feedback we received in the End-of-Course Survey were people thanking us for the course and disclosing a disability that they'd not had to previously disclose because our course was accessible from the start so there was no need for accommodation.

12.5.2.3 Adapt to Your Students During Course Improvement

The third tactic for fostering accessibility is adapting to your students. We recommend dismissing the notion of the "typical student" and making sure you know who is actually in the class. When we first designed the course, we had a generic engineering graduate student in mind (e.g., White, middle-class, heterosexual man from the United States) but soon realized this student only represented a fraction of who we were actually teaching. For example, we quickly realized that most of our students were international and had not grown up in the United States, and many students were in non-thesis Master's programs without substantial research components. We had to reconsider how we could ensure the variety of students was represented in how we taught the course.

12.5.3 Minimizing Harm

The third lesson we learned was the importance of minimizing harm. This lesson included establishing productive norms, disrupting problematic incidents, and implementing feedback from students. Given the nature of the topics covered in our course and the diverse backgrounds of students, we had to balance having difficult conversations with creating an environment that felt safe and respectful. Because it was impossible to predict everything that would come up during class or how every student would respond to the topics, achieving this goal required a continual sensitivity to student feedback.

12.5.3.1 Establish Productive Norms During Course Preparation

The first tactic for minimizing harm was establishing productive norms. We recommend preparing rules of engagement for the space you are creating. For us, that meant creating ground rules which we reminded students about every session. Specifically, our ground rules were: respect each other's personal experiences and humanity; address and refer to everybody according to their personal identity; interpret statements in the most rational way possible; attack the idea, as opposed to the person, if you disagree with someone; avoid disagreeing for the sake of disagreeing; monitor one's own level of engagement to ensure that everyone may contribute; be mindful of other people's privacy. We also established norms around how to respond if you or someone else failed to adhere to a ground rule via the "Oops and Ouch" Communication Recovery technique (Aguilar, 2014). We took this approach because we wanted to model what we considered a more inclusive culture where people could engage in difficult conversations in a respectful way. Our hope was that long taking our course, students would carry certain ideals that would promote equity in the COE.

12.5.3.2 Disrupt Problematic Incidents During the Course Implementation

The second tactic for minimizing harm was disrupting problematic incidents. We recommend interrupting and redirecting conversations as necessary. While we did not want to silence anyone, we also did not want to make space for students to be harmed or experiences to be invalidated because of problematic statements expressed by their peers. This is where establishing productive norms also proved useful. Ground rules enabled us to disrupt someone without making the incident about the individual student; instead, we were interrupting students because their behavior went against the pre-set norms around which we all agreed. Ground rules can serve as guard rails but are only useful when enforced.

12.5.3.3 Implement Feedback During Course Improvement

The third tactic for minimizing harm was implementing feedback. Although tempting to assume instructor-knows-best, we recommend taking suggestions, complaints, or concerns seriously whenever they are brought to your attention. Because students have such different opinions about these kinds of topics, implementing this tactic can be challenging. To minimize harm, we ultimately came up with this guideline: if a minoritized student made a complaint that was grounded in their lived experience about class content related to their minoritized identity(s), we needed to ponder ways to honor their feedback. Otherwise, our instructional team discussed the feedback to discern if it was helpful for the course objectives or coming from a place of defensiveness.

12.6 Conclusion

If the STEM community wants to ensure graduate students can fully participate and reach their potential, new mechanisms for promoting an inclusive culture in STEM graduate education are needed. This chapter describes one effort intended to improve graduate engineering education practices and research cultures. We hope that our lessons will help educators at other universities as they work to address salient equity issues in graduate STEM education and socialize incoming students more intentionally to graduate STEM education.

Acknowledgment We thank Drs. Kirsten A. Davis, Bryan Hanson, John J. Lesko, Aevelina Rahman, and Tremayne O. Waller for their contributions to the development and implementation of this course. Portions of this manuscript were presented at the Collaborative Network for Engineering and Computing Diversity (CoNECD) Virtual Conference, January 24–28, 2021.

References

Aguilar, L. (2014). Ouch! and oops! *The Ouch! Files, 5*(1) Retrieved from https://www.diversity-inclusioncenter.com/archives/ouch_files/Archives/Ouch_Vol5No1.html

American Society for Engineering Education. (2021). *Profiles of engineering and engineering technology*. American Society for Engineering Education, Washington, DC.

Bahnson, M., Satterfield, D., Wyer, M., & Kirn, A. (2022). Interacting with ruling relations: Engineering graduate student experiences of discrimination. *Studies in Engineering Education, 3*(1), 53–78. https://doi.org/10.21061/see.76

Balster, N., Pfund, C., Rediske, R., & Branchaw, J. (2010). Entering research: A course that creates community and structure for beginning undergraduate researchers in the STEM disciplines. *CBE Life Sciences Education, 9*(2), 108–118.

Boyer, E. L. (1990). *Scholarship reconsidered: Priorities of the professoriate*. Princeton University Press.

Branchaw, J. L., Pfund, C., & Rediske, R. (2010). *Entering research facilitator's manual: Workshops for students beginning research in science*. W.H. Freeman.

Branchaw, J. L., Butz, A. R., & Smith, A. R. (2018). *Entering research: A curriculum to support undergraduate & graduate research trainees*. W.H. Freeman.

Burt, B. A. (2017). Learning competencies through engineering research group experiences. *Studies in Graduate and Postdoctoral Education, 8*(1), 48–64. https://doi.org/10.1108/SGPE-05-2017-019

Burt, B. A., McCallum, C. M., Wallace, J. D., Roberson, J. J., Bonanno, A., & Boerman, E. (2021). Moving toward stronger advising practices: How Black males' experiences at HPWIs advance a more caring and wholeness-promoting framework for graduate advising. *Teachers College Record, 123*(10), 31–58. Retrieved from https://tcrecord.org/Content.asp?ContentId=23857

Castania, K. (2003). *The evolving language of diversity*. Cornell Cooperative Extension. Retrieved from https://hdl.handle.net/1813/2665

Crede, E., & Borrego, M. (2012). Learning in graduate engineering research groups of various sizes. *Journal of Engineering Education, 101*(3), 565–589. https://doi.org/10.1002/j.2168-9830.2012.tb00062.x

El-Ghoroury, N. H., Galper, D. I., Sawaqdeh, A., & Bufka, L. F. (2012). Stress, coping, and barriers to wellness among psychology graduate students. *Training and Education in Professional Psychology, 6*(2), 122. https://doi.org/10.1037/a0028768

Fisher, A. J., Mendoza-Denton, R., Patt, C., Young, I., Eppig, A., Garrell, R. L., et al. (2019). Structure and belonging: Pathways to success for underrepresented minority and women PhD students in STEM fields. *PLoS One, 14*(1). https://doi.org/10.1371/journal.pone.0209279

Gardner, S. K. (2009). Student and faculty attributions of attrition in high and low-completing doctoral programs in the United States. *Higher Education, 58*(1), 97–112.

Gardner, S. K., Jansujwicz, J. S., Hutchins, K., Cline, B., & Levesque, V. (2014). Socialization to interdisciplinarity: Faculty and student perspectives. *Higher Education, 67*(3), 255–271. Retrieved from https://digitalcommons.library.umaine.edu/mitchellcenter_pubs/56

Gay, G. (2004). Navigating marginality en route to the professoriate: Graduate students of color learning and living in academia. *International Journal of Qualitative Studies in Education, 17*(2), 265–288. https://doi.org/10.1080/09518390310001653907

Golde, C. M., Bueschell, A. C., Jones, L., & Walker, G. E. (2009). Advocating apprenticeship and intellectual community: Lessons from the Carnegie initiative on the doctorate. In R. G. Ehrenberg & C. V. Kuh (Eds.), *Doctoral education and the faculty of the future* (pp. 53–64). Cornell University Press.

Grote, D., Patrick, A., Lyles, C., Knight, D., Borrego, M., & Alsharif, A. (2021). STEM doctoral students' skill development: Does funding mechanism matter? *International Journal of STEM Education, 8*(1), 1–19.

Holloman, T. K., Lee, W. C., Davis, K. A., London, J. S., Knight, D. B., Waller, T. O. B., & Lesko, J. J. (2021, January). *Work in progress: Preparing graduate students to engage in multicultural environments*. Presented at 2021 CoNECD (virtual). https://doi.org/10.18260/1-2--36146

Jeong, S., Blaney, J. M., & Feldon, D. F. (2019). Identifying faculty and peer interaction patterns of first-year biology doctoral students: A latent class analysis. *CBE—Life Sciences Education, 18*(4), ar59, 1–ar59,13.

Knight, D. B., Lattuca, L. R., Yin, A., Kremer, G., York, T., & Ro, H. K. (2012). An exploration of gender diversity in engineering programs: A curriculum and instruction-based perspective. *Journal of Women and Minorities in Science and Engineering, 18*(1), 55–78. https://doi.org/10.1615/JWomenMinorScienEng.2012003702

Lattuca, L. R., & Stark, J. S. (2015). *Shaping the college curriculum: Academic plans in context* (2nd ed.). Jossey-Bass.

Limas, J. C., Corcoran, L. C., Baker, A. N., Cartaya, A. E., & Ayres, Z. J. (2022). The impact of research culture on mental health & diversity in STEM. *Chemistry—A European Journal, 28*(9), e202102957. https://doi.org/10.1002/chem.202102957

Moss, S. E., & Mahmoudi, M. (2021). STEM the bullying: An empirical investigation of abusive supervision in academic science. *eClinicalMedicine, 40*, 101121. https://doi.org/10.1016/j.eclinm.2021.101121

National Academies of Sciences, Engineering, and Medicine. (2018). *Graduate STEM education for the 21st century*. The National Academies Press. https://doi.org/10.17226/25038

O'Meara, K., Griffin, K. A., Kuvaeva, A., Nyunt, G., & Robinson, T. N. (2017). Sense of belonging and its contributing factors in graduate education. *International Journal of Doctoral Studies, 12*, 251–279. https://doi.org/10.28945/3903

Peters, R. L. (1997). *Getting what you came for: The smart student's guide to earning a master's or a Ph. D.* Farrar, Straus and Giroux.

Primé, D. R., Bernstein, B. L., Wilkins, K. G., & Bekki, J. M. (2015). Measuring the advising alliance for female graduate students in science and engineering: An emerging structure. *Journal of Career Assessment, 23*(1), 64–78.

Rodriguez, S. L., Perez, R. J., & Schulz, J. M. (2022). How STEM lab settings influence graduate school socialization and climate for students of color. *Journal of Diversity in Higher Education, 15*(1), 58. https://doi.org/10.1037/dhe0000361

Sanczyk, A., Merriweather, L. R., Howell, C. D., & Douglas, N. (2021). Cracks in doctoral STEM mentoring relationships: Students' perceptions. *International Journal for Cross-Disciplinary Subjects in Education, 12*(1). https://doi.org/10.20533/ijcdse.2042.6364.2021.0534

Trans Student Educational Resources. (2015). *The gender unicorn*. Retrieved April 6, 2023 from http://www.transstudent.org/gender

Tuma, T. T., Adams, J. D., Hultquist, B. C., & Dolan, E. L. (2021). The dark side of development: A systems characterization of the negative mentoring experiences of doctoral students. *CBE—Life Sciences Education, 20*(2), ar16. https://doi.org/10.1187/cbe.20-10-0231

Twale, D. J., Weidman, J. C., & Bethea, K. (2016). Conceptualizing socialization of graduate students of color: Revisiting the Weidman-Twale-Stein framework. *Western Journal of Black Studies, 40*(2), 80–93.

Virginia Tech Graduate School. (n.d.). *Inclusion and diversity requirement*. Retrieved April 6, 2023, from https://graduateschool.vt.edu/academics/programs/inclusion-diversity-req.html

Weidman, J. C., & Stein, E. L. (2003). Socialization of doctoral students to academic norms. *Research in Higher Education, 44*(6), 641–656.

Chapter 13
Cultural and Systemic Barriers to Graduate Stem Education

Diana Sachmpazidi

Abstract The goal of this chapter is to review the landscape of equity-oriented research in Science, Technology, Engineering, and Mathematics (STEM) graduate education. This chapter will serve as a resource to researchers in understanding the breadth of equity-oriented research in STEM graduate education and support them in identifying existing gaps in the literature where additional work is required. This chapter will also serve as a resource to practitioners and policymakers interested in engaging in institutional-wide efforts to further the quality of graduate education and educational outcomes for their students. In this chapter, I primarily focus on the systemic challenges students from historically excluded social groups face which create barriers to their equitable educational experience. I explore the literature to understand the ways in which the racialized, sexist, homonegative, and ableist structure of STEM graduate education maintains inequities in student experience and educational outcomes. I do so by focusing on discriminatory beliefs, policies, values, and distribution of resources which inform the marginalized nature of STEM culture. Finally, I conclude this chapter by identifying the gaps in the literature and discussing the implications for researchers.

Keywords Graduate education · STEM culture · Systemic barriers and supports · Marginalized students

---

D. Sachmpazidi (✉)
School of Physics and Astronomy, Rochester Institute of Technology, Rochester, NY, USA

Department of Physics, University of Maryland, College Park, MD, USA
e-mail: dxssps@rit.edu

G. L. Cochran, J. D. Adams (eds.), *Equity-Oriented, Discipline-Based STEM Education Research*, Critical Studies of Education 19,
https://doi.org/10.1007/978-981-96-9136-4_13

13.1 Introduction

Marginalization is "the process through which persons are peripheralized based on their identities, associations, experiences, and environment" (Hall et al., 1994, p. 25). Marginalization is tied to the enactment of hegemony and White supremacy, which are functions of institutional structures. Historically excluded social groups (Students of Color, women of all racial and ethnic backgrounds, students who identify with sexual and gender minorities, and students with disabilities) are left with the choice to adapt to the hegemonic values of the dominant culture, navigate the challenges posed by the hostile academic environment, or leave the field (McGee, 2020a).

The representation of students with historically marginalized identities (Black, Latinx, and Native American students of all genders) in Science, Technology, Engineering, and Mathematics (STEM) graduate education has been a concern and area of focus of major national initiatives [i.e., National Science Foundation's (NSF) INCLUDES Alliance: Inclusive Graduate Education Network (IGEN); NSF's Alliance for Graduate Education and the Professoriate (AGEP)] and higher education researchers (e.g., Barthelemy et al., 2016; Posselt et al., 2019; Arnold et al., 2020), especially over the last decade. While structural and systemic shortcomings in academia impact all students and are partly reflected in the high attrition rates (~40% across all STEM fields) (Sowell et al., 2008), students with historically marginalized identities experience additional challenges of institutional racism and sexism.

For many years, researchers approached the phenomenon of unequal educational outcomes from the "student focus," aiming to understand the personal factors that predict academic success. For example, commonly investigated concepts were student self-efficacy, motivation, and identity development (see, Santiago & Einarson, 1998; Allen, 1999; Castillo et al., 2006). While it is well-documented that such personal factors are important predictors of student achievement and retention (i.e., Rockinson-Szapkiw et al., 2022; Allen, 1999; Hernandez et al., 2017), more recent studies approach these factors under the notion that the academic context (e.g., "chilly" climate, faculty members' mindsets and biases) that, for example, inhibits the development of one's self-efficacy. Canning et al. (2019) examined whether STEM faculty members' mindset beliefs about the fixedness or malleability of ability (e.g., "To be honest, students have a certain amount of intelligence, and they really can't do much to change it") link to student motivation and/or course achievement. The authors found that students had lower academic achievement in courses taught by fixed-mindset faculty members than growth mindsets. While students from Black, Latinx, and Native American backgrounds had consistently lower grades than White and Asian American students, the racial achievement gap was nearly twice as large in courses taught by fixed-mindset faculty members. Moreover, faculty members with a fixed-mindset on ability and intelligence tend to emphasize traditional criteria for admission practices that have been found to result in a less diverse student body (Scherr et al., 2017).

The approach of predominantly focusing on the student's personal factors to understand educational achievement gaps is another manifestation of structural racism in STEM as it shifts the focus away from how institutions perpetuate racism and sexism and instead aims to "fix" the student (McGee, 2020a, 2020b). For example, when it comes to increasing graduate retention, one of the critical focus areas where departments strive to improve, students view retention as being tied to systemic change while faculty view attrition as being inevitable when admitting "underprepared" students (Owens et al., 2018; Sachmpazidi, 2021). The STEM culture is dominated by the White male supremacy paradigm that largely governs its policies, practices, and values. For example, STEM culture has traditionally favored norms of success that are based on competitiveness, individualism, and solitary practices, which are stereotypical characteristics of White male scientists and necessitate spaces that favor effects of marginalization for students who do not share these characteristics (Ong et al., 2018; Miller et al., 2021). Explicit and subtle racial and gender discrimination and prejudice in STEM is well-documented (Lee et al., 2020; Harper et al., 2011) and is often manifested as meritocracy, race evasiveness (Brunsma et al., 2017; McGee, 2020a), and ableism (Reinholz & Ridgway, 2021). Institutional actors who espouse the notion of meritocracy view success as a result of one's intelligence, effort, and achievement, ignoring any inherited social privilege. Such meritocratic values generate implications for systemic discrimination, which is reflected, for example, in admission practices.

In this chapter, I discuss the cultural context of STEM graduate education, focusing on admission, mentoring, and academic practices for students with marginalized identities to understand the systemic and cultural factors as barriers to their educational experience. I tie this discussion by also focusing on the student experience more broadly by reviewing studies that explore students' experience of discrimination and microaggressions, mental health, and work–life balance. By focusing on the student experience, I aim to understand more holistically these specific systemic challenges that marginalized students face that hinder equitable access to education and educational outcomes. I conclude this chapter by discussing the gaps in the literature and implications for researchers toward creating safer and more inclusive spaces for marginalized students in STEM graduate education.

13.2 Underrepresentation

Specifically, with women and students of marginalized racial and ethnic backgrounds, their underrepresentation is further exacerbated by the disproportionately higher attrition rates compared to the national average (Sowell et al., 2008, 2015). For example, the 10-year completion rate for White students in the physical sciences and mathematics is 52% while only 37% for African American students (Sowell et al., 2008). Examples of structural and institutional racism and sexism include the notion reflected even among institutional actors' deficit views on the science competency of marginalized students (Bensimon, 2005; Ghee et al., 2016) who are often seen as not belonging to STEM and intellectually inferior to White and Asian American men students (McGee et al., 2022). The culture around science was built by White men, for White men, and set barriers for anyone who does not identify with the dominant scientist stereotype. Marginalization for Students of Color occurs in everyday interactions with faculty and peers. For example, Black women are excluded from participating in peer study groups (Rosa & Mensah, 2016). Similar effects of marginalization for Women and People of Color are evident in countries with a history of colonialism (see Idahosa & Mkhize, 2021).

Students who identify as lesbian, gay, bisexual, transgender, and queer (LGBTQ+) are marginalized in STEM. The LGBTQ+ community remains critically understudied in STEM (Miller et al., 2021). Moreover, there is a lack of adequate collection of demographic data related to sexual orientation and gender identity (Sanchez et al., 2022), which limits researchers' capacity to explore the experience of LGBTQ+ students further. This lack of attention to the issues of sexual orientation and gender identity issues could result in excluding these identities in the STEM workforce (Yoder & Mattheis, 2016). Recent research that documented the experiences of LGBTQ+ scientists found that they often encountered exclusionary behavior (Atherton et al., 2016; Barthelemy et al., 2022). In particular, it was found that trans physicists experience exclusionary behavior at much higher rates than non-trans physicists (Barthelemy et al., 2022).

Students with disabilities have been historically marginalized at all levels of education in the United States. Recent estimates suggest that only about 4.8% of STEM graduate students identify as having a disability (National Science Foundation, 2017), a much lower representation than 11% at the undergraduate level (Snyder et al., 2017). Wells and Kommers (2022) found that students with disabilities were less likely to obtain a STEM degree and had lower GPA grades than their peers. Despite the growing attention to research on equity, disability and access are rarely investigated (Reinholz & Ridgway, 2021). The limited research centering the experiences of students with disabilities suggests that inequities are in place that limit access to graduate education.

13.3 Barriers to and Supports for Accessing Graduate Education

Faculty members who apply the Graduate Record Examination (GRE) score as a decisive admission criterion tend to believe that higher scores reveal one's intelligence and belonging, virtues that they consider to predict success in the field (Posselt, 2016; Posselt et al., 2019). However, research demonstrated that GRE scores are biased against historically marginalized students (Hall et al., 2017; Miller et al., 2019; Miller & Stassun, 2014). More specifically, Miller and Stassun (2014) found that a 700-point cutoff value on quantitative GRE automatically excludes 74% of women applicants and about 95% of applicants from marginalized racial and ethnic backgrounds, as opposed to 27% and 18% of men and applicants from dominant racial and ethnic backgrounds in STEM, respectively. GRE tests and application fees add a significant financial barrier to the student. Due to systemic racism, financial burdens disproportionally affect marginalized students (Cochran et al., 2017; Ong et al., 2011), thus limiting the diversity of the applicant pool. Bias in admission decisions is not limited to heavily relying on GRE scores but also on undergraduate GPA, access to research opportunities, and recommendation letters (Woo et al., 2023; Roberts et al., 2021). Madera et al. (2019) examined 624 letters of recommendation for 174 men and women for gender biases applying for assistant professor positions when controlling for objectives measures from the applicant's curriculum vitae such as the number of first author publications and total publications, honors, postdoc years, courses taught, and ranking of applicant's school. They found that letters for women contained significantly more doubt raisers (negativity, hedging, and faint praise) than letters for men applicants. For example, a doubt raiser for a woman applicant would be something like "Although she has a number of papers in preparation and one under review, I think it would be fair to say that her record on paper would not place her among the top echelon of candidates for first-rate programs." On the other hand, a doubt raiser within the same context for a man candidate would include, "I know that first-author publications are priceless for job applicants. Although [applicant] doesn't have any as of yet, that should not be a concern for you…" Although this study focused on the gender effect in letters of recommendation for assistant professor positions across various disciplines, similar biases are in place in how letters are written for men and women in graduate admissions.

While letters of recommendation are amongst the most highly evaluative criteria for an applicant's capacity to succeed in a given position, there are often neither clear instructions for how to write them nor subjective criteria for evaluating them, which may create opportunities for gender and racial bias and stereotyping where they are more likely to emerge (Morgan et al., 2013). Morgan et al. (2013), in particular, raised their concerns around the ambiguous nature of recommendation letters explaining that in situations of ambiguity or limited information, individuals rely on more automatic, schema-based processing and are more likely to enact stereotypes based on race and gender. A tested and effective method to reduce the effects of stereotyping and bias is the method of elaboration. The authors argued that by elaborating on their answers, evaluators would be "forced to use deliberative, controlled processing to attune to the information most relevant to the given position, and thus reduce automatic, possibly biased thought processes" (p. 2299).

Recent research recommends that admission committees take a holistic approach to review the applicant pool to generate a composite score based on a standardized set of priorities and criteria (personal statements, letters of recommendation, research experience, leadership experience, and community service) (Roberts et al., 2021). Leadership and community service are indicative of one's personal and professional characteristics and, when considered in graduate admissions, may result in a more diverse student body. Structural changes to admission decisions also include removing hard limits on course requirements, admitting students as a cohort instead of faculty sponsors, and diversifying admission committees (Roberts et al., 2021).

While leaders, policymakers, and researchers push toward intensifying the efforts to make STEM disciplines more diverse, about one-third of selective physics graduate programs do not consider the applicants' gender, race, and ethnicity in the admission decisions (Potvin et al., 2017). Even programs that aim to increase gender and racial diversity in graduate admissions are unaware of mechanisms that would effectively foster those efforts (Posselt et al., 2019). As research on equity in STEM graduate education grows, faculty members, especially those serving on admission committees, should consult the literature to learn about recent research recommendations and practices that effectively reduce systemic barriers toward admitting a more diverse pool of students. At the same time, there is a need for more work on how STEM graduate admission committees take up and implement such recommendations. Such studies would be valuable in understanding common pitfalls that enable systemic inequities to persist in STEM graduate education.

13.4 Barriers and Supports for Effective Graduate Mentoring

Mentoring has been denoted as one of the principal factors of the graduate student experience and has been linked to degree completion (Dawson et al., 2015). Lyons et al. (2006) identified three functions that mentors serve in the lives of their mentees: transmitting formal scientific knowledge and technical skills, socializing the mentee into the rules, values, and ethics of the discipline through role modeling, and bolstering mentee's confidence through encouragement and praise. These functions are essential in shaping students' cognitive processes relative to their science identity (i.e., Atkins et al., 2020), self-efficacy (i.e., DeFreitas & Bravo, 2012), and motivation (i.e., Hernandez et al., 2017) and have implications for the success of the graduate student. However, mentoring goes beyond the dual relationship between the mentor and the mentee; it is an academic community of practice designed to provide opportunities for engagement and learning around accepted institutional norms (Howell et al., 2021). The literature has emphasized the connection between mentoring and student science identity. This relationship could become problematic in the absence of cultural responsiveness in mentoring practices in White, androcentric, nondisabled STEM environments. Howell et al. (2021) noted that cultured, gendered, and racialized mentorship experiences could undermine a student's trajectory.

Mentoring and academic practices have significant implications for the success or attrition of marginalized students. In particular, women students cited lack of or poor guidance and support as the primary factors responsible for their decision to leave their programs (Preston, 2004). Similarly, a recent study by Guy and Boards (2019) with eight women students with marginalized racial identities documented how mentoring, research, opportunities, and academia relate to barriers and success factors. As students from historically excluded groups face unique challenges in STEM, they prefer to be mentored by faculty who share similar marginalized identities since they feel that they will benefit from the insights of mentors who have likely experienced similar challenges related to their backgrounds and identities (Baker et al., 2014; Guy & Boards, 2019). Moreover, students see their mentors as successful role models in the field (Griffin et al., 2010). However, given the dearth of STEM faculty from these marginalized communities, most students with marginalized identities will likely be mentored by faculty who are culturally different from them.

Research on mentoring for students with disabilities (SWD) further corroborates the importance of mentors being in a position to understand the unique challenges and experiences of mentees from marginalized groups (i.e., Braun et al., 2017; Hauser, 2013). Braun et al. (2017) developed a survey designed to measure mentoring effectiveness with underrepresented students, mainly focusing on SWD. Results indicated no significant differences for aspects related to Being a Scientist when SWD (deaf students) were paired with deaf mentors or mentors familiar with the deaf community. However, in any other aspects of mentoring related to Deaf Community Capital, Asking for Accommodations, and Communications Access, results were significantly higher when students were paired with Deaf or Deaf-know mentors than when paired with Deaf-unaware mentors. The authors concluded that the cultural awareness element might contribute more to effective mentoring rather than the student–mentor pair with similar backgrounds or characteristics. As a result, the need for attention to culturally responsive mentoring becomes even more critical.

STEM doctoral faculty mentors lack a meaningful understanding of the role of culture in mentoring (Howell et al., 2021; Sanczyk et al., 2021) and the way STEM spaces function. Most faculty mentors described advising practices when discussing what mentoring means to them. As such, mentors emphasize offering academic and research support focusing on research and meeting the graduate school requirements but failing to support students in other areas (Sanczyk et al., 2021). Institutional structures of racism expressed through color evasiveness and deficit perspectives toward students' capabilities are reflected in mentoring practices. For example, Sanczyk et al. (2021) reported that faculty mentors view minoritized students as being ill-prepared for the rigor of graduate school or fail to see race when mentoring students from dominant versus marginalized groups. Similarly, Howell et al. (2021) found that faculty mentors reported lacking confidence in meeting the needs of students from historically excluded racial and ethnic communities. In the same study, in follow-up interviews, a chemistry faculty member from a primarily White institution failed to recognize that domestic doctoral students from these underrepresented groups have unique needs based on their culture and backgrounds as he did for international students. Culturally responsive mentoring training programs for faculty members are essential in helping them more effectively interact and serve the needs of minoritized students (Sanzyk et al., 2021). In such mentoring programs, for example, faculty explore students' worldviews to understand how to develop a meaningful relationship (instead of centering the narrative from the faculty's perspective), share stories about cultural humility, learn about the importance of the racial or ethnic community in the students' lives as a support system, discuss campus-specific recruiting practices, and share external venues for recruitment (Brown et al., 2020). Research demonstrated that traditional mentoring models, primarily those that are institutionally enforced, are less effective for historically underrepresented students (Diekman et al., 2010; Dallaghan et al., 2022) as opposed to those that intentionally pair students with mentors (Dallaghan et al., 2022).

National initiatives have been developed to enhance racial and gender diversity in STEM. One of these programs is the American Physical Society (APS)—Bridge Program (BP) (American Physical Society, 2022). Among others, a key component of the BP was the constellation mentoring model through which students receive mentoring from multiple sources. BP students have noted the importance of being surrounded by others like them who might be faculty members and peers, which helps develop a social network (Sachmpazidi, 2021; Scherr et al., 2020). Research in physics doctoral programs demonstrated that academic and social support (psychosocial support, resources, and support to cope with academic requirements, accessible and caring faculty members) are critical factors influencing student experience (Sachmpazidi, 2021; Abdurrahman, 2021).

From a research perspective, more work needs to be done on developing instruments that will enable programs to collect informative feedback on effective mentoring that structurally includes elements of culturally responsive components. Currently, most mentoring surveys were developed for students from dominant backgrounds (White, nondisabled, straight, and male). As the quality of mentoring shapes the experiences and outcomes of marginalized students, the approach to culturally responsive effective mentoring taken by STEM faculty members becomes critical to understand. Finally, although rich in mentoring models for women in STEM, the literature lacks an exploration of the needs of students who fall outside the gender binary "box" (Dallaghan et al., 2022). Few studies have focused on exploring the climate for LGBTQ+ individuals in STEM (i.e., Patridge et al., 2014; Sanchez et al., 2022). The literature is still far from acknowledging the broad spectrum of challenges faced by LGBTQ+ students and, in turn, understanding the type of support and resources needed. More studies are needed in STEM doctoral education under critical race and critical feminist standpoint theories to unpack the experiences of students who identify with gender and sexual minorities (Rodriguez et al., 2022; Dallaghan et al., 2022).

13.5 Discrimination and Microaggressions in Graduate Education

STEM culture is built on the notion of meritocracy, one of the most shared and widely accepted values in science. The notion of meritocracy manifests that opportunities are distributed based on one's talent, effort, persistence, and achievement.

However, it ignores one's inherited social privilege and its role in one's success. When institutional actors (faculty, mentors, leaders, policymakers) espouse meritocratic values, it results in a hostile environment toward those who are not socially privileged (Bensimon, 2005; Ghee et al., 2016), having implications on who is positioned to belong or not (Holder et al., 2015), thus shaping the landscape of science as we know it today by continuing to nurture the dominant stereotype of White maleness in science (Smith et al., 2019; McGee et al., 2017). STEM programs have become fertile ground for discrimination against those who have historically been denied access to political, economic, and educational participation (Oakes, 1990). In such environments, racism, sexism, and ableism are manifested through microaggressions (i.e., Solorzano et al., 2000), discrimination, and prejudice (Lee et al., 2020; Harper et al., 2011). Discrimination and microaggressions disproportionally affect marginalized students and exacerbate existing inequalities (Boyle et al., 2022).

Racial microaggressions are interactions that degrade individuals based on their membership in a particular racial category (Miles et al., 2020; Pierce et al., 1977). Racial microaggressions are common within academic spaces and result in an adverse racial climate where students develop feelings of frustration and isolation (Solorzano et al., 2000) and doubt their identities as scientists or engineers (Miles et al., 2020). Those microaggressions based on racial stereotypes are present in both faculty–student interactions and among peers and occur within and outside classrooms (McGee et al., 2017; Burt et al., 2019). For example, an African American engineering student described an instance when the professor maintained low expectations of them even when they had high academic achievement (Solorzano et al., 2000, p. 66):

> I was doing really well in the class, like math is one of my strong suits. [...] We took a first quiz and I got like a 95 [...] he [the professor] was like, 'Come into my office. We need to talk,' and I was like, 'Okay.' I just really knew I was gonna be told, 'great job,' but he [said], 'We think you've cheated [...] We just don't know, so we think we're gonna make you [take the exam] again.' And [then] I took it with just the graduate student instructor in the room, and just myself, and I got a 98 on the exam.

Other students in the same study described instances of racial microaggressions (i.e., the professor locking their office door when seeing an African American student walking through the hallway or making assumptions on an African American student's type of scholarship). These microaggressions led to exhaustion from having to negotiate and navigate such hostile and unwelcoming environments (Burt et al., 2019). Moreover, the impact of racial microaggressions is detrimental to students' identities and sense of belonging (Miles et al., 2020; Burt et al., 2019). It can result in a high emotional, psychological, and physical toll on students (McGee, 2020a, 2020b), which can contribute to the development of impostor syndrome (racialized imposter phenomenon) (McGee et al., 2022) and challenge their persistence in the field (Burt et al., 2019; Sachmpazidi, 2021).

Scholarship on the experiences of women graduate students documents gender microaggressions (Barthelemy et al., 2016; Sachmpazidi, 2021). Research on Women of Color shows that Black women experience a more hostile racial climate than women from other racial and ethnic groups (Leath & Chavous, 2018). Similar to the findings on racial microaggressions, gender microaggressions are evident in various ways (sexual objectification, sexist language, assumption of inferiority) in the interactions between women students with faculty and peers. In response to gender and racial microaggressions, marginalized students create counter-spaces to find support and enhance their feeling of belonging. Counter-spaces are safe social spaces where a healthy racial climate can be established and maintained (Ong et al., 2018; Solorzano et al., 2000). A hostile racial climate that shapes negative interpersonal relationships between Students of Color and other institutional actors contributes to students' decision to leave STEM (Brown, 2000; Ong et al., 2011, 2018). The vast majority of work on discrimination and racial microaggressions centered on the experiences of African American students. While it has been documented that African American and Black students experience racist interactions differently than other racial and ethnic groups (Leath & Chavous, 2018), future research should explore the experiences of other Students of Color (Latinx and Native Americans) and the intersection of race with gender, disability status, and sexual orientation to understand better how to disrupt social injustice in its many forms in academic spaces.

Ableism is a structural issue and creates oppressive societal discourses (Reinholz & Ridgway, 2021). The culture of STEM encompasses values predicated on competition, workaholism, and productivity and marginalizes SWD who cannot conform to those standards. A sense of comfort and security needs to be established for disabled students to communicate their needs for accommodations (Pfeifer et al., 2021). SWD in STEM are less likely than their counterparts in non-STEM disciples to use accommodation in their courses (Lee, 2011; Pfeifer et al., 2021). As the nature of microaggressions stems from societal biases ingrained in academic settings, it is reasonable to assume that students with other marginalized identities (disabled and LGBTQ+ students) are also daily victims of such interactions (Platt & Lenzen, 2013). Sue (2010) compiled a typology of homonegative microaggressions against LGBTQ+ people. Those included over-sexualization, homophobia, heterosexist language, sinfulness, assumption of abnormality, denial of individual heterosexism, and endorsement of heteronormative culture. Boyle et al. (2022) documented that LGBTQ+ graduate students face biased treatment and microaggressions, and that such discriminatory barriers are exacerbated for LGBTQ+ People of Color. While there is a plethora of studies on microaggressions centering the experiences of Students of Color and women from all races and ethnicities, there is scarce scholarship on microaggressions against SWD (Olkin et al., 2019) and LGBTQ+ students in STEM (Miller et al., 2021). More research needs to be done in STEM graduate education under a disability justice framework which helps draw attention to how disability intersects with one's other marginalized identities (Reinholz & Ridgway, 2021). As discussed in the following paragraphs, a plethora of research in higher education has demonstrated that systematic exposure to microaggressions is associated with mental health issues (i.e., depression, anxiety, and stress) (see Blume et al., 2012).

13.6 Barriers and Supports for Mental Health and Work–Life Balance in Graduate Education

Discriminatory treatment and microaggressions are commonplace in academic settings and shape negative daily experiences in marginalized students' lives. Research has demonstrated that such negative experiences have detrimental effects on one's mental health (Blume et al., 2012; Boyle et al., 2022) and decreases STEM persistence (Wilkins-Yel et al., 2022). Boyle et al. (2022) found that heterosexual Women of Color and LGBTQ+ People of Color reported significantly greater anxiety 9 months later due to their elevated exposure to microaggressions. Other elements of academic climate, such as lack of interpersonal support, complex interpersonal interactions, and gendered, racialized, and cultural encounters were also sources of adverse mental health for Women of Color (Arnold et al., 2020).

Researchers rang the alarm to the academic community regarding graduate students' mental health and wellbeing since graduate students are significantly more likely than the general population to suffer from mental health issues (i.e., Evans et al., 2018; Cornwall et al., 2019; Yusuf et al., 2020). A lack of work–life balance was associated with poor mental health (Evans et al., 2018), and burnout was found to be one of the sources of academic and psychological struggles (Chisholm et al., 2021). These findings lead us once again to point to the problematic culture of STEM that values competition, workaholism, and productivity as signs of being "gifted" and "fit" for the profession (Abdurrahman, 2021). In studying multiple physics graduate departments, it was found that lack of work–life balance was associated with advisors' mentality around the "ideal" graduate student (Abdurrahman, 2021; Sachmpazidi, 2021). Student interviews revealed that the accepted and encouraged department norm was that students work more than 60 h per week, under the notion that it is an investment for their future in terms of the likelihood of employment in the field. Graduate students attributed their mental health problems to the competitive and workaholic nature of physics with the parallel lack of departmental support (Sachmpazidi, 2021). A slowly growing body of literature has focused on unpacking how racist and sexist microaggressions against marginalized students impact their mental health. However, more work is needed to understand how the hostile academic climate in STEM affects other marginalized students' wellbeing, especially those who identify with multiple marginalized identities.

13.7 Implications

In this chapter, I reviewed the literature on equity-oriented research in STEM graduate education from a cultural, systemic approach focusing on elements related to admission, mentoring, and academic practices. I also discussed the systemic challenges students from historically excluded groups faced of discriminatory treatment and microaggressions and concluded with a discussion on student mental health and wellbeing. This review took a systems approach in studying the sources of long-standing issues in STEM, student underrepresentation and shortcomings in student educational outcomes, under the notion that these issues are deeply rooted in STEM culture. STEM culture is dominated by the White heterosexual male supremacy paradigm that governs its policies, practices, and values and "excludes" people who do not fit within this paradigm. I recognize that there are additional lenses through which unequal achievement and persistence outcomes have been approached in the literature (e.g., student motivation), which were not covered in this review. Instead, by taking a systems approach, I discussed those elements that directly affect STEM culture by changing institutional actors' values and practices to better serve marginalized students' needs and mitigate social injustice in STEM. Potential for cultural change toward equity exists by bringing students into the conversation. Preliminary data from the Access Network participants suggests that students articulated and refined critiques of physical science culture and demonstrated elements of equity ethics in their professional practices (Amezcua et al., 2020).

In this review, I identified several areas that require further research and implications for practitioners (policymakers and faculty members). Here is a summary of the implications for researchers and practitioners for each topic discussed in this chapter.

- *Admissions*: While there are extensive discussions among academics to increase diversity, most admission decisions are made without attention to the mechanisms and practices that are argued to be effective toward enhancing diversity (Potvin et al., 2017; Posselt et al., 2019). As such, a need for training on holistic admission approaches that faculty members, especially those who serve on admissions committees, should consider participating in. From a research standpoint, more work is needed to shed light on how STEM graduate admission committees take up and implement such recommendations in practice. Such studies would be invaluable in understanding common pitfalls that enable systemic inequities to persist in STEM graduate education and, in turn, inform training programs on admissions.
- *Mentoring*: Culturally responsive mentoring training programs for faculty members are necessary to help them better interact and serve the needs of minoritized students (Sanzyk et al., 2021). From a research perspective, there is a need for mentoring survey instruments with culturally responsive components, developed based on the needs of students from historically excluded groups, that will enable programs to collect feedback on effective mentoring. Moreover, more research under the lens of critical race and critical feminist standpoint theories is needed to explore the climate for LGBTQ+ individuals in STEM (i.e., Patridge et al., 2014; Sanchez et al., 2022; Rodriguez et al., 2022; Dallaghan et al., 2022) and document the full range of the challenges faced by students who identify with gender and sexual minorities.

- *Microaggressions:* While there is a plethora of studies on microaggressions centering on the experiences of women and Students of Color, there is a limited body of scholarship on microaggressions against students with disabilities (Olkin et al., 2019) and LGBTQ+ students in STEM (Miller et al., 2021). More research needs to be done in STEM graduate education under a disability justice framework which helps draw attention to how disability intersects with one's other marginalized identities (Reinholz & Ridgway, 2021).
- *Mental health and wellbeing:* More research is needed to understand how the hostile academic climate in STEM impacts the mental health of students who identify with multiple marginalized identities.

The above are some of the existing gaps in equity-oriented scholarship in STEM graduate education. The journey toward mitigating social injustice will be long and difficult. However, we can make incremental and impactful progress toward achieving equity by taking critical steps toward changing and improving our institutions and the mindsets of institutional actors as a community.

Acknowledgments The author acknowledges financial resources as part of the Effective Practices for Physics Programs (EP3) Initiative from the NSF #1821372 that supported her during the time of authoring the chapter. The author is also grateful to the University of Maryland, College Park colleagues, Chandra Turpen, Jennifer Radoff, Fatima Abdurrahman, Lance R. Curtis, Robert P. Dalka, Andrew Elby, and Brandon J. Johnson for their feedback on this chapter.

References

Abdurrahman, F. N. (2021). *Black holes: Finding them, building instruments that help you look, and how systemic racism can prevent you from doing any of that.* UC Berkeley. ProQuest ID: Abdurrahman_berkeley_0028E_20558. Merritt ID: ark:/13030/m5b62s0t. Retrieved from https://escholarship.org/uc/item/1zs693tt

Allen, D. (1999). Desire to finish college: An empirical link between motivation and persistence. *Research in Higher Education, 40*(4), 461–485.

American Physical Society. (2022). *Bridge program.* Retrieved August 23, 2022, from https://www.aps.org/programs/minorities/bridge/index.cfm

Amezcua, F., Quan, G. M., & Turpen, C. A. (2020). Students' exploring and refining their equity ethic within the access network. In *Physics education research conference proceedings* (pp. 17–22). https://doi.org/10.1119/perc.2020.pr.Amezcua

Atherton, T. J., Barthelemy, R. S., Deconinck, W., Falk, M. L., Garmon, S., Long, E., ... & Reeves, K. (2016). LGBT climate in physics: Building an inclusive community. *American Physical Society, 5.*

Arnold, A. C., Wilkins-Yel, K. G., Bekki, J. M., Bernstein, B. L., Natarajan, M., Randall, A. K., Francies, R., & Owku, E. C. (2020). Examining the effects of STEM climate on the mental health of graduate women from diverse racial/ethnic backgrounds. In *Proceedings of the 2020 American Society for engineering education conference.* Retrieved from https://par.nsf.gov/biblio/10164497; https://doi.org/10.18260/1-2%2D%2D34617

Atkins, K., Dougan, B. M., Dromgold-Sermen, M. S., Potter, H., Sathy, V., & Panter, A. T. (2020). "Looking at myself in the future": How mentoring shapes scientific identity for STEM students from underrepresented groups. *International Journal of STEM. Education, 7*(42). https://doi.org/10.1186/s40594-020-00242-3

Baker, V., Pifer, M. J., & Griffin, K. A. (2014). Mentor-protégé fit: Identifying and developing effective mentorship across identities in doctoral education. *International Journal for Researcher Development, 5*(2), 83–98.

Barthelemy, R., McCormick, M., & Henderson, C. (2016). Gender discrimination in physics and astronomy: Graduate student experiences of sexism and gender microaggressions. *Physical Review Physics Education Research, 12*, 020119.

Barthelemy, R. S., Swirtz, M., Garmon, S., Simmons, E. H., Reeves, K., Falk, M. L., et al. (2022). LGBT+ physicists: Harassment, persistence, and uneven support. *Physical Review Physics Education Research, 18*(1), 010124.

Bensimon, E. M. (2005). Closing the achievement gap in higher education: An organizational learning perspective. *New Directions for Higher Education, 2005*, 99–111. https://doi.org/10.1002/he.190

Blume, A. W., Lovato, L. V., Thyken, B. N., & Denny, N. (2012). The relationship of microaggressions with alcohol use and anxiety among ethnic minority college students in a historically white institution. *Cultural Diversity and Ethnic Minority Psychology, 18*(1), 45.

Boyle, K. M., Culatta, E., Turner, J. L., & Sutton, T. E. (2022). Microaggressions and mental health at the intersections of race, gender, and sexual orientation in graduate and law school. *Journal of Women and Gender in Higher Education, 15*(2), 157–180.

Braun, D. C., Gormally, C., & Clark, M. D. (2017). The deaf mentoring survey: A community cultural wealth framework for measuring mentoring effectiveness with underrepresented students. *CBE—Life Sciences Education, 16*(1), ar10.

Brown, S. V. (2000). The preparation of minorities for academic careers in science and engineering: How well are we doing? In G. Campbell, R. Denes, & C. Morrison (Eds.), *Access denied: Race, ethnicity, and the scientific enterprise*. Oxford University Press.

Brown, B., Windchief, S., Komlos, B., & Arouca, R. (2020). Assessing the impact of an indigenous mentoring program on faculty to support american Indian/Alaska native graduate students in STEM. *The Journal of Faculty Development, 34*(3), 27–36.

Brunsma, D. L., Embrick, D. G., & Shin, J. H. (2017). Graduate students of color: Race, racism, and mentoring in the white waters of academia. *Sociology of Race and Ethnicity, 3*(1), 1–13. https://doi.org/10.1177/2332649216681565

Burt, B. A., McKen, A., Burkhart, J., Hormell, J., & Knight, A. (2019). Black men in engineering graduate education: Experiencing racial microaggressions within the Advisor–Advisee relationship. *The Journal of Negro Education, 88*(4), 493–508.

Canning, E. A., Muenks, K., Green, D. J., & Murphy, M. C. (2019). STEM faculty who believe ability is fixed have larger racial achievement gaps and inspire less student motivation in their classes. *Science advances, 5*(2), eaau4734.

Castillo, L. G., Conoley, C. W., Choi-Pearson, C., Archuleta, D. J., Phoummarath, M. J., & Van Landingham, A. (2006). University environment as a mediator of Latino ethnic identity and persistence attitudes. *Journal of counseling psychology, 53*(2), 267.

Chisholm, L. P., Jackson, K. R., Davidson, H. A., Churchwell, A. L., Fleming, A. E., & Drolet, B. C. (2021). Evaluation of racial microaggressions experienced during medical school training and the effect on medical student education and burnout: A validation study. *Journal of the National Medical Association, 113*(3), 310–314.

Cochran, G., Hodapp, T., & Brown, E. (2017). *Identifying barriers to ethnic/racial minority students' participation in graduate physics*. Paper presented at physics education research conference 2017, Cincinnati, OH. Retrieved May 25, 2021, from https://www.compadre.org/Repository/document/ServeFile.cfm?ID=14576&DocID=4753

Cornwall, J., Mayland, E. C., van der Meer, J., Spronken-Smith, R. A., Tustin, C., & Blyth, P. (2019). Stressors in early-stage doctoral students. *Studies in Continuing Education, 41*(3), 363–380. https://doi.org/10.1080/0158037X.2018.1534821

Dallaghan, B. G. L., Coe, C. L., Wright, S. T., & Jordan, S. G. (2022). Mentoring medical education research: Guidelines from a narrative review. *Medical Science Educator, 32*(3), 723–731.

Dawson, A. E., Bernstein, B. L., & Bekki, J. M. (2015). Providing the psychosocial benefits of mentoring to women in STEM: CareerWise as an online solution. *New Directions for Higher Education, 171*, 53–62. https://doi.org/10.1002/he.20142

DeFreitas, S. C., & Bravo, A., Jr. (2012). The influence of involvement with faculty and mentoring on the self-efficacy and academic achievement of African American and Latino college students. *Journal of the Scholarship of Teaching and Learning, 12*(4), 1–11.

Diekman, A. B., Brown, E. R., Johnston, A. M., & Clark, E. K. (2010). Seeking congruity between goals and roles: A new look at why women opt out of science, technology, engineering, and mathematics careers. *Psychological Science, 21*(8), 1051–1057. https://doi.org/10.1177/0956797610377342

Evans, T. M., Bira, L., Gastelum, J. B., Weiss, L. T., & Vanderford, N. L. (2018). Evidence for a mental health crisis in graduate education. *Nature Biotechnology, 36*(3), 282–284. https://doi.org/10.1038/nbt.4089

Ghee, M., Keels, M., Collins, D., Neal-Spence, C., & Baker, E. (2016). Fine-tuning summer research programs to promote underrepresented students' persistence in the STEM pathway. *CBE Life Sciences Education, 15*(3), ar28. https://doi.org/10.1187/cbe.16-01-0046

Griffin, K. A., Pérez, D., II, Holmes, A. P. E., & Mayo, C. E. P. (2010). Investing in the future: The importance of faculty mentoring in the development of students of color in STEM. *New Directions for Institutional Research, 2010*, 95–103. Retrieved from https://doi-org.libproxy.library.wmich.edu/10.1002/ir.365

Guy, B., & Boards, A. (2019). A seat at the table: Exploring the experiences of underrepresented minority women in STEM graduate programs. *Journal of Prevention and Intervention in the Community, 47*(4), 354–365.

Hall, J. M., Stevens, P. E., & Meleis, A. I. (1994). Marginalization: A guiding concept for valuing diversity in nursing knowledge development. *Advances in Nursing Science, 16*(4), 23–41.

Hall, J. D., O'Connell, A. B., & Cook, J. G. (2017). Predictors of student productivity in biomedical graduate school applications. *PLoS One, 12*, e0169121.

Harper, S. R., Davis, R. J., Jones, D. E., McGowan, B. L., Ingram, T. N., & Platt, C. S. (2011). Race and racism in the experiences of black male resident assistants at predominantly white universities. *Journal of College Student Development, 52*(2), 180–200. https://doi.org/10.1353/csd.2011.0025

Hauser, P. C. (2013, July). Deaf scientists need deaf mentors. In *Proceedings of the 6th international deaf academics and researchers conference*. Lisbon.

Hernandez, P. R., Bloodhart, B., Barnes, R. T., Adams, A. S., Clinton, S. M., Pollack, I., et al. (2017). Promoting professional identity, motivation, and persistence: Benefits of an informal mentoring program for female undergraduate students. *PLoS One, 12*(11), e0187531.

Holder, A. M. B., Jackson, M. A., & Ponterotto, J. G. (2015). Racial microaggression experiences and coping strategies of black women in corporate leadership. *Qualitative Psychology, 2*(2), 164–180.

Howell, C., Villanueva, K., Douglas, N., Merriweather, L., & Sanczyk, A. (2021). Doctoral mentoring relationships in STEM programs: Faculty perspectives. *The Chronicle of Mentoring and Coaching, 5*(14), 435–441.

Idahosa, G. E., & Mkhize, Z. (2021). Intersectional experiences of Black South African female doctoral students in STEM: Participation, success and retention. *Agenda, 35*(2), 110–122. https://doi.org/10.1080/10130950.2021.1919533

Leath, S., & Chavous, T. (2018). Black Women's experiences of campus racial climate and stigma at predominantly white institutions: Insights from a comparative and within-group approach for STEM and non-STEM majors. *The Journal of Negro Education, 87*(2), 125–139.

Lee, A. (2011). A comparison of postsecondary science, technology, engineering, and mathematics (STEM) enrollment for students with and without disabilities. *Career Development for Exceptional Individuals, 34*(2), 72–82.

Lee, M. J., Collins, J. D., Harwood, S. A., Mendenhall, R., & Huntt, M. B. (2020). "If you aren't White, Asian or Indian, you aren't an engineer": Racial microaggressions in STEM education. *International Journal of STEM Education, 7*, 48. https://doi.org/10.1186/s40594-020-00241-4

Lyons, W., Scroggins, D., & Rule, P. B. (2006). The mentor in graduate education. *Studies in Higher Education, 15*(3), 277–285. https://doi.org/10.1080/03075079012331377400

Madera, J. M., Hebl, M. R., Dial, H., Martin, R., & Valian, V. (2019). Raising doubt in letters of recommendation for academia: Gender differences and their impact. *Journal of Business and Psychology, 34*, 287–303. https://doi.org/10.1007/s10869-018-9541-1

McGee, E. O. (2020a). Interrogating structural racism in STEM higher education. *Educational Researcher, 49*(9), 633–644. https://doi.org/10.3102/0013189X20972718

McGee, E. O. (2020b). *Black, brown, bruised: How racialized STEM education stifles innovation*. Harvard Education Press.

McGee, E. O., Thakore, B. K., & LaBlance, S. S. (2017). The burden of being "model": Racialized experiences of Asian STEM college students. *Journal of Diversity in Higher Education, 10*(3), 253–270. https://doi.org/10.1037/dhe0000022

McGee, E. O., Botchway, P. K., Naphan-Kingery, D. E., Brockman, A. J., Houston, S., II, & White, D. T. (2022). Racism camouflaged as impostorism and the impact on Black STEM doctoral students. *Race Ethnicity and Education, 25*(4), 487–507. https://doi.org/10.1080/13613324.2021.1924137

Miles, M. L., Brockman, A. J., & Naphan-Kingery, D. E. (2020). Invalidated identities: The disconfirming effects of racial microaggressions on Black doctoral students in STEM. *Journal of Research in Science Teaching, 57*, 1608–1631. Retrieved from https://doi-org.libproxy.library.wmich.edu/10.1002/tea.21646

Miller, C., & Stassun, K. (2014). A test that fails. *Nature, 510*, 303–304. https://doi.org/10.1038/nj7504-303a

Miller, C. W., Zwickl, B. M., Posselt, J. R., Silvestrini, R. T., & Hodapp, T. (2019). Typical physics Ph. D. admissions criteria limit access to underrepresented groups but fail to predict doctoral completion. *Science Advances, 5*(1), eaat7550.

Miller, R. A., Vaccaro, A., Kimball, E. W., & Forester, R. (2021). "It's dude culture": Students with minoritized identities of sexuality and/or gender navigating STEM majors. *Journal of Diversity in Higher Education, 14*(3), 340.

Morgan, W. B., Elder, K. B., & King, E. B. (2013). The emergence and reduction of bias in letters of recommendation. *Journal of Applied Social Psychology, 43*(11), 2297–2306. https://doi.org/10.1111/jasp.12179

National Science Foundation. (2017). *Women, minorities, and persons with disabilities in science and engineering: 2017* (Special Report No. 17-310). Author. Retrieved from http://www.nsf.gov/statistics/wmpd

Oakes, J. (1990). Multiplying inequalities: The effects of race, social class, and tracking on opportunities to learn mathematics and science.

Olkin, R., Hayward, H. S., Abbene, M. S., & VanHeel, G. (2019). The experiences of microaggressions against women with visible and invisible disabilities. *Journal of Social Issues, 75*(3), 757–785.

Ong, M., Wright, C., Espinosa, L., & Orfield, G. (2011). Inside the double bind: A synthesis of empirical research on undergraduate and graduate women of color in science, technology, engineering, and mathematics. *Harvard Educational Review, 81*(2), 172–208.

Ong, M., Smith, J. M., & Ko, L. T. (2018). Counterspaces for women of color in STEM higher education: Marginal and central spaces for persistence and success. *Journal for Research in Science Teaching, 55*, 206–245. Retrieved from https://doi-org.libproxy.library.wmich.edu/10.1002/tea.21417

Owens, L., Zwickl, B. M., Franklin, S. V., & Miller, C. W. (2018, August 1–2). Misaligned visions for improving graduate diversity: Student characteristics vs. systemic/cultural factors. In A. Traxler, Y. Cao, & S. Wolf (Eds.), *PERC proceedings*. https://doi.org/10.1119/perc.2018.pr.Owens

Patridge, E. V., Barthelemy, R., & Rankin, S. R. (2014). Factors impacting the academic climate for LGBQ STEM faculty. *Journal of Women and Minorities in Science and Engineering, 20*(1). https://doi.org/10.1615/JWomenMinorScienEng.2014007429

Pfeifer, M. A., Reiter, E. M., Cordero, J. J., & Stanton, J. D. (2021). Inside and out: Factors that support and hinder the self-advocacy of undergraduates with ADHD and/or specific learning disabilities in STEM. *CBE—Life Sciences Education, 20*(2), ar17.

Pierce, C. M., Carew, J. V., Pierce-Gonzalez, D., & Wills, D. (1977). An experiment in racism: TV commercials. *Education and Urban Society, 10*(1), 61–87.

Platt, L. F., & Lenzen, A. L. (2013). Sexual orientation microaggressions and the experience of sexual minorities. *Journal of Homosexuality, 60*(7), 1011–1034.

Posselt, J. R. (2016). *Inside graduate admissions: Merit, diversity, and faculty gatekeeping*. Harvard University Press.

Posselt, J. R., Hernandez, T. E., Cochran, G. L., & Miller, C. W. (2019). Metrics first, diversity later? Making the short list and getting admitted to physics PhD programs. *Journal of Women and Minorities in Science and Engineering, 25*(4), 283–306. https://doi.org/10.1615/JWomenMinorScienEng.201902786

Potvin, G., Chari, D., & Hodapp, T. (2017). Investigating approaches to diversity in a national survey of physics doctoral degree programs: The graduate admissions landscape. *Physical Review Physics Education Research, 13*(2), 020142.

Preston, A. E. (2004). *Leaving science: Occupational exit from science careers*. Russell Sage Foundation.

Reinholz, D. L., & Ridgway, S. W. (2021). Access needs: Centering students and disrupting ableist norms in STEM. *CBE—Life Sciences Education, 20*(3), es8.

Roberts, S. F., Pyfrom, E., Hoffman, J. A., Pai, C., Reagan, E. K., & Light, A. E. (2021). Review of racially equitable admissions practices in STEM doctoral programs. *Education Sciences, 11*, 270. https://doi.org/10.3390/educsci11060270

Rockinson-Szapkiw, A. J., Sharpe, K., & Wendt, J. (2022). Promoting self-efficacy, mentoring competencies, and persistence in STEM: A case study evaluating racial and ethnic minority women's learning experiences in a virtual STEM peer mentor training. *Journal of Science Education and Technology, 31*, 386–402.

Rodriguez, M., Barthelemy, R., & McCormick, M. (2022). Critical race and feminist standpoint theories in physics education research: A historical review and potential applications. *Physical Review Physics Education Research, 22*(1), 013101. Retrieved from https://link.aps.org/doi/10.1103/PhysRevPhysEducRes.18.013101

Rosa, K., & Mensah, F. M. (2016). Educational pathways of Black women physicists: Stories of experiencing and overcoming obstacles in life. *Physical Review Physics Education Research, 12*(2), 020113.

Sachmpazidi, D. (2021). *Investigating the relationship between departmental support structures, self-efficacy and intention to persist: An examination of students' experience in 19 physics graduate programs across the United States*. Dissertations 3771. Retrieved from https://scholarworks.wmich.edu/dissertations/3771

Sanczyk, A., Merriweather, L. R., Howell, C. D., & Douglas, N. C. (2021). STEM doctoral mentoring: a call for a conscious, culturally responsive journey. *International Journal of Mentoring and Coaching in Education*.

Sanchez, D., Guerra, R. C., & Farrell, S. (2022). *Demographic data collection of LGBTQ+ identities: Barriers and motivations*. In 2022 ASEE annual conference proceedings.

Sanczyk, A., Merriweather, L. R., Howell, C. D., & Douglas, N. C. (2021). STEM doctoral mentoring: A call for a conscious, culturally responsive journey. *International Journal of Mentoring and Coaching in Education, 10*(3), 284–297.

Santiago, A. M., & Einarson, M. K. (1998). Background characteristics as predictors of academic self-confidence and academic self-efficacy among graduate science and engineering students. *Research in Higher Education, 39*(2), 163–198. Retrieved from http://www.jstor.org/stable/40196292.

Scherr, R. E., Plisch, M., Gray, K. E., Potvin, G., & Hodapp, T. (2017). Fixed and growth mindsets in physics graduate admissions. *Physical Review Physics Education Research, 13*(2), 020133.

Scherr, R. E., Lopez, M. A., & Rosario-Franco, M. (2020). Isolation and connectedness among Black and Latinx physics graduate students. *Physical Review Physics Education Research, 16*(2), 020132. https://doi.org/10.1103/PhysRevPhysEducRes.16.020132

Smith, K. C., Boakye, B., Williams, D., & Fleming, L. (2019). The exploration of how identity intersectionality strengthens STEM identity for Black female undergraduates attending a historically Black college and university (HBCU). *The Journal of Negro Education, 88*(3), 407–418. Retrieved from https://www.muse.jhu.edu/article/802629

Snyder, T. D., de Bray, C., & Dillow, S. A. (2017). *Digest of education statistics, 2016 (No. NCES 2017–094)*. National Center for Education Statistics, Institute of Education Sciences, U.S. Department of Education.

Solorzano, D., Ceja, M., & Yosso, T. (2000). Critical race theory, racial microaggressions, and campus racial climate: The experiences of African American college students. *Journal of Negro Education, 69*(1/2), 60.

Sowell, R. S., Zhang, T., & Redd, T. (2008). *Ph.D. completion and attrition: Analysis of baseline demographic data from the Ph.D. completion project*. Council of Graduate Schools.

Sowell, R., Allum, J., & Okahana, H. (2015). *Doctoral initiative on minority attrition and completion*. Council of Graduate Schools.

Sue, D. W. (2010). *Microaggressions in everyday life: Race, gender, and sexual orientation*. Wiley.

Wells, R., & Kommers, S. (2022). Graduate and professional education for students with disabilities: Examining access to STEM, legal, and health fields in the United States. *International Journal of Disability, Development and Education, 69*(2), 672–686. https://doi.org/10.1080/1034912X.2020.1726299

Wilkins-Yel, K. G., Arnold, A., Bekki, J., Natarajan, M., Bernstein, B., & Randall, A. K. (2022). "I can't push off my own mental health": Chilly STEM climates, mental health, and STEM persistence among Black, Latina, and White graduate women. *Sex Roles, 86*(3), 208–232.

Woo, S. E., LeBreton, J. M., Keith, M. G., & Tay, L. (2023). Bias, fairness, and validity in graduate-school admissions: A psychometric perspective. *Perspectives on Psychological Science, 18*(1), 3–31.

Yoder, J. B., & Mattheis, A. (2016). Queer in STEM: Workplace experiences reported in a national survey of LGBTQA individuals in science, technology, engineering, and mathematics careers. *Journal of Homosexuality, 63*(1), 1–27. https://doi.org/10.1080/00918369.2015.1078632

Yusuf, J., Saitgalina, M., & Chapman, D. W. (2020). Work-life balance and well-being of graduate students. *Journal of Public Affairs Education, 26*(4), 458–483. https://doi.org/10.1080/15236803.2020.1771990

Chapter 14
Theoretical and Methodological Innovative Futures for Physics in Postsecondary Contexts

Eleanor S. Armstrong, Allison J. Gonsalves, Anna T. Danielsson, and Emma Mckay

Abstract The aim of the chapter is to highlight promising theoretical and methodological innovations that work to understand and address inequities in postsecondary physics contexts. In particular, we move beyond a focus on the individual and identity as the main lens. Instead, our focus on relationships pulls a range of new research sites into consideration as vital sites of postsecondary physics education, where new, existing, and non-members of the field learn to navigate the profession. We show how relations between people, communities, and institutions can unsettle what is focused on in physics education research (PER).

Keywords Physics education research · Postsecondary physics · Communities · Theoretical innovations · Methodological innovations

14.1 Introduction and Overview

The aim of the chapter is to highlight promising theoretical and methodological innovations that work to understand and address inequities in postsecondary physics contexts. In particular, we move beyond a focus on the individual and identity as the

E. S. Armstrong (✉)
Institute for Space Research, University of Leicester, Leicester, United Kingdom
e-mail: ea377@leicester.ac.uk

A. J. Gonsalves · E. Mckay
McGill University, Montreal, QC, Canada
e-mail: allison.gonsalves@mcgill.ca; emma.m.mckay@mail.mcgill.ca

A. T. Danielsson
Stockholm University, Stockholm, Sweden
e-mail: anna.t.danielsson@su.se

G. L. Cochran, J. D. Adams (eds.), *Equity-Oriented, Discipline-Based STEM Education Research*, Critical Studies of Education 19,
https://doi.org/10.1007/978-981-96-9136-4_14

main lens. We instead show how a focus on relations between people, communities, and institutions can unsettle physics education research (PER) and offer new avenues for scholarly investigation. In particular, if PER researches education, then its hegemonic contexts, interests, goals, and methods play a role in defining what education *means*. The broad overview we provide in this chapter of questions and re-orientations for all of these pieces of PER pushes then toward reconceptualizing what counts as education within PER, moving toward relationships, community, and critical critique as core to the idea of education.

This chapter takes up and responds to two important contexts. First, PER and physics equity work more broadly has primarily been conducted in ways which support seeing science education as a process in service of developing capital (see, for example, Pierce, 2012) and, consequently, have focused on "diversity" and "inclusion" issues like binarized gender inequalities. Second and more specifically, PER has traditionally focused on a genre of research that prioritizes learning standard physics knowledge and practices in the university classroom and student laboratories—a context which has only recently started to challenge the colonial, enlightenment values of physics research at large, instead continuing to train and socialize individuals into these practices. In this chapter, then, we focus on how attending to efforts that grapple with "politicized" questions and themes, and the social backgrounds to the research in the community, can broaden the field of (postsecondary) physics education research. We look at examples of action and learning in a variety of contexts, which responds to the call to pluralize the typical focus of PER on classroom contexts.

14.1.1 Positionality

As researchers, we acknowledge that research is a subjective endeavor that is situated in our varied positionalities. All four authors bring varying relations with ancestors, the lands we live on, and physics as a subject, to our scholarship on physics education. We recognize that we have White racialized identities, and are/have always been settlers; all four of us living and working in colonial states. We understand that engaging in research means also recognizing those privileges and colonized perspectives in our research and work. Our epistemological positionalities are aligned with research that seeks to identify and challenge issues of power, equity, and social justice in physics contexts, particularly in academia. Some of us position ourselves as insiders to physics, while others are outsider to physics but benefit from positions as insiders to postsecondary physics.

14.1.2 Research Context for Postsecondary Physics Education

Often motivating physics education research is the goal to have more (and more diverse) people in STEM. Why do we need more people working in STEM? An answer to that question involves complex, entangling ideas about scientists and engineers as contributing to economic growth with considerations about the significance of STEM professionals for a country's defense capabilities and national security (Fensham, 1985; Lövheim, 2016, 2021). To complicate matters further, utilitarian arguments concerning a perceived need for more STEM professionals are often interspersed with arguments drawing on equity and social justice (DeBoer, 2013). This emphasis on "getting more qualified professionals into STEM fields" has led to research on barriers that under-represented groups face in STEM, and solutions to support their trajectories into the "STEM pipeline," rather than questions about the STEM pipeline itself. In physics education research in particular, the issue of the uneven distribution of women and men in physical sciences has received considerable attention. Looking back, notable works include Laura McCullough's research on gender differences in physics assessments (e.g., McCullough, 2004), the work on highlighting and reducing gender gaps in higher education physics (e.g., Pollock et al., 2007), and research in secondary schools investigating "physics identity" and strategies to encourage girls to enter university-level physics programs (e.g., Hazari et al., 2013). Such work has been pivotal in uncovering binarized gender inequities in physics, and in postsecondary physics education especially.

Casting the net wider, feminist critiques of science (e.g., Fox, 1985; Prescod-Weinstein, 2020) and studies of physics communities from anthropological points of view (e.g., Traweek, 1988; Hasse, 2002) have provided complementary perspectives on the intersection between gender and the epistemology and practices of physics. This contrasts to much equity work in physics education oriented work in the context of postsecondary physics that has, implicitly or explicitly, relied on notions of gender as something static and binary (Traxler et al., 2016). Lately, this work has diversified to include both a recognition of a wider set of social categories (e.g., Barthelemy's (2020) work on LGBT+ physicists) and expanding theoretical vantage points, such as critical race theory (e.g., Hyater-Adams et al., 2018) and intersectionality (e.g., Avraamidou, 2020a). Such identity-oriented perspectives have lately gained a lot of traction in physics education research (Gonsalves & Danielsson, 2020) for understanding connections between gender and who is recognized as a particular kind of physics student or physicist. As useful as this work has been in uncovering nuances of exclusionary practices in physics, we cannot but wonder where the perspectives' limits are in terms of reimagining teaching, learning, and research practices in physics. Therefore, in this chapter, rather than give an overview of equity-oriented research in postsecondary physics education research to date, we look forward toward future possibilities. We seek to identify strands in the contemporary research that we see as auspicious for engaging a diversity of learners and disrupting harmful norms and practices in postsecondary physics contexts.

14.1.3 Grounding Lens and Showcasing Innovations

In this chapter, we cover three innovations: theoretical innovations; methods and practices; and rethinking impact. In these sections, we draw into PER context developments that have taken place elsewhere in academic scholarship; and demonstrate their use and future potential uses within postsecondary physics DBER. We have selected some examples from different frameworks which are themselves expansive, and we recommend that the references we include here can act as a guide for those interested in reading and thinking with these theoretical frameworks at large.

Our view of postsecondary contexts will include not just higher education programs and university students' experiences in physics, but also learning in professional communities, informal learning spaces in postsecondary physics institutions, and community engaged physics research. Learning and education have long been conceived of expansively (e.g., Sandlin et al., 2011) to include extra-institutional contexts; de-institutionalizing the conception of teaching and learning and liberating knowledge development. In expanding our focus beyond traditional learning institutions, we draw inspiration from Herrenkohl et al. (2019) who argue for a "learning humanities" that puts relationships at the center of inquiry. Thus, we move away from examining the cognitive dimensions of physics learning, and from the examination of practices shaping identity work, to instead explore how relationships are centered in (in)formal physics practices, and how practices which place relationships at the center can transform forms of engagement in physics. Thus, we no longer wish to answer questions about how to get more qualified people into physics. Instead, we seek to learn from research and practices that ask questions about obligations vis-a-vis physics practices, responsibilities that physicists and physics students have to act in the world (in relation to human and more-than-human actors), and mechanisms that retain those who are qualified in the profession. Our focus on relationships pulls informal learning contexts into consideration as vital sites of postsecondary physics education, where new, existing, and non-members of the field learn to navigate the profession.

Thus, In Sect. 14.2, we cover theoretical innovations that engage with critical race and critical Whiteness theories and queer theory, as well as exploring expansively other theoretical innovations elsewhere in science learning contexts that we see innovative futures in for postsecondary physics education research. In Sect. 14.3, we demonstrate how mentorship communities and professional development networks offer novel methods and practices toward futures in postsecondary physics communities that are not structured simply around "including" new folks into existing communities of practice. Finally, in Sect. 14.4, we pull to the fore how innovation in what university outreach is and how it is studied in PER contexts, and in the practices and research of citizen science exercise can expand postsecondary physics education. We conclude the chapter looking forward to innovation in futures in postsecondary PER.

14.2 Theoretical Innovations Adopted Recently in PER

In this section, we show how innovations in application of theoretical lenses offer critical perspectives on PER and innovative futures toward which we can expand physics education (and the research that studies these practices expansively).

14.2.1 Critical Race Theory and Critical Whiteness

Recent developments in postsecondary PER have seen an uptake of critical perspectives on race in physics education, drawing on critical race theory, intersectionality, and critical Whiteness (Rosa & Mensah, 2016; Hyater-Adams et al., 2019; Avraamidou, 2020b, 2022; Robertson & Hairston, 2022). Critical race theory (CRT) describes racism as an everyday ordinary occurrence, understanding that race structures social interactions, and that racism is normalized in social systems, social institutions, and social norms. This normalization of racism is an important principle that CRT seeks to make visible as a structuring feature of society (Delgado, 1995). As race, through the lens of CRT, is produced in social interactions, to understand racism research must center the voices of People of Color (Delgado & Stefancic, 2017). Rosa and Mensah (2016) documented the voices of Black women in physics, by utilizing one of the key principles of CRT: counterstories. In centering the stories of Black women in physics, telling counterstories confronts the dominant stories of those who are in power in physics. These counterstories revealed obstacles that Black women face in physics, such as ordinary racism and microaggressions, and frames these obstacles, such as feelings of isolation and decreased self-efficacy, mechanisms of exclusion which compromised their learning and performance in physics. Rosa and Mensah argue that traditional narratives in research about persistence in physics overplays "academic achievement…and downplays the role of the racial social climate in STEM fields" (p. 11).

The counterstories enabled by the CRT lens reveal how racism functions specifically in students' performance and persistence in physics, and challenges the dominant narratives that place student deficits as a key feature that interventions in the field should seek to address. In this study, Hyater-Adams et al. (2019) adopt a CRT lens to investigate how Black folks identify as physicists, developing a framework (critical physics identity, CPI) to describe such identities. The framework draws on the basic tenets of science identity framings (featuring the importance of recognition, performance, and competence) but also combines these with Nasir's (2011) model of racialized identities that foregrounds the importance of relational, ideational, and material resources in the construction of identity. The findings showed that all participants strongly identified with physics, but experienced difficulties performing their physics identities due to similar structural barriers as those described by Rosa and Mensah. The value of this framework which places race at the analytical centre is that it provides a means to capture the counterstories of Black participants, while identifying the structural obstacles that impact Black students' performances in physics.

Drawing on intersectionality theory, Avraamidou (2020b, 2022) challenges the prioritizing of the "politics of recognition" in physics identity research, asking researchers to consider where the power resides in research that seeks to learn about how recognition supports insider identity work in physics. Avraamidou's use of intersectionality—which develops an understanding of how the way people experience society is shaped by the overlap of their different identities that are inextricably interlinked and not separable (Nash, 2011)—in this research highlights the often problematic ways that the concept of recognition is mobilized in identity research (in PER and science education research more broadly). Recognition has emerged as a strong analytical concept in identity research investigating the underrepresentation of women and racialized people in physics (e.g., Hazari et al., 2010). However, Avraamidou (2022) points out, recognition has been largely mobilized as a binary concept (either one is recognized as a physics person, or not), which neglects the multiple, intersecting ways that identity positions can yield various forms of recognition, non-recognition, or misrecognition that are not fixed between different contexts. Thus, Avraamidou argues, there is a need to attend to the politics of recognition in identity research, and especially to attend to the ways that current conceptualizations of recognition "leaves open the possibility of women in physics seeking recognition from the very people and institutions that are responsible for their underrepresentation" (p. 62).

Robertson and Hairston (2022) operationalize Critical Whiteness theory to examine how White supremacist ideology can structure interactions in physics classrooms. The authors draw on CRT and literature from Critical Whiteness studies to identify "markers for whiteness" and to analyze how these show up in introductory physics classrooms. This research examines micro-interactions between students, a teacher, a whiteboard, and a physics concept to analyze the interactions and the teacher's attention. Mobilizing several concepts from CWT, the authors demonstrate how interactions in the classroom center Whiteness in a way that elevates dominant voices, ideas, and ideologies while marginalizing others. These interactions, and those described in follow-up interviews about the class, demonstrate how meritocracy is intertwined with Whiteness in ways that are valued among students and the teacher. These ideologies are likely to be silencing to many of those participating in the class, yet they are reproduced by students in interactions. This new research in physics education is a radical departure for the field. The response to its publication signals the troubling relationship physics education has with race and racial ideology. Shortly after this paper's publication, the editors of Physical Review Physics Education journal published an editorial supporting the research and their choice to publish it, and admonishing the "harassing, intimidating, and diminishing tweets and emails in response to the publication of the article" (Henderson & Thoennessen, 2022). The authors have since published subsequent work employing a CWT framework that demonstrates how faculty members teaching in mainstream physics programs will adopt "race evasive frames" (e.g., grit or equal opportunity) to discuss representation and success in ways that maintain race-neutrality and the minimization of racism in physics (Robertson et al., 2023).

14.2.2 LGBT+ Research and Queer Theorising

Where early equity and diversity work in PER was almost exclusively focused on binarized gender, recent work has instead expanded to consider the experiences of LGBT+[1] physicists and the constructions of gender and sexuality. Rather than only accounting for the experience of cis women and cis men, this research reports on the experiences of trans and non-binary physicists. A notable series of papers by Ramón S Barthelemy and colleagues (Barthelemy, 2020; Barthelemy et al., 2022a, 2022b) report from an online quantitative and qualitative survey of LGBTQ+ physicists, exploring LGBTQ+ physicists and physics students experiences of exclusionary behavior and harassment. While those respondents who could be out about their LGBTQ+ identity were more comfortable with their institution (Barthelemy et al., 2022a), in the 71 qualitative responses that reported on experiences of exclusionary behavior and harassment for cis and trans physicists, Barthelemy (2020) found that the majority were based on non-normative gender expression, including for cis physicists, and being a woman. Their data demonstrate that nearly half of the transgender respondents experienced the most exclusionary behavior (with one fifth reporting at least one experience of this in the past year), and over one third of transgender respondents considering leaving their workplace. This finding is echoed in a survey of astronomy and planetary scientists, where Richey et al. (2019) found that although LGBTQPAN women and gender minorities were more likely to feel unsafe at work due to their race, sexual orientation, and gender identity than cisgender, straight women.

To date, this line of research has focused on the experiences of LGBT+ students and scientists, and contributes to broadening the discussion about diversity and inclusion in PER beyond binary gender; demonstrating the need to take into account the intersection of different axes of power. There have been, for example, vital calls to adapt survey practices and institutional policies to take a complex approach to gender, specifically moving beyond gender as something binary, discrete, or that can be easily discernible through names or other external characteristics (Rasmussen et al., 2019; Strauss et al., 2020). Similarly, raising the profile in PER projects of both LGBTQ+ physics students and physicists through developing resources about prominent LGBTQ+ folks (e.g., 500 Queer Scientists, n.d.), or of popular culture

[1] We use the term LGBT+ in our writing, but where we refer to findings from other papers, we use the language in the original text. This is to honor the ways that the researchers have worked to categorize their participant experiences, and highlights the plurality of approaches in the scholarship to date.

figures engaging with physics such as Amrou Al-Kadhi describing quantum mechanics' non-binary state as "beautiful, strange and glorious" (Al-Kadhi, 2024) has been a vital step toward developing both representation and identity-confirmation in postsecondary physics education.

However, we might distinguish this work on LGBTQ+ visibility and recognition, to practices that embed queer theory and queer approaches to both PER and postsecondary physics education contexts. In postsecondary informal science education contexts, Armstrong (2022) has demonstrated the use of queer theorizing to unpack "hidden" narratives that both discuss LGBTQ+ identities and moved to apply critical queer perspectives on physics and other science research for tour guide attendees in a museum. This demonstrates the potential of queer theorizing in postsecondary physics education where it seeks to deconstruct dominant narratives around gender and sexuality, dismantling binary constructions such as heterosexual/homosexual and masculine/feminine. Elsewhere, work to theorize queer epistemologies of identities in physics contexts is taking shape (see, for example, Armstrong & Danielsson, 2023, and Orthia & Roberson (2023)). In a recent anthology, Letts and Fifield (2019), who have collected STEM education writings engaging with queer theories over two decades, state that "Queer theorizing uncovers and incites paradoxical relations of the normal and strange in ways of knowing, feeling, and being ...[and attends to] processes that normalize and naturalize (and abnormalize and denaturalize) particular manifestations of human subjects, society, culture, and nature." (p. 4). This can be seen in Helene Götschel's (2019) physics university classroom. She seeks to teach queering of physics by, for example, "scrutinizing what is supposed to be normal and what is invisible or silenced, making the familiar strange, reflecting on the discursive production of physical knowledge" (p. 126). In this way, Götschel also discusses how our understanding and storying of the world through physics sometimes relies on binary, heteronormative gender systems; reflective of Barad's work to use queer to rethink the very ontology of physics (Barad, 2011).

14.2.3 Theories Used More Widely in Other Science Learning Contexts

In addition to these two theoretical vantage points, we note that there are a range of innovations that have been taken up elsewhere in both non-physics postsecondary science education and physics education at other levels, and here we gesture toward the possibilities for postsecondary physics education research by engaging with these rich bodies of work. Collectively these (along with the queer and critical race theory and Whiteness sections) share a willingness to look at physics as a system of knowledge that is part of larger social systems and to critically to engage with social justice in the context of physics education (Cochran et al., 2021).

Well integrated into research on early-years learning contexts, posthumanism and other theories which include matter offer theoretical innovation for postsecondary PER by emphasizing the co-emergence of learning between learners and matter. Introducing this into theorizing of postsecondary offers mechanisms to engage with how, in contexts beyond the classroom, engagement and experience are often a priority. This is in direct contrast to the formal classroom where "often concepts are afforded more ontological power and are considered more real than the lifeworld experiences of the learner" (Milne, 2019). Particularly rich in theorizing around elements of the natural environment, for example, in exploration of water by Elkin Postila (2022), or engaging learners in ethical education in science (e.g., Scantlebury & Milne, 2020), these approaches offer the potential to draw attention to dimensions of physics education practice that are currently under-theorized in the postsecondary domain. Such work directs our attention to the formation of identity through intra-action with matter (Barad, 2007), and in the social engagement with the nonhuman world. However, an emphasis on the genealogies of posthumanism as *the* mode of engaging with the interaction of human and non-human worlds in other science learning overlooks and erases theorizing by Indigenous scholars and with Indigenous knowers (Todd, 2016). For example, the Turtle Island Institute (founded by Melanie Goodchild (Goodchild 2023)) supports a Virtual Teaching Lodge that engages in Indigenous-led teaching of science specifically beyond the university classroom; Aboriginal and Torres Strait Islander Astronomy project works with teacher education in the postsecondary context at the University of Melbourne (*Aboriginal and Torres Strait Islander Astronomy*, 2022).

Agitating for greater engagement in the implications of physics being part of a settler colonial project in the education that is provided about physics is part of activist work within the field (see Sect. 14.3: Methods and Practices for more details). Deliberate engagement with, and challenge to, entitlement to Indigenous land and access for physics research can be seen in relation to large astronomical science operations (e.g., Casumbal-Salazar, 2017; Neilson, 2019; Goodyear-Ka'ōpua, 2020; Maile, 2021). But, thus far, the impacts of including such critiques in physics education, for example, in the content of classroom work or ethical skills for engaging communities of practice, has been undertheorized in PER. Such work could unpack what work postsecondary learning does or could do to (re)shape beliefs that physics as a body of knowledge is modern, contemporary, progressive, future-oriented *and* that those attributes are good, desirable, and/or inevitable; highlighting where or how learners both within and the beyond the field are socialized to believe this to be the case. In this way, we call for innovative futures in physics education that move research past the support of colonial states and assimilatory equity that currently dominates the field. All of these critiques encourage us to ask: What will it take in PER to interrogate how colonial, capitalist, and extractivist practices, systems, and theories exist in physics?

14.3 Methods and Practices of Physics Education

We highlight postsecondary contexts that expand the concept of "physics education" to include learning beyond teaching in the classroom. Our contribution here is to move a focus on methods and practices of physics education in the university classroom context to more varied social contexts. We exemplify by one practice in the context of under-graduate students and one in the context of professional learning, but recognize that these practices also are powerful across contexts and that the distinction between "professional" and "student" is porous and not clearly defined.

14.3.1 Physics Learners: Mentorship and Training Programs

Socialization into the field of physics has been an important and long-standing focus of research in PER (e.g., Danielsson, 2012; Gonsalves, 2014), theorizing at how learners learn to belong in the research field. Socialization takes place in many venues. Beyond learning through university classrooms, mentorship and training programs can provide such socialisation events. A wave of programs focus on providing opportunities for more inclusive socialization for those who come from minoritized backgrounds into the field. However, even well-intentioned mentorship provided by people outside the community from which the junior participants come from, and especially that undertaken by White scholars, can potentially be alienating. Counter-narrative led research by Martinez-Cola (2020) demonstrates that Students of Color are subject to imbalance mentoring relationships that, especially in their worst cases, do not see challenging of community norms in developing a place within the community for the mentees, instead trying to fit those outside into existing norms. While there are some White mentors who act as "Allies" (as Martinez-Cola describes them) to disrupt community norms through inviting mentees to events, introducing them to their community, co-authoring with them, she makes the point that these are infrequent persons in the academy. Rather, many mentors "collect" students (then extractively utilize mentees for personal gain) or "nightlight" mentees (by shining guiding lights for mentees without recognizing systemic challenges) in mentoring relationships which socialize students in harmful ways.

To support physics learners, developing communities can be vital. Community change can be supported, convened, led, and populated by the minoritized groups that are historically excluded. These projects offer a different idea about what it means to be socialized into the field, in what we might call "by-with-for PER projects," led by community members, for community members, with community members as active agents in their development (an adaptation of Simon, 2016). For example, at the California Academy of Sciences, a summer intern program for queer youth, "STARS" (Striving for Trans Inclusion and Anti-Racism in Science Learning) lead by trans and/or BIPOC museum workers to forge paths that see queer identities as part of postsecondary science fields. As an intern on the program, Eli Ramos (2021) gathered the experiences of their colleagues on this program. Ramos notes that many colleagues were united in their "ostracization by race or gender" from science community spaces prior to the program. However, STARS interns described how development "alongside other people of color and LGBTQ+ people has been instrumental to us" allowing them not to have "to explain the basics of my identity and justifying my personhood"—which "was a relief" (Ramos, 2021). Elsewhere, participants at Space Science in Context reported their experience of not having to explain or justify themselves in a conference setting (based on racialized, ableist, or misogynistic judgements) as being integral to finding a community in space sciences that encouraged them to see they belonged in the field (Persaud & Armstrong, 2022). Such experiences describe the relief of not having to do what Murray et al. have theorized as the additional "imposter work" that minoritized folks often undertake to survive and thrive in postsecondary science education contexts (Murray et al., 2022) by being outside the social conditions that make them outsiders to the science community at large. This shift in focus from the individual to the system of relations that maintain the internalization of "feeling like a fraud despite one's achievements" (p. 7) rejects the need for individualized success narratives that often belie the privileges that are required to experience overcoming such exclusions. As Hodari et al. (2022) urge in their paper on socialization in physics training it "costs us so little to counter those experiences with cues that tell them we are happy they [minoritized students] are there" (p. 21).

14.3.2 *Physics Professionals and Communities: Reshaping Professionalism Through Community Building and Action*

Other postsecondary physics learning spaces (re)develop norms within professional STEM communities, which are by members of the profession, for their peers, in community with them. Although the distinction is somewhat arbitrary from the section (what does "to be within the field" mean as opposed to becoming part of the field, for instance), in this section, we focus on professional development and community building projects and programs. In their paper, Hodari et al. (2022) note while much research focuses on undergraduate contexts, late-postgraduate and postdoctoral positions in physics fields are key times where researchers are socialized into the field. It is during these periods that "historically marginalized scientists are often excluded or kept out of these professional networks that are tied to increased productivity, which in turn affects their professional success" (p. 4).

Thus, the development of space to reshape social norms of the field from within are vital, which can lead to relearning what participation within the professional physics field could be. These take a wide range of guises from labs within institutions, to trans-national digital communities. From digital community building of projects such as #BlackinSTEM, Particles for Justice (n.d.), Pride in STEM (n.d.), Shutdown STEM (n.d.), VanguardSTEM (2021) leading to off-line action, to projects such as the *Astronomers for Planet Earth* which encourages climate conscious research and conference participation (Burtscher et al., 2022); or to local institutional LGBTQ+ STEM community action across science departments at UCL (Lock et al., 2025); it is possible to expand the focus on professional community change. We unite these projects in their work to support relearning behaviors within the postsecondary physics community, through supporting, validating, and focusing change in the community and institutions at large through reshaping ideas about how physicists can be within the community. In their work on the experiences on Women of Color in STEM, Ong et al. (2018) describe how such communities can be theorized as counter spaces for Women of Color, instrumental for "facilitating their educational preparation and readying them for careers and beyond" (p. 216).

14.4 Rethinking the Impacts of Informal Postsecondary Learning Contexts

While existing literature in the field addresses the impact of informal learning primarily on the participants in such projects who are from "the public," we here make an intervention to draw attention to the work that is being done by those developing informal learning practices and the opportunities that this provides to reshape their understandings and knowledge of physics. We focus particularly on reconceptualizing *who* is learning the practices through the example of University outreach and *what* is being learned through exploring innovative citizen science projects.

14.4.1 University "Outreach"

Engagement with informal learning in postsecondary physics contexts has been shown to impact university students' identity work in their physics programs in ways which expand the sense of what physics identity is or ought to be. These programs can be meaningful opportunities for students to build community and find a sense of belonging within their university programs. Research investigating the impacts of university students' experiences as facilitators of physics outreach projects suggests that informal physics learning experiences can have positive impacts on students' identity work outside of the constraints that are usually found in academic physics settings (Fracchiolla et al., 2020). Participating in a physics outreach community of practice provides opportunities for students to be recognized as members in the physics community in new ways, and expanding their participation beyond the normative practices of university physics (Prefontaine et al., 2021). This is not limited to university students. Outreach practices of researchers also provide new spaces to negotiate their science identities through creating and sustaining multiple science identities theorized as giving researchers the opportunity to experiment with developing shifting and contingent identity performances in university science outreach (Davies, 2021). Additionally, research specifically investigating women's involvement in facilitating informal physics programs found that participants gained increased opportunities for internal and external recognition, thus providing opportunities for them to "curate their own sense of expertise" which improved their confidence in their chosen major (Randolph et al., 2021, p. 363).

These programs also provide resources for youth who are engaged in learning that facilitates their own socialization into physics. For example, research into the significance of "university adjacent" informal physics learning programs suggests that university outreach programs provide numerous resources that contribute to youths'—especially young women's—identity work vis-a-vis physics (Gonsalves et al., 2022). Programs identified as university-adjacent (such as outreach programs, internships, summer schools, and public lectures) can provide opportunities for youth to begin to see themselves as physics people, and to learn about the lives of physicists beyond the classroom or even the lab. While we acknowledge the importance of these practices to contribute to pluralizing the people who participate in physics, we also wish to point to these as often elite and difficult to access programs for many minoritized students.

14.4.2 Citizen Science

"Citizen science" designates a wide range of different activities involving public participation in science. This can include, for example, amateur naturalists collecting observational data about birds (Parrish et al., 2019), people classifying images of galaxies (Reed et al., 2013; Lingard et al., 2020), or monitoring of radioactivity in food (Reiher, 2016; Andersson-Sundén et al., 2019). Today, citizen science is typified by distributed fieldwork or crowd-sourcing of data analysis (McQuillan, 2014), with the common goal of "educating" the public about science. Research in PER related to these activities have hegemonically, thus, looked at the "success" of various projects against such goals. But citizen science has an epistemically transformative potential: challenging the very foundation of the modern regime of scientific knowledge production where the work of professional scientists is separated from the lay public, who are understood as—at best—consumers of science. The involvement of lay people in science not only changes who participates in the knowledge production, it can also transform how this is done as well as where and why—including lay people in experimental research not just in collection of observations (Strasser et al., 2019). Strasser and colleagues instead attend to "striking" importance of the perceptive and affective qualities employed by participants during citizen science projects, despite "knowing that [these attributes'] exclusion was a key element in the formation of modern science, based on the ideal of objective, rational, and disinterested knowledge" (2019, p. 65).

McQuillan (2014) takes this potential further, arguing for citizen science as nomadic science focused on the local, the situated, and the culturally reflective. In this nomadic science also lies a potential for community activism, to enact science as a bottom-up practice, attentive to local needs and practices (McQuillan, 2014). For example, the Public Lab (n.d.) seeks to pursue environmental justice through community science and open technology. Through low-cost tools, people are able to participate in the building of community knowledge about environmental health based on physics experimental techniques (e.g., do-it-yourself spectrometers used to detect oil pollution). Another example of citizen science within physics is the Original Research By Young Twinkle Students (ORBYTS) program, where secondary school students have been involved in original research on exoplanets, supervised by PhD students and postdocs (Dunn, 2021). Part of the data used comes from data obtained by citizen astronomers, thereby linking junior research, school students, and citizen scientists. One aim with the ORBYTS program is to by involving school students in state-of-the-art research empower them to participate in STEM and challenge ideas about who can be a producer of scientific knowledge (Sousa-Silva et al., 2017).

14.5 Conclusions, Contributions to the Field, and a Call to Innovative Futures

Throughout this chapter, one of our overarching contributions has been to (re)conceptualize what counts as *education* within the field of physics education research by introducing and highlighting theories, methods, and contexts that are a departure from hegemonic PER. In our selection of cases, we particularly ask what it means to think of *education* beyond the normative postsecondary contexts of the university classroom, with a greater focus on *learning* in its expansiveness to capture experiences, knowledge transfers, skills, and behaviors. Not only does this transformation open a range of new practices and sites of subjectivities to the purview of PER, it also moves beyond a conceptualisation of postsecondary education that is narrowly focused on those for whom physics remains a central part of their lives. Instead, this view allows exploration of the value of understanding what physics learning might mean for those beyond the field. Thus, we call for innovative moves in postsecondary PER: from the individual to networks of relationships, from a specifically university classroom context to more varied social contexts, and from support of socializing people into physics more efficiently into critiquing colonial physics at its core. This transformation of physics education research and practices is a call to think critically not only about the sites of physics learning that are researched, but also the goals of the research practice. We call for innovative futures in PER that resist reproducing oppressive, colonialist, extractive cultures, or the goals of improving "diversity and inclusion" in physics, for example, and instead invest in imagining sociocultural change beyond increasing representation.

References

#ShutDownAcademia #ShutDownSTEM. (n.d.). Retrieved October 19, 2022, from https://www.shutdownstem.com/

Al-Kadhi, Amrou (2018). What quantum physics taught me about queer identity. BBC https://www.bbc.co.uk/videos/c51yldgd2r7o (Accessed 12 October 2023)

Andersson-Sundén, E., Gustavsson, C., Hjalmarsson, A., Jacewicz, M., Lantz, M., Marciniewski, P., et al. (2019). Citizen science and radioactivity. *Nuclear Physics News, 29*(2), 25–28.

Armstrong, E. S. (2022). Towards queer tours of science and technology museums. *Museums and Society, 20*(2). https://doi.org/10.29311/mas.v20i2.3941

Armstrong, E. S., & Danielsson, A. T. (2023). Science Butch Blues. *Queer-feminist science and technology studies forum, 8*, 11–30.

Australian Indigenous Astronomy. (2022, July 8). *Aboriginal and Torres strait islander astronomy*. Retrieved October 19, 2022, from http://www.aboriginalastronomy.com.au/

Avraamidou, L. (2020a). Science identity as a landscape of becoming: Rethinking recognition and emotions through an intersectional lens. *Cultural Studies of Science Education, 15*(2), 323–345. https://doi.org/10.1007/s11422-019-09954-7

Avraamidou, L. (2020b). "I am a young immigrant woman doing physics and on top of that I am Muslim": Identities, intersections, and negotiations. *Journal of Research in Science Teaching, 57*(3), 311–341.

Avraamidou, L. (2022). Identities in/out of physics and the politics of recognition. *Journal of Research in Science Teaching, 59*(1), 58–94.

Barad, K. (2007). *Meeting the universe halfway: Quantum physics and the entanglement of matter and meaning*. Duke University Press.

Barad, K. (2011). Nature's queer performativity. *Qui Parle: Critical Humanities and Social Sciences, 19*(2), 121–158.

Barthelemy, R. S. (2020). LGBT+ physicists qualitative experiences of exclusionary behavior and harassment. *European Journal of Physics, 41*(6), 065703.

Barthelemy, R. S., Swirtz, M., Garmon, S., Simmons, E. H., Reeves, K., Falk, M. L., et al. (2022a). LGBT+ physicists: Harassment, persistence, and uneven support. *Physical Review Physics Education Research, 18*(1), 010124.

Barthelemy, R. S., Hughes, B. E., Swirtz, M., Mikota, M., & Atherton, T. J. (2022b). Workplace climate for LGBT+ physicists: A view from students and professional physicists. *Physical Review Physics Education Research, 18*(1), 010147. Retrieved from https://www.bbc.co.uk/ideas/videos/what-quantum-physics-taught-me-about-queer-identit/p06ltbj1

Burtscher, L., Balaguer-Núñez, L., D'Orazi, V., Barret, D., Beuchert, T., Janiuk, A., et al. (2022). Astronomy organizations should lead in our battle against the climate crisis. *Nature Astronomy, 6*(7), 764–764.

Casumbal-Salazar, I. (2017). A fictive kinship: Making "modernity," "ancient Hawaiians," and the telescopes on Mauna Kea. *Native American and Indigenous Studies, 4*(2), 1–30.

Cochran, G. L., Hyater-Adams, S., Alvarado, C., Prescod-Weinstein, C., & Daane, A. R. (2021). Social justice and physics education. In *Teaching and learning for social justice and equity in higher education* (pp. 125–147). Palgrave Macmillan.

Danielsson, A. T. (2012). Exploring woman university physics students 'doing gender' and 'doing physics'. *Gender and Education, 24*(1), 25–39.

Davies, S. R. (2021). Performing science in public: Science communication and scientific identity. *Community and Identity in Contemporary Technosciences, 31*, 207.

DeBoer, G. E. (2013). Science for all: Historical perspectives on policy for science education reform. In *Moving the equity agenda forward* (pp. 5–20). Springer.

Delgado, R. (1995). The Rodrigo chronicles: *Conversations about America and race*. NYU Press.

Delgado, R., & Stefancic, J. (2017). *Critical race theory: An introduction* (Vol. 20). NYU Press.

Dunn, W. (2021). A Lecture in 2 Parts: 1. ORBYTS: Inspirational Research Partnerships for School Students from Historically-Excluded Groups and 2. Extraordinary X-rays from the Outer Planets. In EGU General Assembly Conference Abstracts (pp. EGU21-11236).

Elkin Postila, T. (2022). Stories of water: Preschool children's engagement with water purification. *Cultural Studies of Science Education, 17*(2), 277–299.

Fensham, P. J. (1985). Science for all: A reflective essay. *Journal of Curriculum Studies, 17*(4), 415–435.

Fox, K. E. (1985). *Reflections on gender and science*. Yale University Press.

Fracchiolla, C., Prefontaine, B., & Hinko, K. (2020). Community of practice approach for understanding identity development within informal physics programs. *Physical Review Physics Education Research, 16*(2), 020115.

Gonsalves, A. (2014). "Physics and the girly girl—there is a contradiction somewhere": Doctoral students' positioning around discourses of gender and competence in physics. *Cultural Studies of Science Education, 9*, 503–521. https://doi.org/10.1007/s11422-012-9447-6

Gonsalves, A., & Danielsson, A. (2020). *Physics education and gender: Identity as an analytic lens for research.* Springer Press (183 pp.). Retrieved from https://link-springer-com.proxy3.library.mcgill.ca/book/10.1007/978-3-030-41933-2

Gonsalves, A., Danielsson, A., Johansson, A., & Nyström, A.-S. (2022). Other spaces for youths' identity-work in physics. *Physical Review Physics Education Research, 18*, 020118. (online).

Goodchild, M. (2023). *Niigani Miinigowiziiwin (we Give These Gifts to the Future)* (Doctoral dissertation, University of Waterloo).

Goodyear-Ka'ōpua, N. (2020). On the cattle guard. *Biography, 43*(3), 527–529.

Götschel, H. (2019). Teaching queering physics: An agenda for research and practice. In *STEM of desire* (pp. 125–145). Brill Sense.

Hasse, C. (2002). Gender diversity in play with physics: The problem of premises for participation in activities. *Mind, Culture, and Activity, 9*, 250–269.

Hazari, Z., Sonnert, G., Sadler, P. M., & Shanahan, M. C. (2010). Connecting high school physics experiences, outcome expectations, physics identity, and physics career choice: A gender study. *Journal of Research in Science Teaching, 47*(8), 978–1003.

Hazari, Z., Potvin, G., Lock, R. M., Lung, F., Sonnert, G., & Sadler, P. M. (2013). Factors that affect the physical science career interest of female students: Testing five common hypotheses. *Physical Review Special Topics-Physics Education Research, 9*(2), 020115.

Henderson, C., & Thoennessen, M. (2022). Research on advancing equity is critical for physics. *Physical Review Physics Education Research, 18*(1), 010001.

Herrenkohl, L. R., Lee, J., Kong, F., Nakamura, S., Imani, K., Nasu, K., et al. (2019). Learning in community for STEM undergraduates: Connecting a learning sciences and a learning humanities approach in higher education. *Cognition and Instruction, 37*(3), 327–348.

Hodari, A. K., Krammes, S. B., Prescod-Weinstein, C., Nord, B. D., Esquivel, J. N., & Assamagan, K. A. (2022). Informal socialization in physics training. *Proceedings of the US Community Study on the Future of Particle Physics (Snowmass 2021)* Preprint available: arXiv:2203.11518v2

Hyater-Adams, S., Fracchiolla, C., Finkelstein, N., & Hinko, K. (2018). Critical look at physics identity: An operationalized framework for examining race and physics identity. *Physical Review Physics Education Research, 14*(1), 010132.

Hyater-Adams, S., Fracchiolla, C., Williams, T., Finkelstein, N., & Hinko, K. (2019). Deconstructing Black physics identity: Linking individual and social constructs using the critical physics identity framework. *Physical Review Physics Education Research, 15*(2), 020115.

Isler, J. C., Berryman, N. V., Harriot, A., Vilfranc, C. L., Carrington, L. J., & Lee, D. N. (2021). Defining the Flow—Using an Intersectional Scientific Methodology to Construct a VanguardSTEM Hyperspace. *Genealogy, 5*(1), 8. Retrieved from https://www.mdpi.com/2313-5778/5/1/8

Letts, W., & Fifield, S. (2019). *STEM of desire: Queer theories and science education.* Brill Sense.

Lingard, T. K., Masters, K. L., Krawczyk, C., Lintott, C., Kruk, S., Simmons, B., et al. (2020). Galaxy Zoo builder: Four-component photometric decomposition of spiral galaxies guided by citizen science. *The Astrophysical Journal, 900*(2), 178.

Lock, S. J., Jones, E., Hellyer, R. J., & Armstrong, E. (2025). LGBTQ+ staff experiences at UCL: teaching, research and institutional culture. UCL, LESG.

Lövheim, D. (2016). *Naturvetarna, ingenjörerna och valfrihetens samhälle: Rekrytering till teknik och naturvetenskap under svensk efterkrigstid* (p. 260). Kriterium.

Lövheim, D. (2021). Cold War fostering of scientific elites: International Youth Olympiads in chemistry and physics 1967–1984. *History of Education, 50*(5), 685–703.

Maile, D. U. (2021). On being late: Cruising mauna kea and unsettling technoscientific conquest in Hawai'i. *American Indian Culture and Research Journal, 45*(1), 95–122. https://doi.org/10.17953/aicrj.45.1.maile

Martinez-Cola, M. (2020). Collectors, nightlights, and allies, oh my. *Understanding and Dismantling Privilege, 10*(1), 61–82.

McCullough, L. (2004). Gender, context, and physics assessment. *Journal of International Women's Studies, 5*(4), 20–30.

McQuillan, D. (2014). The countercultural potential of citizen science. *M/C Journal, 17*(6). https://doi.org/10.5204/mcj.919

Milne, C. (2019). Intra-actions that matter: Building for practice in a liberal arts science course. In *Material practice and materiality: Too long ignored in Science Education* (pp. 81–100). Springer.

Murray, Ó. M., Chiu, Y. L. T., Wong, B., & Horsburgh, J. (2022). Deindividualising imposter syndrome: Imposter work amongst marginalised STEMM undergraduates in the UK. *Sociology, 57*(1), 00380385221117 3.

Nash, J. C. (2011). Practicing love: Black feminism, love-politics, and post-intersectionality. *Meridians, 11*(2), 1–24.

Nasir, N. I. (2011). *Racialized identities: Race and achievement among African American youth.* Stanford University Press.

Neilson, H. R. (2019). Astronomy must respect rights of Indigenous peoples. *Nature, 572*, 312.

Orthia, L. A., & Roberson, T. eds., (2023). Queering science communication: representations, theory, and practice. Bristol University Press.

Ong, M., Smith, J. M., & Ko, L. T. (2018). Counterspaces for women of color in STEM higher education: Marginal and central spaces for persistence and success. *Journal of Research in Science Teaching, 55*(2), 206–245.

Parrish, J. K., Jones, T., Burgess, H. K., He, Y., Fortson, L., & Cavalier, D. (2019). Hoping for optimality or designing for inclusion: Persistence, learning, and the social network of citizen science. *Proceedings of the National Academy of Sciences, 116*(6), 1894–1901.

Particles for Justice. (n.d.). Retrieved October 19, 2022, from https://www.particlesforjustice.org/

Persaud, D. M., & Armstrong, E. S. (2022). Space science in context: Lessons learned and recommendations for IDEA practice and beyond. *LPI Contributions, 2679*, 2067.

Pierce, C. (2012). *Education in the age of biocapitalism: Optimizing educational life for a flat world.* Springer.

Pollock, S. J., Finkelstein, N. D., & Kost, L. E. (2007). Reducing the gender gap in the physics classroom: How sufficient is interactive engagement? *Physical Review Special Topics-Physics Education Research, 3*(1), 010107.

Prefontaine, B., Mullen, C., Güven, J. J., Rispler, C., Rethman, C., Bergin, S. D., et al. (2021). Informal physics programs as communities of practice: How can programs support university students' identities? *Physical Review Physics Education Research, 17*(2), 020134.

Prescod-Weinstein, C. (2020). Making Black women scientists under white empiricism: The racialization of epistemology in physics. *Signs: Journal of Women in Culture and Society, 45*(2), 421–447.

Pride in STEM. (n.d.). Retrieved October 19, 2022, from https://prideinstem.org/

Public Lab: A DIY Environmental Science Community. (n.d.). Retrieved October 19, 2022, from https://publiclab.org/

Queer Scientists (n.d.) 500 Queer Scientists Visibility Campaign. https://500queerscientists.com/ (Accessed 12 October 2023).

Ramos, E. (2021). *Towards museum inclusion.* Museum education roundtable towards museum inclusion reflections from Exploratorium STARS internship program comments. Retrieved October 19, 2022, from https://www.museumedu.org/towards-museum-inclusion-reflections-from-exploratorium-stars-internship-program/

Randolph, J., Hay, E., Rethman, C., Erukhimova, T., Donaldson, J. P., & Perry, J. (2021, August). Impact of informal physics programs on female university students. In *Proceedings of the physics education research conference (PERC)* (pp. 359–364).

Rasmussen, K. C., Maier, E., Strauss, B. E., Durbin, M., Riesbeck, L., Wallach, A., et al. (2019). The nonbinary fraction: Looking towards the future of gender equity in astronomy. *APC white paper submitted to the Astro2020 Decadal Survey* Preprint available at arXiv:1907.04893.

Reed, M. J., Lardner, R. A., & Carney, K. (2013). *An exploratory factor analysis of motivations for participating in Zooniverse, a collection of virtual citizen science projects.* 2013 46th Hawaii International Conference on System Sciences (pp. 610–619). https://doi.org/10.1109/HICSS.2013.85

Reiher, C. (2016). Lay people and experts in citizen science: Monitoring radioactively contaminated food in post-Fukushima Japan. *Asien, 140*, 56–73.

Richey, C. R., Lee, K. M. N., Rodgers, E., & Clancy, K. B. (2019). Gender and sexual minorities in astronomy and planetary science face increased risks of harassment and assault. *Bulletin of the American Astronomical Society, 51*(4), 0206.

Robertson, A. D., & Hairston, W. T. (2022). Observing whiteness in introductory physics: A case study. *Physical Review Physics Education Research, 18*(1), 010119.

Robertson, A. D., Vélez, V., Hairston, W. T., & Bonilla-Silva, E. (2023). Race-evasive frames in physics and physics education: Results from an interview study. *Physical Review Physics Education Research, 19*(1), 010115.

Rosa, K., & Mensah, F. M. (2016). Educational pathways of Black women physicists: Stories of experiencing and overcoming obstacles in life. *Physical Review Physics Education Research, 12*(2), 020113.

Sandlin, J. A., O'Malley, M. P., & Burdick, J. (2011). Mapping the complexity of public pedagogy scholarship: 1894–2010. *Review of Educational Research, 81*(3), 338–375.

Scantlebury, K., & Milne, C. (2020). Beyond dichotomies/binaries: Twenty-first century post humanities ethics for science education using a Baradian perspective. In *Examining ethics in contemporary science education research* (pp. 159–176). Springer.

Simon, N. (2016). *The Art of Relevance* Museum 2.0.

Sousa-Silva, C., McKemmish, L. K., Chubb, K. L., Gorman, M. N., Baker, J. S., Barton, E. J., et al. (2017). Original Research By Young Twinkle Students (ORBYTS): When can students start performing original research? *Physics Education, 53*(1), 015020.

Strasser, B., Baudry, J., Mahr, D., Sanchez, G., & Tancoigne, E. (2019). "Citizen science"? Rethinking science and public participation. *Science and Technology Studies, 32*, 52–76.

Strauss, B. E., Borges, S. R., Faridani, T., Grier, J. A., Kiihne, A., Maier, E. R., et al.. (2020). Nonbinary systems: Looking towards the future of gender equity in planetary science. *A State of the Profession White Paper for the Planetary and Astrobiology Decadal Survey 2023-2032.* Preprint arXiv:2009.08247*v1*.

Todd, Z. (2016). An indigenous feminist's take on the ontological turn: 'Ontology' is just another word for colonialism. *Journal of Historical Sociology, 29*(1), 4–22.

Traweek, S. (1988). *Beamtimes and lifetimes.* Harvard University Press.

Traxler, A. L., Cid, X. C., Blue, J., & Barthelemy, R. (2016). Enriching gender in physics education research: A binary past and a complex future. *Physical Review Physics Education Research, 12*(2), 020114.

FSC
www.fsc.org
MIX
Papier aus verantwortungsvollen Quellen
Paper from responsible sources
FSC® C105338

If you have any concerns about our products,
you can contact us on
ProductSafety@springernature.com

In case Publisher is established outside the EU,
the EU authorized representative is:
Springer Nature Customer Service Center GmbH
Europaplatz 3, 69115 Heidelberg, Germany

Printed by Libri Plureos GmbH
in Hamburg, Germany